Natural History Society of Montreal

**The Canadian naturalist and geologist**

Natural History Society of Montreal

**The Canadian naturalist and geologist**

ISBN/EAN: 9783337208516

Printed in Europe, USA, Canada, Australia, Japan

Cover: Foto ©berggeist007 / pixelio.de

More available books at **www.hansebooks.com**

THE

# CANADIAN NATURALIST

## AND GEOLOGIST:

### A Bi-Monthly Journal of Natural Science,

CONDUCTED BY A COMMITTEE OF THE NATURAL
HISTORY SOCIETY OF MONTREAL.

**NEW SERIES —Vol. 3.**

(WITH THREE PLATES.)

MONTREAL:
DAWSON BROTHERS, GREAT ST. JAMES STREET.
1868.

MONTREAL: PRINTED BY THE MONTREAL PRINTING AND PUBLISHING COMPANY.

# CONTENTS.

iv

THE

# CANADIAN NATURALIST.

## SECOND SERIES.

## ON THE ROCKS AND CUPRIFEROUS BEDS OF PORTAGE LAKE, MICHIGAN.

### By Thomas Macfarlane.

During the summer of 1865 I was employed on the Geological Survey of Canada in making certain explorations on the north and east shores of Lake Superior. I had instructions to visit also the mines of the south shore, in order to acquire some idea of the experience there gained in mining the deposits of native copper, it being anticipated that such might be advantageously applied in explorations on the Canadian side of the lake. The observations which I made on the south shore, although sufficiently interesting, could not well find a place in a report having reference to Canadian territory, and, Sir William Logan having kindly consented, I have made them the subject of the following paper.

One of the most conspicuous geographical features of the south shore of Lake Superior, is Keweenaw Point. Like the rocks constituting it, it strikes out into the lake in a north-easterly direction for a distance of fifty miles. Portage Lake is situated near its base, and together with Sturgeon River, which flows into Keweenaw Bay, almost severs the point from the main land. The north-western part of Portage Lake intersects the various strata of trap and other rocks which run along the whole length of Keweenaw Point. While to the north-eastward, at Eagle River and elsewhere, the mines of greatest note are generally situated upon veins crossing the strike of the trap, those in the neighbourhood of

Portage Lake are worked almost exclusively upon beds, the strike and dip of which are parallel with that of the enclosing rocks. Such beds are not, however, altogether absent in other districts of the copper region, where they have been called ' ash beds,' but it is in the Portage Lake district that they occur most frequently, and are mined most successfully. The rocks with which they are interstratified are principally what are called traps and greenstones, together with conglomerates and sandstones. They maintain a general strike of N. 20° to N. 40° E., and have a dip of 50° to 60° north-westward.

In attempting to describe these rocks more minutely, I shall begin with those lying immediately west of the great cupriferous bed on which the Quincy, Pewabic and Franklin mines are situated, and proceed then to notice those lying to the eastward, which are, geologically, lower lying rocks.

The rock which is observed at the side of the road leading past the Quincy mine to the Pewabic, and which lies several hundred feet west of the cupriferous bed, is distinctly of a compound nature, but all its constituent minerals are not large enough to be accurately determined. Conspicuous among them is a dark green chloritic mineral, the grains of which vary from the smallest size to one fourth of an inch in diameter. In the latter case they are irregularly shaped, with rounded angles, but they are never quite round or amygdaloidal. They frequently consist in the centre of dark green laminæ. The mineral is very soft and has a light greenish-grey streak. It fuses readily before the blow-pipe to a black magnetic glass, and it would seem to be the preponderating mineral in the rock. The other constituents are in very fine grains, and consist of a reddish-grey feldspathic mineral, with distinct cleavage planes, and closely resembling it, light greenish-grey particles but whether of a feldspathic, pyroxenic or hornblendic nature could not be determined. The prevailing colour of the rock is dark greyish-green. Hydrochloric acid produces no effervescense with it, even when in a state of fine powder. Its specific gravity is 2.83, and the magnet attracts a very small quantity of magnetite from its powder. The colour of the powder when very fine is light greenish-grey. When ignited it loses 3.09 per cent. of its weight and changes to a light brown colour. When digested with nitric acid, and then afterwards with a weak solution of caustic potash (to remove free silica) it experiences, including the loss by ignition, a loss of 46.36 per cent. This consists of

Silica.............................. 14.73
Alumina........................... 7.17
Peroxide of iron.................. 14.87
Lime .............................. 4.47
Magnesia........................... 2.03
Water.............................. 3.09

———

46.36

In the undecomposed residue light red and dark coloured particles are discernible. On digesting it with hydrochloric acid and subsequently with a weak solution of potash, it sustains a further loss of 10.6 per cent., which consists of

Silica ............................... 3.48
Alumina............................. 3.03
Peroxide of iron.................... 1.98
Lime ................................ 1.76
Magnesia ............................ .35

———

The undecomposed residue was still found to consist of a light red and a dark coloured constituent. The latter was the heavier, and an approximate separation was accomplished by washing. The dark coloured particles, which could not however be freed wholly from the light coloured felspathic constituent, fused readily to a dark brown glass. To judge from its gravity and fusibility it would not appear unreasonable to regard it as either pyroxene or hornblende. In quantity, however, it did not exceed one-eighth of the felspar. The latter fused easily before the blow-pipe to a colourless glass, tinging the flame strongly yellow. It would therefore seem to be of the nature of labradorite, although it is only slightly decomposed by hydrochloric acid. Since, according to Girard, neither labradorite, nor pyroxene nor magnetite are decomposable by nitric acid, it may reasonably be concluded that the constituents removed by the nitric acid are those of the chloritic mineral. On treating the rock, previous to ignition, with hydrochloric acid, much of the iron is removed as protoxide. Although some peroxide is also possibly present, I have calculated the whole of the iron as protoxide, and have moreover added the difference of weight between it and the iron as peroxide, to the loss sustained by ignition, and put it down as water. In this way the composition of the chloritic mineral calculated to 100 parts, would be

| | |
|---|---|
| Silica................................. | 31.78 |
| Alumina............................. | 15.47 |
| Protoxide of iron................... | 28.87 |
| Lime................................. | 9.64 |
| Magnesia............................ | 4.37 |
| Water .............................. | 9.87 |

$$100.00$$

In these figures the quantity of iron is much greater, and that of magnesia much less than in ordinary chlorite. In its composition, and in being easily decomposed by acids, the mineral most closely resembles the ferruginous chlorite of Delesse,* (the delessite of Naumann), but differs from it in containing a considerable amount of lime, and in being readily fused before the blow-pipe. Assuming, nevertheless, that the chloritic constituent is delessite, and that one half of the iron removed by hydrochloric acid belongs to the magnetite, then the rock would be composed mineralogically of

| | |
|---|---|
| Delessite ........................... | 46.36 |
| Labradorite......................... | 47.43 |
| Pyroxene or hornblende........... | 5.26 |
| Magnetite .......................... | 0.95 |

$$100.00$$

The next rock to the eastward, to which I paid some attention, is that which constitutes the hanging wall of the Quincy Mine. It is a fine-grained mixture of reddish-grey feldspar, and dark green delessite, the former predominating. In this mixture larger crystals of feldspar and larger rounded grains of the ferruginous chlorite are occasionally discernible. Its sp. gr. is 2.83. The powder is of a reddish-grey tint, and the magnet shews the presence in it of a trace of magnetite. On ignition it changes to light brown,

---

* The following is the composition of ferruginous chlorite according to Delesse's analysis :

| | |
|---|---|
| Silica...................... | 31.07 |
| Alumina.................... | 15.47 |
| Peroxide of iron............ | 22.21 |
| Protoxide of manganese...... | traces |
| Lime....................... | 0.46 |
| Magnesia................... | 19.14 |
| Water...................... | 11.55 |

Bischof : Chemical and Physical Geology, III, 228.

sustaining at the same time a loss of 1.32 p. c.   No effervescence is produced by hydrochloric acid, which dissolves out from the rock 32.44 per cent. of bases, consisting of

| | |
|---|---:|
| Alumina | 7.52 |
| Peroxide of iron | 15.04 |
| Lime | 4.34 |
| Magnesia | 5.54 |

which, doubtless, principally belong to the chloritic mineral.   The residue contains a very small quantity of the heavier and darker constituent which was found in the rock first described.   The residue is not decomposed by concentrated sulphuric acid.

Next, in downward succession, comes the cupriferous bed generally known as the ' Pewabic Lode,' although it possesses none of the characters of a vein.   It has a thickness of about 12 feet, and in places resembles the rock which constitutes the foot-wall of the mine, into which it seems to graduate.   In its characteristic varieties it differs, however, completely from that rock.   It is a reddish-brown or chocolate coloured uncrystalline rock with amygdaloidal structure and uneven, almost earthy fracture.   The matrix sometimes contains some small amygdules, which are not always completely filled, and thus render the rock porous.   The matrix is fusible to a black, slightly magnetic glass.   It is in places impregnated with grains of metallic copper, from the minutest size to those having a diameter of a tenth of an inch. Those of a still larger size very generally project from the matrix into the amygdules, or form rounded particles lying entirely within these cavities, and filling them.   The copper is here accompanied by a mineral of a light green colour, very soft, and separable from the rock as a green powder.   It fuses before the blowpipe to a black slightly magnetic glass.   On ignition it changes to a light yellow colour losing 0.4 p. c. of its weight.   It is decomposed by hydrochloric acid and the resulting solution contains protoxide as well as peroxide of iron.   On analysis, it gave the following results, in which all the iron is calculated as protoxide, and the difference between it and peroxide put down as water

| | |
|---|---:|
| Silica | 46.48 |
| Alumina | 17.71 |
| Protoxide of iron | 21.17 |
| Lime | 9.89 |
| Magnesia | trace |

Alkalies ........................ 1.97 by difference.
Water ........................ 2.78
                              ———
                              100

It is probably a variety of green-earth. Some of the amygdules are altogether filled with it, in which case it frequently contains small isolated grains of metallic copper. Sometimes calcspar is found along with the green-earth, the two minerals generally occupying separate parts of the cavity. Very frequently the green mineral merely lines the cavities, and the rest is filled up with calcspar. The foregoing description is of a specimen of the bed exceedingly rich in copper. At other places the matrix is more compact and darker coloured, and the amygdules are exclusively filled with calcspar, without any enclosing film of green-earth. Sometimes quartz, delessite, laumontite and prehnite occur filling the cavities. In many parts of the bed, large irregular patches and veins of calcspar are seen, through which and through the adjoining rock, run huge irregular masses of copper frequently weighing several tons, with which small quantities of native silver are associated. Epidote is also often met with in the bed, generally unconnected with the amygdules, and forming small irregular masses in the chocolate-coloured rock. The foregoing description applies equally to the cupriferous bed as developed in the Pewabic and Franklin mines. These are situated on the north side of Portage Lake. The continuation of the bed to the southeast was sought for a long time fruitlessly, until at last it was discovered accidentally at a distance of about four miles south-west of Portage Lake. At this point, on the property of the South Pewabic Mining Company, it is being opened and presents the following characters. The rock is of the same colour as on the Quincy Mine, but it is finer grained, and in places a conchoidal fracture is even observable. The amygdules are smaller, and the metallic copper seems altogether confined to them, forming solid rounded pellets. It is accompanied by delessite, calcspar, laumontite and prehnite, which minerals also occur in the cavities alone. The matrix of this bed is also fusible to a black magnetic glass.

The rock which underlies the copper-bearing bed of the Quincy Mine is distinctly amygdaloidal. The matrix is fine grained, but it is crystalline and is seen to consist of different constituents. Its colour is dark reddish-grey, and it is fusible to a black glass. The cavities, which seldom exceed the size of a pea, are filled with what appears to be the same chloritic mineral which occurs as a constituent in the first two rocks above described. It is very soft and may be cut into small, slightly coherent slices. These fuse readily to a black glass, which is slightly magnetic. In fine powder its colour is light greenish grey, and by ignition it turns dark brown, losing 5.85 p. c. of its weight. Hydrochloric acid decomposes it readily. On analysis, and calculation as above described, it gave.

Silica.................................. 30.59
Alumina ........................... 26.07
Protoxide of iron.................. 22.01
Lime................................. 1.92
Magnesia........................... 12.36
Water............................... 7.23
_____
100.18

It will be observed that these results correspond much more closely with the composition of delessite than that calculated from the constituents dissolved by nitric acid from the rock first described. The specific gravity of the rock, including the amygdules, is 2.78. The colour of the fine powder is dark reddish-grey. On ignition it turns brown and loses 2.33. Nitric acid dissolves 25.67, and hydrochloric acid 34.12 of its weight. In the residue from treatment with the latter acid, no heavy dark coloured constituent could be detected. From the above particulars the following mineralogical composition is deducible.

Delessite in amygdules and grains... 38.
Labradorite ........................... 62.
_____
100

An occasional crystal of foldspar is met with in the rock, which seems to be identical with that occurring in the matrix, and is only partially decomposed by hydrochloric acid.

The various bands of rock which underlie the Pewabic lode have been intersected by a cross-cut, more than five hundred feet in length, from the seventy fathoms level of the Pewabic mine. This working has passed through the following rocks, the local names and thicknesses (horizontally) of which are as follows :

Trap.............................. 137 feet.
Old Pewabic lode.............. 34   ".
Trap.............................. 85   "

Green amygdaloid vein.......... 19 feet.
Trap ............................ 98 "
Albany and Boston vein......... 7 "
Trap ............................ 45 "
Epidote or Mesnard vein....... 23 "
Trap ............................ 20 "
Fluckan ........................ 1 "
Conglomerate ..................... 31 "
Sandstone........................ 6 "

———

506 feet.

The general strike of these strata is N. 38° E. and the dip 55° northwestward. The two beds above denominated as the Green amygdaloid vein and the Mesnard vein are also found on the Quincy property, where the first named bears a general resemblance to the rock of the Pewabic lode. The matrix is perhaps darker coloured, and contains grains and crystals of feldspar as well as amygdules of green-earth and calcspar, the latter containing copper in fine grains. The rock of the Mesnard vein is dark brown, with a bluish tint. The minerals of the amygdules are principally green-earth, quartz and metallic copper. This bed is also called the Epidote vein but the green-earth has probably been mistaken for epidote.

The trap which overlies the conglomerate in the Albany and Boston Mine is a fine grained mixture of dark green delessite, (in grains less distinctly isolated than in the rocks already described) greenish-grey feldspar, and reddish-brown mica, some of the laminæ of the latter shewing ruby-red reflections. Its sp. gr. is 2.81, and the smallest trace only of its powder is attracted by the magnet. The colour of the powder is greenish-grey, which changes on ignition to brown, a loss of 4.19 being sustained. Nitric acid dissolves from it 24.52 p. c., which consist of

Alumina ........................... 5.96
Peroxide of iron................... 14.78
Lime.............................. 3.41
Magnesia........................... 0.37

These figures agree pretty closely with the quantities of bases dissolved from the rocks already described, but the quantities of lime and magnesia are a little smaller. The residue consists of a dark coloured, heavier, and a reddish-white coloured lighter part, the latter about twice as large in quantity as the former. The dark coloured portion consisted probably in greater part of mica, and to judge from the comparatively low specific gravity of the rock, little or no pyroxene or hornblende could be present. The mineralogical composition of this trap is therefore probably as follows :

| | |
|---|---|
| Delessite | 40 |
| Mica | 20 |
| Labradorite | 40 |
| | 100 |

The ' Fluckan ' which underlies the trap last described is separated from it by a small seam of clay. The fluckan itself is a fine grained, dark-red shaly rock in which pieces of a greenish blue colour are sometimes seen. Both substances are fusible before the blow-pipe and contain occasionally small grains and flakes of copper. It resembles the old *Thonstein* (claystone) of the Germans, now more properly named Felsite tuff.

The conglomerate upon which the foregoing rock rests, has acquired some celebrity on account of its being mined for copper on the property of the Albany and Boston Mining Company. The boulders and pebbles consist of various species of porphyry. One of them has a dark brown matrix with small white crystals of feldspar ; another has a matrix of the same colour but with larger crystals of orthoclase, while a third variety consists principally of a fine grained mass of orthoclase with which a small quantity of a dark coloured mineral occurs in particles too small for determination. The matrix consists of a coarse grained sand of porphyritic material, impregnated with calcareous matter. In many places the interstices are not at all filled up, in others calcspar is the matrix, and very often in the lower part of the bed the matrix is almost pure metallic copper. Sometimes the metal completely fills the whole space between the pebbles, sometimes it is accompanied by calcspar, but much more frequently it is disseminated in fine particles through the coarse grained matrix. Sometimes a pebble is found quite saturated with copper, but it seems to have been of a more porous nature than the others and an amygdaloidal structure may be detected in it.

As above mentioned, a bed of sandstone underlies the conglomerate. It shews traces of stratification, is of a dark-red colour, and evidently consists of the same material as the conglomerate pebbles but in finer particles.

The trap which underlies this sandstone is amygdaloidal, but becomes more compact at a distance from the sandstone. In the adit which is being driven across the strata on the Quincy property, and which, so far as it has yet gone, is in the trap underlying the conglomerate, the rock much resembles the one first described as occurring on the road passing the Quincy mine. The grains of delessite are however smaller, seldom exceeding one tenth of an inch in diameter. An occasional crystal of feldspar is also observable in the fine grained mass of the rock. This mineral is in places reddish-grey, and in others greenish-grey, fuses readily to a colourless blebby glass and colours the blow-pipe flame strongly yellow. The sp. gr. of the rock is 2.89, and the colour of the powder light greenish-grey, but somewhat darker than that of the rock first described. It changes like that to a light brown on ignition, losing at the same time 2.77 p. c. On being treated with nitric acid and caustic potash the following substances are removed from it :

| | | |
|---|---|---|
| Silica | 12.41 | per cent. |
| Alumina | 5.96 | " |
| Peroxide of iron | 15.85 | " |
| Lime | 3.77 | " |
| Magnesia | 1.84 | " |

39.83 per cent.

These substances, together with the water lost on ignition, calculated in the same manner as in the case of the rock first described, for 100 parts give

| | |
|---|---|
| Silica | 29.52 |
| Alumina | 14.00 |
| Protoxide of iron | 33.47 |
| Lime | 8.80 |
| Magnesia | 4.29 |
| Water | 9.92 |

100.00

The residue from this treatment, which amounts to 57.17 per cent. of the original rock, on being digested in hydrochloric acid lost 6.7 p. c. additional, consisting of

| | |
|---|---|
| Alumina | 2.38 |
| Peroxide of iron | 2.45 |

Lime ................................. 1.57
Magnesia ........................... .30

The residue consisted of the same dark and light coloured parts as in the case of the rock first described. Calculated in the same manner as it, the mineralogical composition of this rock from the Quincy adit would be

Delessite ........................... 42.60
Labradorite ........................ 50.69
Pyroxene or hornblende.......... 5.62
Magnetite ........................... 1.09
—————
100.00

From the particulars above given, it would seem that the constituents of the traps of the Portage Lake district are principally feldspar of the labradorite species, and chlorite of a species allied to delessite, with which are found occasionally mica, small quantities of magnetite and perhaps of augite or hornblende. Similar results are given in Foster and Whitney's Lake Superior Report II, 87; but the relative proportions of the constituents are not given, nor is the peculiar nature of the chlorite referred to. The name of greenstone would seem altogether inapplicable to these rocks, because augite or hornblende only occurs in them occasionally if at all, and then in comparatively small quantity. As to the name of trap, the rocks previously so called have been by the best lithological authorities subdivided into two families, Melaphyre and Basalt.* The latter family which includes dolerite, anamesite and common basalt is distinguished by the dark, mostly black or greyish-black colour, the high specific gravity, and the richness in augite and magnetite of its rocks, and by the frequent occurrence in them of olivine and zeolites. The melaphyres on the other hand are characterised by their apparent want of augite, by their comparatively low specific gravity, by their colour of reddish-grey mixed with green and black, and their frequent development as amygdaloidal varieties; in which case quartz, calcspar and delessite fill the cavities more frequently than zeolites. The traps above described would seem to belong to the class of melaphyres, and to resemble especially those of Mansfeld described by Freiesleben, of Saxony,† and that of Faucogney described by Delesse.

* Naumann; Lehrbuchder Geognosie i, 599; Senft. Classification und Beschreibung der Felsarten, pp. 262 & 272.

† Geognostische Beschreibung des Konigreiches Sachsen ii, 447.

It is in the latter locality that the ferruginous chlorite, of which the analysis is quoted above, is found. It not only occurs in the amygdaloidal varieties of other localities, but, according to Naumann, it is also a constituent of many compact melaphyres. The following translation is from Naumann's Lehrbuch (I, 600) and is descriptive of the peculiarities of the melaphyres. It will be seen at once that it in every particular applies to the melaphyres of Portage Lake. "The principal characteristic of these rocks is " founded, on the one hand, on the decided nature of the felspa- " thic constituent, which when distinctly developed, has always " been recognized as labradorite, and on the other hand on the cir- " cumstance that pyroxene is very seldom present in recognizable " crystals, or grains, and usually cannot be determined mineralogi- " cally. The melaphyres generally appear as micro- or crypto- " crystalline rocks and only sometimes have arrived at a distinctly " granular developement. A third peculiarity is recognizable in " the tendency which these rocks have to the formation of air- " cavities and amygdaloidal structure, on which account the mela- " phyres are very frequently developed as amygdaloids or spilites. " In the amygdules, which sometimes reach a considerable size, " and then appear as geodes of varied constitution, the following " minerals are mostly found:—calcspar or brown-spar, and many " varieties of the species quartz (chalcedony, carnelian, jasper, " quartz, amethyst, agate) as also a mineral resembling chlorite or " green-earth which usually forms the periphery of the amygdules " like a shell or rind. *A similar, soft and green-coloured mineral* " is also often disseminated in the rock in grains and indistinct " crystals. The zeolites which are so frequent in the amygdaloidal " basalts, belong to the more rare occurrences in melaphyres pro- " perly so called. If we now add to these characters the com- " plete absence of quartz in the form of a rock constituent, the " predominating reddish-brown to reddish-grey colour of the mass " of the rock, which sometimes runs into greenish-grey, dark- " green and black, and the frequent occurrence of rubellan or " mica, we shall have tolerably exhausted the general petrographical " peculiarities of the melaphyres." Dr. T. Sterry Hunt, in his valuable paper on lithology, refers to this class of rocks as requiring a distinctive name, but he seems unwilling to adopt that of melaphyre. Since, however, Von Buch, Naumann and Senft *

* My objection to retaining the name of melaphyre is based upon the fact that these authors apply the name to different rocks. Brongnart, who invented it,

favor its adoption, and the science of lithology is already well stocked with terms of by no means general adoption, it would seem advisable to retain the word melaphyre to denote such rocks as those above described. With regard to the copper-bearing beds, the fusibility of the rock, and its transition in places into the neighbouring rock connects it distinctly with the melaphyres. This, together with the total absence of crystalline structure, and its apparently detrital character in places, would lead one to suppose that these beds are melaphyre tuffs, bearing the same relation to melaphyre, which volcanic tuffs bear to trachytes and basalts. The trap of the Portage Lake District might therefore be properly termed granular melaphyre when it is small-grained and crystalline; amygdaloidal melaphyre when cavities are present in a crystalline matrix; compact melaphyre when the rock is fine-grained and crystalline; and tufaceous melaphyre when the matrix is destitute of crystalline structure.

The rocks which occur to the eastward of the trap last described, I had no opportunity of examining minutely. They consist probably however of the same rocks as those above mentioned, alternating with each other for about one and a quarter miles, which is the distance across the strata from the conglomerate bed of the Albany and Boston property to the so called vein explored by the Isle Royale, and other mines.

About 260 feet west of the ‘Isle Royale Vein,’ the bed occurs upon which the Grand Portage mine is situated. The colour of the matrix is light-green, thus differing greatly from the beds hitherto described. It has an uneven earthy fracture, is non-crystalline, with small white spots here and there through it. It is fusible and gives water when heated in a glass tube. The amygdules are all of a dark-green colour, and frequently consists exclusively of delessite. Quite as frequently, however, they consist of that mineral, with a kernel of quartz, or much more seldom of calcspar. The copper is found oftener in the amygdules than in the matrix. As in the other beds larger aggregations of crystal-

gave it to black porphyries holding hornblende; Von Buch and d'Halloy use the name as synonymous with an augite-porphyry, while finally Naumann and Senft restrict the term to rocks which contain neither hornblende nor augite, and are not black in color, as the name melaphyre would imply. Hence I agree with Bernhard Cotta in rejecting the name, while admitting at the same time that some term is requisite to designate the important class of anothosite rocks in which a hydrous mineral (ferruginous chlorite) takes the place of hornblende or augite.—T. S. H.—(EDITOR'S NOTE.)

line minerals occur, in which quartz generally preponderates, associated with calcspar, prehnite and native copper.   Some specks of
native silver sometimes occur in this veinstone.   The strike of the
bed is N. 30° E., and the dip about 52° north-westward.

Between the Grand Portage and Isle Royale Veins the  trap is
of the usual character, reddish-grey coloured, with dark-green
grains and spots of delessite impregnating it.

The cupriferous bed of the Isle Royale mine is often of a dark-
chocolate colour similar to that of the Pewabic lode.   In other
places it has the character of the Portage lode, being light-green
coloured, non-crystalline and with an uneven fracture, but it is
comparatively free from amygdules.

Trap, as usual, underlies the Isle Royale Vein, and, with other
rocks, fills up the space between it and Mabb's vein which lies
about a mile to the south-eastward.   One of these is a conglomerate resembling that of the Albany and Boston mine, so far as the
nature of the pebbles is concerned.   The matrix is very porous,
and in coarse grains, which are in places cemented together by
quartz as well as calcspar.

Mabb's Vein, upon which mining has also been commenced by
the Isle Royale Co., has a matrix of a much more crystalline character than any of the cupriferous beds already described.   It is
of a dark-green colour, and is impregnated with grains and irregular spots (but not amygdules) of quartz, which is accompanied
by epidote and metallic copper.   Sometimes, however, an approach
to the light-green earthy rock of the Isle Royale vein is noticeable.

A short distance to the east of Mabb's vein another conglomerate bed is found.   The pebbles are porphyritic here also, but contain crystals of quartz as well as of felspar, and the paste is difficultly fusible before the blow-pipe, fine splinters of it only becoming
glazed.   The pebbles do not seem to be so well rounded as in the
other beds.

I had no opportunity of examining any of the rocks further
eastward, which form the base of the trap formation, but since those
already described form part of a series of strata having a vertical
thickness of about 10,000 feet, it may be supposed that they afford
good average specimens of the whole.

There is probably no one point, even in Europe, where within a
limited area, there are to be found such a number of mines, many
of them rich, well appointed and well managed ; such a display of
beautiful mining  machinery ; or such magnificent stamp-works as

are to be found within say five miles of the towns of Hancock and Houghton on Portage Lake. Even the professional visitor, who has given previous attention to the subject, cannot but be astonished as he rounds the point beneath these towns, and sails up to them, at the scene of life and activity which suddenly opens up before him. Having only spent ten days in the district, it would be impossible for me to attempt to describe with a moderate degree of minuteness even its principal mines. There are at least twelve mines in operation within a short distance of the lake, and of these the majority are producing copper in quantity varying from 20 to 120 tons of the pure metal monthly. The mines which have the largest production are those of the Pewabic lode, and it will be sufficient to refer briefly to their mining and dressing operations.

In exploring the cupriferous bed in the Quincy mine, as in following the other beds in the district, the miner has only its lithological character to guide him, there being no distinct joints or walls on either side. The shafts, levels and winzes of the mine are all opened within the bed so that the amount of *dead work* done is the very least possible. At the 100-fathom level the strike is N. 30° E., and the dip 70° north-westward. The shafts on the Quincy mine are from 200 to 300 feet apart, and the levels from 72 to 75 feet beneath each other on the incline of the bed, and 60 feet perpendicularly. The width of the bed is from 6 to 30 feet and the average thickness ten feet. According to the general experience at the mine, the thicker the bed the richer is the rock in copper. About two-thirds of the area of the bed is removed as remunerative; the other third, although it may contain some copper, is left standing, as unworthy of excavation. The amount of ingot copper yielded by the ground actually removed in 1864 was 562 lbs. per cubic fathom. Assuming the sp. gr. of the rock of the lode to be 2.7, it thus yielded 1.4 per cent. Of course the copper was unequally distributed through the bed rock, and the true per centage would be at many places above, and at others below that just mentioned. The bed is excavated by a very judicious combination of over-hand and under-hand stopping. The rock is removed to the shafts in waggons containing about one ton each, and hoisted in *skips* or waggons of a peculiar shape, running on tracks in the inclined shafts. The contrivance whereby these skips are emptied on their reaching the surface is without doubt the simplest and most beautiful anywhere in use. There are six shafts; the deepest, No. 4, is 660 feet vertically, and about 800 feet on the incline of the bed, below the surface. The pumps have a six-inch bore with a seven-inch column, but they only work three hours in twenty-four, so little is the mine troubled with water. On reaching the surface the bed-rock under-goes a sorting and about one-third is set aside as worthless. The other two-thirds are roasted in huge heaps much in the same manner as iron-stone. The object of this operation is to render the rock more easily pulverized. After roasting, the larger masses of copper are sorted out and sent directly to the furnace, where they yield about 60 per cent. The remainder is forwarded in waggons, on an inclined tram-way (where the full waggons in descending pull up the empty ones) to the stamp-work situated close to the lake, below the village of Hancock. Here Wayne's stamps, Shiermann's jiggers and ordinary Cornish buddles are employed in concentra-ting the ore. Each stamp weighs 900 lbs., and has 16 inches lift. The stamped rock passes through a sieve made of boiler plate, $\frac{1}{4}$ inch thick. The holes are $\frac{1}{4}$ inch in diameter, and have a slight diminishing taper towards the stamps. The latter are stopped every eleven hours in order that the larger pieces of copper may be removed from the stamp-box. The stamped ore is discharged into a shallow run which has an inclination of a half inch in a foot. From this it comes on to a sieve which is constantly in mo-tion, has $\frac{1}{8}$ inch holes, and separates it into coarse and fine work for the jigger. The fine work in passing down into the jigging sieve meets an upward current of water which carries away the slimes from it. The jigging machine, in which the sieve is stationary, apparently cleans the ore very effectually. A sample of the coarse ragging from it was given me which assayed 98.6 per cent., while the *skimpings* or refuse contained only 0.6 per cent. The fine *ragging* from the same machine assayed 89.3 per cent. and the refuse 0.73 p. c. The product from washing the finer stuff on the buddles assayed 78.6 per cent. while the *tailings* from the same operation gave 0.46 per cent. The whole of the refuse products of the stamp-work are, however, passed through an adjoining building, and some part of them worked over. The yield of the rock treated in the 'stamp-work was, during 1864, 2.96 per cent. I make no attempt to describe the magnificent machinery of the Pewabic and Franklin stamp-works where Ball's patent stamps and washers are employed. To judge, however, from the per-centage of copper in the refuse products, the work is not so well done here as in the Quincy stamp-works. With the permission of the superintendent of the Franklin stamp-work, I took several samples from various parts of the run-house, and from the waste heap outside, which assayed as follows:

| | | |
|---|---|---|
| From head of run........................ | 4.93 per cent. | |
| " middle of do ...................... | 3. | " |
| " end of do ...................... | 3.13 | " |
| " a heap immediately outside of run house ............................ | 0.66 | " |
| " sand bank............................ | 1.00 | " |

When it is recollected that the yield of the rock treated in the Franklin stamp-work is only 1.69 per cent. the loss in the refuse products would appear to be very large. At the stamp-works of the Albany and Boston Mining Co., Gates's revolving stamps and Collom's jiggers are employed. This is also the case at the Huron stamp-work. (The Huron mine is on the Isle Royale bed.) It appears to be as yet uncertain as to which system of dressing is the most advantageous, but in view of the experience which is being acquired in the district almost daily, this cannot long remain a matter of doubt. It is, however, singular that in a district where such an enormous amount of capital is invested in mines and stamp-works, there should be no provision made for testing accurately, by the wet process, the various refuse and other products of the ore-dressing operations. It would seem difficult without such means, to come to a reliable result as to which method of concentration is the best.

The system of dividing the lands into small sections seems to have contributed not a little to the rapid developement of the mines of the Portage Lake district. The sections contain one square mile of 640 acres, and each of these is subdivided into four quarters. Some of the best of the mines have no more length of lode to work upon than may be contained in a quarter section. As a consequence, the attention and energies of the mining companies, and their managers, are, on the discovery of a cupriferous bed, at once turned to exploring and mining in depth. The opposite system, which prevails on the north shore of the lake, of having very large mining locations is as detrimental to the progress of the country as it is to the interests of the owners. The explorations are carried on over too great an area, they are desultory, are not easily superintended, and seldom yield any definite result.

In concluding this paper, I venture to hope that some of the facts which it relates concerning the mines of Portage Lake will be found useful in detecting the presence of remunerative cupriferous beds on the Canadian shore of the lake.    The existence of such there can scarcely be doubted, and it is equally certain that if the same energy, intelligence and capital were employed in their developement as on those of Portage Lake, the north shore, now a wilderness, would soon become studded with towns as flourishing and populous as those which now ornament the south shore.

Acton Vale, C. E., January 3, 1866.

# NATURAL HISTORY SOCIETY.
## MONTHLY MEETINGS.

At the first monthly meeting convened at the rooms of the Society on Monday evening September 25, and at the second held Monday evening October 30, only routine business was done. The following donations were announced :—

### TO THE LIBRARY.

The Statutes of Canada, for 1865 : from the Provincial Government.

Journal of Education, L. C. ; from the Superintendent.

United States Coast Survey Report; from the Superintendent.

Report of the Smithsonian Institute ; from the Director.

Statistics of U. S. Commerce ; from Secretary Chase.

Notes on *Selandria Cerasi* ; from Prof. Winchell.

Report on the Geological Survey of the Province of Canterbury, by Julius Haast, F. G. S. ; from the author.

Animals of N. A., by H. B. Small, (2 copies) ; from the author.

Journal of Prison Discipline, Philadelphia.

Diagnosis of new Gasteropods, by Dr. Stimpson ; from the author.

Report of the Northern Home for friendless children, Philadelphia.

Calendar of the University of St. Andrews, Scotland.

Pre-Historic Man, by Dr. Wilson ; from the author.

Descriptions of new fossils, by Prof. Winchell ; from the author.

Pennsylvania School Report for 1865.

Report on the Geology of New Brunswick; from Prof. Hind.

Défenses des Colonies, par Joachim Barrande; from the author.

*And in exchange for the Canadian Naturalist.*

Journal of the Society of Arts, London.

Geological Magazine, London.

Quarterly Journal of Science, London.

Journal of the Geological Society, London.

Technologist, London.

Popular Science Review, London.

Journal of the Board of Arts for U. C.

Transactions of the Lit. and Hist. Society of Quebec.

Journal of the Franklin Institute, Philadelphia, Pa.

Proceedings of the Academy of Sciences, Philadelphia, Pa.

Proceedings of the Essex Institute, Salem, Mass.

Silliman's Journal, New Haven, Conn.

Annals of the Lyceum of Nat. Hist., New York.

Proceedings of the Society of Nat. History, Boston, Mass.

---

The third monthly meeting was held Monday evening November 27; the President Dr. Smallwood in the chair.

The following donations were announced, and the Society's thanks voted to the donors:—

### TO THE MUSEUM.

A young specimen of the white variety of the Canadian Deer (*Cervus Virginianus*) from Mr. W. S. Macfarlane; Sword, Powder-horn and Pouch, made by the Mandingoes, from Sierra Leone, from Commissary General Winter; Stone Hatchet, &c., found in New Jersey, from Mr. J. M. Brown; White-footed Mouse, (*Mus leucopus*, Raff.), from the Cabinet Keeper.

### NEW MEMBERS.

Dr. Daniel Wilson, Toronto, and Mr. Westwood, Professor of Zoology, University of Oxford, were elected honorary members; Mr. G. F. Angas, of London, a corresponding member; and Messrs. Thomas Watson and Thomas Robinson, ordinary members.

### PROCEEDINGS.

Mr. Alfred Rimmer read a paper on certain proposed alterations of the Game Laws. A discussion ensuing, the subject was referred to a Committee consisting of Messrs. Drummond, Rimmer and Watt, when the meeting adjourned.

---

The fourth monthly meeting was held at the society's rooms on Monday evening, December 18; the President, Dr. Smallwood, in the chair.

The following donations were announced and thanks voted to the donors :—

### TO THE MUSEUM.

A fine specimen of the American deer (*Cervus Virginianus*), from Mr. W. S. Macfarlane; seven specimens of Central American birds from Mr. Haig, through Mr. Leeming; specimen of a South American turtle-dove from Mr. Struthers; nine specimens of Devonian fossil fishes from Orkney, Scotland, from Mr. Barnston.

### PROCEEDINGS.

A paper on the natural history of *Sanguinaria Canadensis* or Canada blood-root, by Dr. Gibb, of London, was read by the Secretary.

Principal Dawson afterwards exhibited a number of specimens of flint implements and fossils from St. Acheul, near Amiens, and made some observations on the mode of their occurrence in the 'high level gravel,' in the valley of the Somme. He referred to the investigations of Boucher-de-Perthese, Lyell, and Prestwich, and quoted a portion of the description of the locality by the latter geologist. He stated that he had come to the following conclusions, derived from an examination of the locality and of the specimens, more especially those in the collection of Mr. Prestwich:

1. The implements cannot be considered so much as characteristic of a particular age as of particular work. They are not spears, or arrows, or hatchets, but picks and diggers, adapted for digging in the earth, or hollowing wooden canoes. A consideration of the implements of the American stone age renders it in the highest degree improbable that the makers of these tools did not possess also stone arrows, spears, knives, and other implements. The application of the idea of an older and ruder stone age to such implements is gratuitous, and contradicted by the evidence afforded by American antiquities.

2. There are some reasons which induce the belief that these implements have been used in burrowing small horizontal adits into the gravel beds of St. Acheul, in search of flints. In this case they may not be of great antiquity, though certainly older than the Roman occupation of Gaul.

3. They may have been deposited with the gravel. In this case they belong historically to a very ancient period, though geologically modern; and at the time when they were so deposited the climate of France must have been more severe than at present, its level different, its surface covered with dense forests, inhabited by several great quadrupeds now extinct, and the River Somme must have been much larger than at present, and must have spread its waters over a wide plain, in which the St. Acheul gravel constituted a bank or point, inundated in times of flood, and perhaps resorted to by the aborigines as a place for making canoes.

4. Before either of the two theories above stated can be finally accepted, much more thorough investigations must be made, and also careful topographical surveys of the whole district. In event of the view last mentioned being sustained, the question of the absolute time required will still be difficult to determine, since the causes of erosion and deposition in operation at the period in question must have been very dissimilar from those now in action; and other unknown causes, whether sudden or gradual in their operation, must have intervened to produce the present state of the country. In this case, however, there would be a strong probability that the *Rhinoceros tichorhinus* and the Mammoth had continued to exist in Europe down to the period of the implement making.

It is much to be desired that a series of systematic excavations in these gravels, and a geological and topographical survey of the whole basin of the Somme should be undertaken by some scientific body in France or England, as it may require many years to enable individual explorers to obtain the data required to settle the questions that have been raised in connection with these deposits.

The society's thanks were voted to Dr. Gibb and to Dr. Dawson. and the meeting thereafter adjourned.

---

The fifth monthly meeting of the Society was held Monday evening, January 29; The President in the chair.

### NEW MEMBER.

Mr. Alexander Agassiz, of Cambridge, U. S., was elected a corresponding member.

### PROCEEDINGS.

It was resolved to hold the Annual Conversazione on Thursday evening, March 1, and a committee was appointed to make the necessary arrangements.

Dr. Dawson moved the adoption of the following new by-law (of which he had given due notice) which was unanimously carried :—

" That ordinary members not resident in Montreal shall be required to pay an annual subscription of $3, and shall be entitled to received the *Canadian Naturalist* for each year; the said contribution to be paid in advance, and such members to be designated non-resident ordinary members."

Mr. Rimmer made some remarks on the proposed amendments to the Game Laws and read the draft of a report. His views had not the support of the committee and the discussion was therefore adjourned till next meeting.

Mr. H. G. Vennor presented a catalogue of the birds noted on the Great Manitoulin Islands, and accompanied it with a few observations on its physical features. Having given a brief topographical description of the Island and a sketch of its geology, some of the silicified fossils of the Clinton group from the neighborhood of Lake Manitou were exhibited; also photographs of glacial groovings and scratchings on rocks on the south shore of the island. The following are extracts from the notes then read :

" From the village of Manitouaning, a fair portage road or trail leads off to the first and largest lake on the Island, Lake Manitou, or the Lake of the Great Spirit. The portage is about three miles in length and runs through fine open woods, comparatively free from under-brush. For the information of any who may hereafter visit the Great Manitoulin, I may state that no canoes are to be had on any of the interior lakes of the island, and that it is not unusual to paddle for days on these, without even meeting with an Indian family. Consequently all canoes and Indians required have to be procured either at Little Current or Manitouaning.  *  *  *  *  *  Manitouaning Bay is ten miles long, and reaches to within two and one-half miles of South Bay, on the South side of the Island, thus nearly cutting off the unceded portion of the Island.

" The waters of Lake Manitou are beautifully clear, and abound in fine fish—such as Black-bass, Salmon and Brook-trout, White-fish, and Perch.

" At the extreme Western end of this lake the Indians cross by a portage to another large lake called ' Mindemooya' or ' Old Woman's Lake'; here canoes have also to be portaged.

" The whole of this portage is strewn over with very fine Clinton fossils. The cliffs around this lake lie at some distance from the shores, so that we were not much surprised at finding a belt of good and well timbered land, between these cliffs and the shores. On such land we noticed large crops of corn and potatoes. From the middle of the lake rises Mindemooya Island, which is said to be much infested by snakes. Farther westward we have another large lake called Kagaweng, and numerous smaller ones generally distributed over the island.

" Oil wells were being successfully worked at Wequemakong by the Great Manitoulin Oil Company. The oil from this locality is of the finest description. An office has been opened in Montreal in connection with this Company.

" On the interior lakes the bald-eagle and fish-hawk were very numerous; the former bird apparently living by the toiling of the latter species. Ruffed-grouse, Spruce-partridge and Wild-pigeons were very numerous all through the interior of the island. The islands in the lakes swarmed with the Silvery and Black-backed gulls, while the waters resounded with the cries of the Loon. The Whip-poor-will might always be heard along the rocky shores and particularly near the mouth of rivers."

On the whole, the reader remarked that the Great Manitoulin presented many advantages to the settler; for although perhaps one third of the island was of a rocky and consequently barren character, the remaining two-thirds contained land of the finest description, covered at present either by Indian crops, or splendid hard-wood forests, which last yielded large quantities of maple sugar—generally at the rate of 1,000 lbs per acre. Mr. Vennor concluded by expressing a hope that ere long we might be able to hear of this great Manitoulin Island as being the home of the white settler, where he might be seen surrounded by waving fields of grain, and possessing not only the comforts, but also the luxuries of life.

----

The sixth monthly meeting of the Society was held at the rooms of the Society, on Monday evening, February 26; the President, Dr. Smallwood, in the chair.

PROCEEDINGS.

" The Committee on the Game Laws submitted the following Report :—

The Committee on the Game Laws has the honour to report the following recommendations :

1. That all game legislation be consolidated into one general act.

2. That the following be the close-terms for the whole Province.

Woodcock and Snipe;—March 1, to August 1.

Ptarmigan and all kinds of Duck ;—March 1, to September 1.

Deer of all kinds ;—February 1, to September 1.

Turkey, Pheasant, Partridge, and Grouse of all kinds ;—February 1, to September 1.

Quail ;—February 1, to October 1.

Fur bearing animals ;—April 1, to November 1.

Your Committee does not consider these dates to be absolutely the best, but rather as compromise close-terms such as would probably unite different interests.

3. That egging and bird-nesting be prohibited, save on the North-shore east of the Saguenay, and on the Islands of the Gulf, where it shall be legal up to June 1 as at present.

4. That there should be no close-term for birds within these limits.   [Except for Eider-ducks.]

5. That this Report be sent to the Fish and Game Club with a view to a joint effort being made to procure the necessary legislation.

The Committee is of opinion that no action is needed in the matter of fish, inasmuch as the administration of the Fisheries Department has been judicious, and the operation of the new Fishery Act (in itself greatly in advance of similar enactments in the mother country) promises to be on the whole satisfactory.

Respectfully submitted.

GEORGE A. DRUMMOND.<br>DAVID A. P. WATT.

The Report having been received was thereafter unanimously adopted, excepting the last clause relating to fish, which was reserved for discussion.

Mr. A. RIMMER believed that fishing by means of fixed-engines should be made illegal; and contended that all such were destructive of fish and ruinous to salmon grounds.   Since they had been suppressed in England, the yield of salmon had been increased immensely.   He remarked on the demoralizing effects of such nets, killing and maiming the fish by night and by day ; and asserted that the destruction of salmon in Upper Canada was owing to these nets, as the fish were thereby driven off their proper breeding

grounds. Formerly salmon abounded in the rivers to the west of
Montreal, and formed a staple article of food for the inhabitants;
but they had long since ceased to exist, and for many years none
had been seen. One solitary fish found its way to the St. Regis
river last season, but the Indians who killed it were unable to tell
its name and looked on it as a sort of *lusus naturæ*. He objected
to any fixed obstacles being placed in the way of fish going to their
spawning grounds, and said that since these had been abolished in
England, salmon could there be purchased cheaper than in Canada.
As to using seines for catching fish, they were used in England,
and our Canadian rivers were much better adapted to their use.
One river that he knew, the Jacques Cartier, in which salmon had
been exterminated, now abounded with these fish, the result of care
and of allowing a free passage to the spawning ground; the Murray
river too formerly abounded with salmon, but they had been exterminated by brush-weirs, and now a single fish was the season's
catch. The owners of brush-weirs at Murray Bay had told him
that formerly they took herrings by means of them in such abundance that they had to use them for manure; while now they got
very few herrings or fish of any kind, a result not to be wondered
at as he had found these weirs full of herring-fry and other small
fish; in one brush-weir upwards of five thousand smolts had been
killed in one tide.

Mr. DRUMMOND maintained that the only question at issue was
how to catch for the market, at the smallest expense, the greatest
weight of salmon, making sure to leave in the rivers, as well an ample
supply for keeping up the breed as all the immature fish. He argued
that these ends could most easily be attained by means of fixed-
engines in the salt-water (where seining was practically out of the
question), and had in fact, to a considerable extent, been already
attained by the Canadian nets hitherto used, inasmuch as the
numbers of fish in our salmon-rivers had of late years vastly
increased. He asserted that the British modes of fishing were
much more destructive than the Canadian, and quoted statements
to prove that salmon had not increased in the United Kingdom
under recent legislation and that they were very much dearer there
than here.

Dr. DAWSON said that the chief objection which he saw could
be urged against brush-weirs was their inefficiency; they captured
too few fish, and were rude clumsy implements which fish soon
learned to avoid. He thought a good deal of misapprehension

existed as to the kinds of fish caught in them, his observations led him to believe that no salmon- or herring-fry nor other immature fish were taken by them; at least he had never seen such though he had examined several weirs.

MR. WATT stated that the Fisheries Act left the Commissioner of Crown Lands free to allow or to disallow any sort of net or combination of nets, and that he and his subordinates might be supposed to understand their own business better than amateurs and to have the interests of the fisheries as much at heart. He said that so far from fixed-engines being abolished in Britain it was perfectly lawful to use them even in fresh-water and for salmon, and quoted official advertisements approved at the Home Office in January last, containing regulations for the guidance of salmon fishermen using stake-nets, bag-nets, stake-weirs and fly-nets, authorizing meshes much smaller than ours, and netting five weeks later; he averred that the modes of salmon fishing pursued in Britain were much more destructive than that pursued here, and would, owing (among other causes) to the different physical conformation of Lower Canada, empty our rivers in a few years if practised by us. He denied that the salmon nets now in use were in any way responsible for the evils complained of. His observations on brush-weirs coincided with those of Dr. Dawson. Having examined many such he had found neither smolts nor immature fish of any kind; their contents consisted chiefly of tomcods, sand-launce, caplin, sardines, and smelts—some of which fish had often been confounded with salmon-fry. As regulated by the Act, Mr. Watt considered these weirs should be harmless enough modes of fishing.

MR. A. MURRAY (President of the Game Protection Club) said that as this matter had been taken up by the Society, it was important that its decision should be a correct one and based on a sufficient knowledge of the subject. In the Game Protection Societies of Montreal and Quebec the opinion was almost unanimous against fishing by means of fixed-engines. He had with him a number of authorities on the subject and was prepared to enter upon it, but as the discussion was not likely to be a short one, he preferred to adopt the suggestion already thrown out and allow the matter to lie over until next meeting. The report of the Montreal Fish and Game Club would be issued in course of a week or two; it would discuss the subject at some length and he would see that a copy of it was placed in the hands of each member of the Society.

Further discussion was accordingly adjourned.

' Mr. J. F. Whiteaves then made a communication " On certain new additions to the Society's museum."

He remarked that the few statements which he had been requested to make would refer only to the collection he had brought from England during the summer of 1865, and that he did not wish that any remarks he might offer concerning the specimens should be looked upon as the result of original investigation, or that they had any claim to novelty.

The following is a list of the donations in question, which have not previously been recorded :—

| | |
|---|---|
| Prof. ROLLESTON, Oxford University. | Skin of the grey headed kalong or flying-fox (*Pteropus poliocephalus*, Temminck). |
| | Cast of the head of the dodo, from the specimen in the Oxford University Museum. |
| | Two cuttle-fishes, (*Loligo vulgaris*), in spirits. |
| From the late Rev. F. W. HOPE, through Professor WESTWOOD, Oxford University. | Three cases of crustaceans from the Mediterranean (mostly brachyurous decapods) consisting of forty-five specimens, of twenty-six species. |
| | Two cases of exotic insects, mostly coleoptera, some of them from Central Africa, as follows : |

Coleoptera,     84 species.
Hymenoptera,   1     "
Orthoptera,     4     "
Hemiptera,     15     "

| | |
|---|---|
| MR. G. F. ANGAS, London. | Seven species of shells, two of bryozoa, three of annelida, three of echinodermata, four of corals, and four of sponges; all from Southern Australia. |
| | One lepas from California. |
| Prof. TENNANT, King's College, London. | Six species of fossils from the Upper Chalk of Gravesend, Kent. |

| | |
|---|---|
| Mr. Jas. Parker, Jr., Oxford. | Six species of fossils from the English Upper Silurian, and five from the Purbeck beds of Dorsetshire. |
| Mr. W. E. Jameson, London. | Fossils from the Oxfordshire oolites, the greensand of Farringdon (Berkshire) and from the Norfolk Crag; in all eight species. |
| Mr. R. S. Standen, London. | Fossils from the Great Oolite of Minchinhampton, Gloucestershire; and from the Inferior Oolite of the neighbourhood of Cheltenham. Altogether ten species. |
| Mr. J. F. Whiteaves. | Two specimens of the Sagouin, (*Jacchus vulgaris* Geoffroy St. Hilaire, *Hapales jacchus* Illiger,) from Brazil. |

One skin of the Malabar squirrel; (*Sciurus maximus*).

Six species of exotic shells.

Five species of Echinodermata.

One coral.

A fine specimen of the *Balanus tulipa*, from Australia.

A number of European fossils, including a series of fishes from the Old Red Sandstone of Scotland: the Carboniferous deposits of Staffordshire, &c.: the Permian of Durham: the Lias at Lyme Regis: the Oolites of Oxfordshire: the Chalk of Kent: the Eocene of Monte Bolca, near Verona: and the Crag of Norfolk.

Estimate of this collection.

| | | |
|---|---|---|
| Upper Silurian, | 11 | species. |
| Devonian, | 1 | " |
| Carboniferous, | 14 | " |
| Permian, | 1 | " |
| Lias, | 12 | " |
| Oolites, | 59 | " |
| Chalk, | 27 | " |
| Tertiary, | 40 | " |
| Post-tertiary, | 1 | " |
| | | |
| Altogether | 166 | species. |

From Principal DAWSON, (in exchange for duplicate specimens brought from England).

Fourteen species of Echinodermata from Norway.

A series of Tertiary fossils, consisting of forty-one species from the Eocene and Miocene of Paris; of eight species from the Eocene, Miocene, and Pliocene of the United States ; and five from the English Pliocene.

Specimen of *Dictyonema Websteri*, from the Upper Silurian shales of Nova Scotia.

Mr. Whiteaves said :— In the few remarks which I propose making on these specimens, I shall adopt the ordinary zoological classification.

A pair of specimens of the Sagouin, *Jacchus vulgaris* of Geoffroy St. Hilaire, *Hapales jacchus* of Illiger, were exhibited. They were stated to belong to the order Quadrumana, a group which includes the Baboons, the Apes, the Monkeys generally, and the Lemurs. The Sagouin is one of the American or Platyrhine monkeys, a group peculiar to the New World, and one which is characterized by the flatness and broadness of the nose, and the width of its septum, which makes the nostrils appear far apart from each other on each side of the nose. The species in question has received several popular names. It is the Sagouin or Sanglin of Edwards and of other authors; the Ouistiti of Buffon and of French naturalists; the striated monkey of Pennant; while by some it is called loosely the Marmoset. It is a small species, not much larger than some squirrels, and is very squirrel like in its habits. It inhabits the forests of Guiana and Brazil, to some extent is omnivorous in its habits, but its favourite food, in a wild state, is said to be the banana. It has two tufts of hair round the ears, its tail is long but not prehensile.

The grey-headed flying-fox, (*Pteropus poliocephalus*) belongs to the order Cheiroptera, which includes the Bats, the Vampyres, &c. The ordinary bats are for the most part insectivorous in their habits, while the flying-foxes, from the blunt tubercular crowns of their molars, were supposed to be essentially frugivorous. All the members of the order, however, are more or less omnivorous, and it was found that the Pteropus in confinement fed readily on the flesh of birds. They derive their name of flying-foxes from the resemblance of the head to that of a fox. Their jaws are more elongated than are those of the bats and vampyres. Probably the idea of the harpy was derived from animals of this order, and it has been thought likely by some writers that the bat of the Bible was a species of Pteropus.

The Malabar squirrel (*Sciurus Maximus*) is a true squirrel and belongs to the genus Sciurus as restricted by modern zoological writers. It inhabits the Malabar coast, and is chiefly remarkable for the peculiar colouring of its fur. It is said to prefer living among palm trees, and to be very fond of the milky juice of the cocoa-nut, as well as of the more solid part of the fruit.

A cast of the skull of the Dodo was exhibited, taken from the specimen in the Oxford University Museum. The species, of which only a few fragments of the skeleton, &c., are preserved, formerly inhabited the Mauritius, and is supposed to have been extinct for about 200 years. Considerable discussion has taken place amongst naturalists as to its supposed affinities; some have thought that it should be classed in the order Raptores, and placed near the Vultures; others again have regarded it as belonging to the group Cursores, on account of the rudimentary character of its wings. Messrs. Strickland and Melville, in a comparatively recent treatise on this bird, have placed it among the pigeons, and consider that its nearest living ally is the Didunculus of the Navigator's Islands, a bird which, however, can fly tolerably well. Bones of three other species of large wingless birds from the Island of Rodriguez, an island east of those of Bourbon and of the Mauritius, are in the possession of the Zoological Society of London. As these last three birds, and the Dodo, could hardly pass from one island to another, being provided with rudimentary wings only, it has been supposed by some naturalists, that the islands of Bourbon, of the Mauritius, and of Rodriguez, at one time formed part of a great continent, which is now submerged beneath the waves of the Indian Ocean.

Two letters were read from Mr. G. A. Rowell, the Assistant Curator of the Oxford University Museum, in which a contribution of skins of mammals and birds was promised by the professors of geology and zoology, in the spring of 1866.

Several species of South Australian Mollusca have been presented by Mr. G. F. Angas, and some miscellaneous exotic species by Mr. Whiteaves. One of the S. Australian shells is a Solemya (*S. Australis*) interesting as closely resembling a species (*S. velum*, Say,) found on the Atlantic coast of the United States.

Two species (four specimens) of Bryozoa have also been received from Mr. Angas, who collected them in S. Australia, two of them belonging to the genus Retepora. The difference between the hardparts of a bryozoon and those of a true coral was explained ; and it was shewn that the stony cells of bryozoa are destitute of the radiating calcareous partitions usually seen in the cells of corals.

An interesting named series of crustaceans from the Mediterranean, has been received from Mr. Westwood, the Professor of Zoology in the University of Oxford. They formed a part of the fine collection presented by the late Rev. F. W. Hope to that University. All of them belong to the order Decapoda, in which order all the stalk-eyed crustacea of which it is composed have the whole of the thoracic segments united, with the head, into a single mass, " incased in a common shell, with no traces of segmentary division." Their branchial organs are inclosed within a cavity on each side of the cephalo-thorax, and their true thoracic legs are nearly always ten in number, whence the name of the order. One of the species, *Scyllarus Arctus*, belongs to the macrurous or long-tailed division of the Decapods, a division to which Shrimps, Prawns and Lobsters belong. The remainder of the twenty-six species are brachyurous or short-tailed decapods, and are mostly peculiar kinds of crab.

A beautiful series of exotic insects has been presented by Prof. Westwood. Among the most noticeable of the beetles are seven species from Tropical Africa, collected by some of the members of Dr. Livingstone's expedition. Of these, *Texius Megerlei* is a fine large carnivorous ground beetle, belonging to the family Carabidæ. A fine pair of the rare *Dynastes taurus* has been received, a genus which is allied to the well known Hercules beetle of Brazil, and belongs to the family Dynastidæ of the lamellicornes. Other examples of the lamellicorn beetles from Tropical Africa are a pair of the large Rhinoceros beetle, *Oryctes boas*, and of *Copris gigas*, an insect not very dissimilar to the sacred beetle (*Ateuchus sacer*) of the Egyptians. The *Gnathocera Iris*, a brilliant green cockchafer, and the *G. suturalis*, another cockchafer with black longitudinal stripes on a light olive green ground, are also representatives of the lamellicornes of Tropical Africa. There remain two specimens of a Calandra, a large and curious weevil, also *Tetrognatha gigas*, from the same country, which is a large longicorn species. Attention was called to a series of Buprestidæ,

from India, New Holland, Brazil, &c. These beetles surpass all
others of their class in the beauty of their metallic colouring, and
are used at the present day as jewelry. Other curious forms are
the Goliath beetles, *Goliathus (Ceratorrhina) guttata* and *G. aurata*,
a pair of each of which, from Cape Palmas, have been received
from Prof. Westwood. The two species indicated are not however among the larger forms of the group, but are remarkable for
beauty of colour. These insects, like the crustaceans, were part of
the Rev. F. W. Hope's collection, presented by him to the University of Oxford. Mr. Angas has kindly presented a series of
annelida, echinodermata, corals and sponges, from S. Australia;
Mr. Whiteaves, several interesting exotic echinodermata and corals;
and Principal Dawson a collection of Norwegian echinodermata.
The Society's collection of fossils previously consisted of a little
more than 300 species, and was very deficient in fossil fishes. Pains
have been taken to supply this deficiency, and with some success,
twenty-six species, from rocks of various ages, having been added to
the collection. The latest classification of recent fishes was briefly
explained, and specimens of fossil fishes, from Palæozoic, Mesozoic and tertiary rocks, were exhibited, and their affinities described. It was shewn that the Palæozoic fishes in point of organization, belong to a very high order among fishes, a fact which
by Hugh Miller and others has been thought to militate against
Mr. Darwin's views as to the origin of species. Some of the
Palæozoic fishes have many reptilian characteristics. Throughout
the Palæozoic and in the older mesozoic age, ganoids, and sharks
(selachians with placoid scales) were the dominant race of fishes,
and true bony fishes (teleosts), which are the prevalent forms now
in existence, do not date farther back than the cretaceous period. A
number of miscellaneous European fossils were exhibited, and some
of the more interesting were explained verbally somewhat in detail.
It was stated that about 250 species had been added to the Society's
collection of fossils, the result of last summer's collecting in England.

A special vote of thanks was unanimously voted by the Society
to each of the donors of the specimens referred to, also a vote of
thanks to Mr. Whiteaves for his zeal in collecting.

# COMPARISONS OF THE ICEBERGS OF BELLE-ISLE
## WITH THE GLACIERS OF MONT BLANC,
## WITH REFERENCE TO THE BOULDER-CLAY OF CANADA.

### By J. W. Dawson, LL.D., F.R.S., F.G.S., Principal of McGill College.

The snow-clad hills of Greenland send down to the sea great glaciers, which in the bays and fiords of that inhospitable region, form at their extremities huge cliffs of everlasting ice, and annually ' calve,' as the seamen say, or give off a great progeny of ice islands which, slowly drifted to the southward by the Arctic current, pass along the American coast, diffusing a cold and bleak atmosphere, until they melt in the warm waters of the Gulf stream. Many of these bergs enter the Straits of Belle-Isle, for the Arctic current clings closely to the coast, and a part of it seems to be deflected into the Gulf of St. Lawrence through this passage, carrying with it many large bergs.

Mr. Vaughan, late superintendent of the Light-house at Belle-Isle, has kept a register of icebergs for several years. He states that for ten which enter the straits, fifty drift to the southward, and that most of those which enter pass inward on the north side of the island, drift toward the western end of the straits, and then pass out on the south of the island, so that the straits seem to be merely a sort of eddy in the course of the bergs. The number in the straits varies much in different seasons of the year. The greatest number are seen in spring, especially in May and June ; and toward autumn and in the winter very few remain. Those which remain until autumn, are reduced to mere skeletons ; but if they survive until winter, they again grow in dimensions, owing to the accumulations upon them of snow and new ice. Those that we saw early in July were large and massive in their proportions. The few that remained when we returned in September, were smaller in size and cut into fantastic and toppling pinnacles. Vaughan records that on the 30th of May, 1858, he counted in the Straits of Belle-Isle 496 bergs, the least of them sixty feet in height, some of them half a mile long and two hundred feet high. Only one-eighth of the volume of floating ice appears above water, and many of these great bergs may thus touch the ground in a depth of thirty fathoms or more, so that if we imagine four hundred of them moving up and down under the influence of the current, oscillating slowly with the motion of the sea, and grinding on the rocks and stone-covered bottom at all depths from the centre of the channel, we may form some conception of the effects of these huge polishers of the sea-floor.

Of the bergs which pass outside of the straits, many ground on the banks off Belle-Isle. Vaughan has seen a hundred large bergs aground at one time on the banks, and they ground on various parts of the banks of Newfoundland, and all along the coast of that island. As they are borne by the deep-seated cold current, and are scarcely at all affected by the wind, they move somewhat uniformly in a direction from N. E. to S. W., and when they touch the bottom the striation or grooving which they produce must be in that direction.

In passing through the straits in July, we saw a great number of bergs, some were low and flat-topped with perpendicular sides, others were concave or roof-shaped like great tents pitched on the sea ; others were rounded in outline or rose into towers and pinnacles. Most of them were of a pure dead white like loaf sugar, shaded with pale bluish green in the great rents and recent fractures. One of them seemed as if it had grounded and then overturned, presenting a flat and scored surface covered with sand and earthy matter.

At present we wish to regard the icebergs of Belle-Isle in their character of geological agents. Viewed in this aspect, they are in the first place parts of the cosmical arrangements for equalizing temperature, and for dispersing the great accumulations of ice in the Arctic regions, which might otherwise unsettle the climatic and even the static equilibrium of our globe, as they are believed by some imaginative physicists and geologists to have done in the so-called glacial period. If the ice islands in the Atlantic, like lumps of ice in a pitcher of water, chill our climate in spring, they are at the same time agents in preventing a still more serious secular chilling which might result from the growth without limit of the Arctic snow and ice. They are also constantly employed in wearing down the Arctic land, and aided by the great northern current from Davis's Straits, in scattering its debris of stones, boulders and sand over the banks along the American coast. Incidentally to this work, they smooth and level the higher parts of the sea bottom, and mark it with furrows and striæ indicative of the direction of their own motion.

When we examine a chart of the American coast, and observe the deep channel and hollow submarine valleys of the Arctic current, and the sand-banks which extend parallel to this channel from the great bank of Newfoundland to Cape Cod, we cannot avoid the conclusion that the Arctic current and its ice have great power both of excavation and deposition.   On the one hand, deep hollows are cut out where the current flows over the bottom, and on the other, great banks are heaped up where the ice thaws and the force of the current is abated.   I have been much struck with the worn and abraded appearance of stones and dead shells taken up from the banks off the American coast, and am convinced that an erosive power comparable to that of a river carrying sand over its bed, and materially aided by the grinding action of ice, is constantly in action under the waters of the Arctic current. The unequal pressure resulting from this deposition and abrasion, is not improbably connected with the slight earthquakes experienced in Eastern America, and also with the slow depression of the coast ; and if we go back to that earliest of all geological periods when the Laurentian rocks of Sir Wm. Logan, constituting the Labrador Coast and the Laurentide Hills, were alone above water, we may even attribute in no small degree to the Arctic current of that old time the heaping up of those thousands of feet of deposits which now constitute the great range of the Alleghany and Appalachian mountains, and form the breast-bone of the American continent.

But such large speculations might soon carry us far from Belle-Isle, and to bring us back to the American coast and to the domain of common things, we may note that a vast variety of marine life exists in the cold waters of the Arctic current, and that this is one of the reasons of the great and valuable fisheries of Labrador, Newfoundland and Nova Scotia, regions in which the sea thus becomes the harvest field of much of the human population.   On the Arctic current and its ice also floats to the southward the game of the sealers of St. John and the whalers of Gaspé.   The distance that some of these creatures come, is shown by the fact that I once found upon the skin of a whale killed by the Gaspé fishermen, a species of acorn-shell (*Coronula reginæ*, Darwin,) supposed to be peculiar to the Pacific, an evidence that the creature had navigated the Arctic channels from Behring's Straits to be slain in the gulf of Saint Lawrence.

We may now proceed to connect these statements as to the distri-

bution of icebergs, with the glaciated condition of our continents,
with the remarkable fact that the same effects now produced by
the ice and the Arctic current in the strait of Belle-Isle and the
deep-current channel off the American coast, are visible all over
the North American and European land north of forty degrees of
latitude, and that there is evidence that the St. Lawrence valley
itself was once a gigantic Belle-Isle, in which thousands of bergs
worked perhaps for thousands of years, grinding and striating its
rocks, cutting out its deeper parts and heaping up in it quantities
of northern debris. Out of this fact of the so-called glaciated
condition of the surface of our continents, has however arisen one
of the great controversies of modern geology. While all admit
the action of ice in distributing and arranging the materials which
constitute the last coating which has been laid upon the surface
of our continents, some maintain that land glaciers have done the
work, others that sea-borne icebergs have been the agents employed.
As in some other controversies, the truth seems to lie between the
extremes. Glaciers are slow, inactive and limited in their sphere.
Icebergs are locomotive and far-travelled, extending their action
to great distances from their sources. So far, the advantages are
in favor of the iceberg. But the work which the glacier does is
done thoroughly, and time and facilities being given, it may be
done over wide areas. Again, the iceberg is the child of the
glacier, and therefore the agency of the one is indirectly that of
the other. Thus, in any view we must plough with both of these
geological oxen, and the controversy becomes like that old one of
the Neptunists and Plutonists, which has been settled by admitting
both water and heat to have been instrumental in the formation
of rocks.

Our country is one of those which have been most thoroughly
glaciated, and in the midst of these controversies a geologist
resident here should have some certain doctrine as to the
question whether at that period, geologically recent, which we
call the Post-pliocene period, Canada was raised to a great height
above the sea, and covered like Greenland with a mantle of perpetual ice, or whether it was, like the strait of Belle-Isle and the
banks of Newfoundland, under water, and annually ground over
by icebergs. A great advocate of the glacier theory has said that
we cannot properly appreciate his view without exploring
thoroughly the present glaciers of Greenland and ascertaining
their effects. This I have not had opportunity to do, but I have

endeavoured to do the next best thing by passing as rapidly as possible from the icebergs of Belle-Isle to the glaciers of Mont Blanc, and by asking the question whether Canada was in the post-pliocene period like the present Belle-Isle or the present Mont Blanc, or whether it partook of the character of both ?

Transporting ourselves then to the monarch of the Alps, let us suppose we stand upon the Flegere, a spur of the mountains fronting Mont Blanc, and commanding a view of the entire group. From this point the western end of the range presents the rounded summit of Mont Blanc proper, flanked by the lower eminences of the Dome and Aiguille de Gouté, which rise from a broad and uneven plateau of nevé or hard snow, sending down to the plain two great glaciers or streams of ice, the Bossons and Tacony glaciers. Eastward of Mont Blanc the nevé or snow plateau is penetrated by a series of sharp points of rock or aiguilles, which stretch along in a row of serried peaks, and then give place to a deep notch through which flows the greatest of all the glaciers of this side of Mont Blanc, the celebrated Mer de Glace, directly in front of our stand-point. To the left of this is another mass of aiguilles, culminating in the Aiguille Verte, only recently ascended by Mr. Whymper, of melancholy notoriety in connection with the fatal ascent of the Metterhorn. This second group of needles descends into the long and narrow Glacier of Argentiere, and beyond this we see in the distance the Glacier and Aiguille de Tour. As seen from this point it is evident that the whole system of the Mont Blanc glaciers originates in a vast mantle of snow capping the ridge of the chain, and extending about twenty miles in length with a breadth of about five miles. This mass of snow being above the limits of perpetual frost, would go on increasing from year to year, except so far as it might be diminished by the fall of avalanches from its sides, were it not that its plasticity is sufficient to enable the frozen mass to glide slowly down the valleys, changing in its progress into an icy stream, which descending to the plain melts at its base and discharges itself in a torrent of white muddy water. The Mont Blanc chain sends forth about a dozen of large glaciers of this kind, besides many smaller ones. Crossing the valley of Chamouni, and ascending the Montanvert to a height of about 6,000 feet, let us look more particularly at one of these glaciers, the Mer de Glace. It is a long valley with steep sides, about half a mile wide and filled with ice, which presents a general level or slightly inclined surface, traversed with innumerable transverse cracks or crevasses, penetrating apparently to the bottom of the glacier, and with slippery sloping edges of moist ice threatening at every step to plunge the traveller into the depths below. Still the treacherous surface is daily crossed by parties of travellers apparently without any accident. The whole of the ice is moving steadily along the slope on which it rests, at the rate of eight to ten miles daily ; the rate of motion is less in winter and greater in summer ; and farther down, where the glacier goes by the name of the Glacier du Bois, and descends a steeper slope, its rapidity is greater ; and at the same time by the opening of immense crevasses its surface projects in fantastic ridges and pinnacles. The movements and changes in the ice of these glaciers are in truth very remarkable, and show a mobility and plasticity which at first sight we should not have been prepared to expect in a solid like ice. The crevasses become open or closed, curved upwards or downwards, perpendicular or inclined, according to the surface upon which the glacier is moving, and the whole mass is crushed upward or flattens out, its particles evidently moving on each other with much the same result as would take place in a mass of thick mud similarly moving. On the surface of the ice there are a few boulders and many stones, and in places these accumulate in long irregular bands indicating the lines of junction of the minor ice streams coming in from above to join the main glacier. At the sides are two great mounds of rubbish, much higher than the present surface of the glacier. They are called the lateral moraines, and consist of boulders, stones, gravel and sand, confusely intermingled, and for the most part retaining their sharp angles. This mass of rubbish is moved downward by the glacier, and with the stones constituting the central moraine, is discharged at the lower end, accumulating there in the mass of detritus known as the terminal moraine.

Glaciers have been termed rivers of ice ; but there is one respect in which they differ remarkably from rivers. They are broad above and narrow below, or rather their width above corresponds to the drainage area of a river. This is well seen in a map of the Mer de Glace. From its termination in the Glacier du Bois to the top of the Mer de Glace proper, a distance of about three and a half miles, its breadth does not exceed half a mile, but above this point it spreads out into three great glaciers, the Geant, the du Chaud and the Talefre, the aggregate width of which is six or seven miles. The snow and ice of a large interior table-land or series of wide valleys are thus emptied into one narrow ravine, and pour their whole accumulations through the Mer de Glace.   Leaving however the many interesting phenomena connected with the motion of glaciers, and which have been so well interpreted by Saussure, Agassiz, Forbes, Hopkins, Tyndall and others, we may consider their effects on the mountain valleys in which they operate—

1.—They carry quantities of debris from the hill-tops and mountain valleys downward into the plains.   From every peak, cliff and ridge, the frost and thaw are constantly loosening stones and other matters which are swept by avalanches to the surface of the glacier, and constitute lateral moraines.   When two or more glaciers unite into one, these become medial moraines, and at length are spread over and through the whole mass of the ice ; eventually all this material, including stones of immense size, as well as fine sand and mud, is deposited in the terminal moraine or carried off by the streams.

2.—They are mills for grinding and triturating rock.   The pieces of rock in the moraine are, in the course of their movement, crushed against one another and the sides of the valley, and are cracked and ground as if in a crushing-mill.   Farther the stones on the surface of the glacier are ever falling into crevasses, and thus reach the bottom of the ice, where they are further ground against one another and the floor of rock.   In the movement of the glacier these stones seem in some cases to come again to the surface, and their remains are finally discharged in the terminal moraine, which is the waste-heap of this great mill. The fine material which has been produced, the flour of the mill, so to speak, becomes diffused in the water which is constantly flowing from beneath the glacier, and for this reason all the streams flowing from glaciers are turbid with whitish sand and mud.

The Arve which drains the glaciers of the north side of Mont Blanc, carries its burden of mud into the Rhone, which sweeps it with the similar material of many other Alpine streams into the Mediterranean, to aid in filling up the bottom of that sea, whose blue waters it discolours for miles from the shore, and to increase its own ever enlarging delta which encroaches on the sea at the rate of about half a mile per century.   The upper waters of the Rhone, laden with similar material, are filling up the Lake of Geneva ; and the great deposit of ' loess ' in the alluvial plain of

the Rhine, about which Gaul and German have contended since
the dawn of European history, is of similar origin. The mass of
material which has thus been carried off from the Alps, would
suffice to build up a great mountain chain. Thus by the action
of ice and water—

> " The mountain falling cometh to naught
> And the rock is removed out of its place."

Many observers who have commented on these facts have taken
it for granted that the mud thus sent off from glaciers, and which
is so much greater in amount than the matter remaining in their
moraines, must be ground from the bottom of the glacier valleys,
and hence have attributed to these glaciers great power of cutting
out and deepening their valleys. But this is evidently an error,
just as it would be an error to suppose the flour of a grist-mill
ground out of the mill stones. Glaciers it is true groove and
striate and polish the rocks over which they move, and especially
those of projecting points and slight elevations in their beds,
but the material which they grind up is principally derived from
the exposed frost-bitten rocks above them, and the rocky floor
under the glacier is merely the nether mill-stone against which
these loose stones are crushed. The glaciers in short can scarcely
be regarded as cutting agents at all, in so far as the sides and
bottoms of their beds are concerned, and in the valleys which the
old glaciers have abandoned, it is evident that the torrents which
have succeeded them have far greater cutting power.

The glaciers have their periods of advance and of recession.
A series of wet and cool summers causes them to advance and
encroach on the plains, pushing before them their moraines, and
even forests and human habitations. In dry and warm summers
they shrink and recede. Such changes seem to have occurred in
by-gone times on a gigantic scale. All the valleys below the present glaciers, present traces of former glacier action. Even the
Jura mountains seem at one time to have had glaciers. Large
blocks from the Alps have been carried across the intervening
valley and lodged at great heights on the slopes of the Jura, leading the majority of the Swiss and Italian geologists to believe that
even this great valley and the basin of Lake Leman were once
filled with glacier ice. But unless we can suppose that the Alps
were then vastly higher than at present, this seems scarcely to be
physically possible, and it seems more likely that the conditions
were just the reverse of those supposed, namely, that the low land

was submerged and that the valley of Lake Leman was a strait like Belle-Isle, traversed by powerful currents and receiving icebergs from both Jurassic and Alpine glaciers, and probably from further north. One or other supposition is required to account for the appearances, which may be explained on either view. The European hills may have been higher and colder, and changes of level elsewhere may have combined with this to give a cold climate; or on the other hand, a great submergence may have left the hills as islands, and may have so reduced the temperature by the influx of Arctic currents and ice, as to enable the Alpine glaciers to descend to the level of the sea. Now we have evidence of such submergence in the beds of sea-shells and travelled boulders scattered over Europe, while we also have evidence of contemporaneous glaciers in their traces on the hills of Wales and Scotland and elsewhere, where they do not now occur.

I have long maintained that in America all the observed facts imply a climate no colder than that which would have resulted from the subsidence which we know to have occurred in the temperate latitudes in the post-pliocene period, and though I would not desire to speak so positively about Europe, I confess to a strong impression that the same is the case there, and that the casing of glacier ice imagined by many geologists, as well as the various hypotheses which have been devised to account for it and to avoid the mechanical, meteorological and astronomical difficulties attending it, are alike gratuitous and chimerical, as not being at all required to account for observed, facts and being contradictory, when carefully considered, to known physical laws as well as geological phenomena.*

Carrying with me a knowledge of the phenomena of the glacial drift as they exist in North America, and of the modern ice drift or its shores, I was continually asking myself the question—To what extent do the phenomena of glacier drift and erosion resemble these? and standing on the moraine of the Bosson glacier, which struck me as more like boulder clay than anything else I saw in the Alps, with the exception of some recent avalanches, I jotted down what appeared to me to be the most important points of difference. They stand thus :—

1.—Glaciers heap up their debris in abrupt ridges. Floating

---

* Canadian Naturalist, Vols. viii and ix. Geological Magazine, December, 1865.

ice sometimes does this, but more usually spreads its load in a more or less uniform sheet.

2.—The material of moraines is all local. Icebergs carry their deposits often to great distances from their sources.

3.—The stones carried by glaciers are mostly angular, except where they have been acted on by torrents. Those moved by floating ice are more often rounded, being acted on by the waves and by the abrading action of sand drifted by currents.

4.—In the marine glacial deposits mud is mixed with stones and boulders. In the case of land glaciers most of this mud is carried off by streams and deposited elsewhere.

5.—The deposits from floating ice may contain marine shells. Those of glaciers cannot, except where, as in Greenland and Spitzbergen, glaciers push their moraines out into the sea.

6.—It is of the nature of glaciers to flow in the deepest ravines they can find, and such ravines drain the ice of extensive areas of mountain land. Icebergs on the contrary act with greatest ease on flat surfaces or slight elevations in the sea bottom.

7.—Glaciers must descend slopes and must be backed by large supplies of perennial snow. Icebergs act independently, and being water-borne may work up slopes and on level sufaces.

8.—Glaciers striate the sides and bottoms of their ravines very unequally, acting with great force and effect only on those places where their weight impinges most heavily. Icebergs on the contrary being carried by constant currents and over comparatively flat surfaces, must striate and grind more regularly over large areas, and with less reference to local inequalities of surface.

9.—The direction of the striæ and grooves produced by glaciers depends on the direction of valleys. That of icebergs on the contrary depends upon the direction of marine currents, which is not determined by the outline of the surface, but is influenced by the large and wide depressions of the sea bottom.

10.—When subsidence of the land is in progress, floating ice may carry boulders from lower to higher levels. Glaciers cannot do this under any circumstances, though in their progress they may leave blocks perched on the tops of peaks and ridges.

I believe that in all these points of difference the boulder clay and drift of Canada and other parts of North America, correspond rather with the action of floating ice than of land ice. More especially is this the case in the character of the striated surfaces, the bedded distribution of the deposits, the transport of material up the natural slope, the presence of marine shells, and the mechanical and chemical character of the boulder clay. In short, those who regard the Canadian boulder clay as a glacier deposit, can only do so by overlooking essential points of difference between it and modern accumulations of this kind.

In conclusion, I would wish it to be distinctly understood, that I do not doubt that at the time of the greatest post-pliocene submergence of Eastern America, at which time I believe the greater part of the boulder clay was formed, and the more important striation effected, the higher hills then standing as islands would be capped with perpetual snow, and through a great part of the year surrounded with heavy field and barrier ice, and that in these hills there might be glaciers of greater or less extent. Further it should be understood that I regard the boulder clays of the St. Lawrence valley as of different ages, ranging from the early post-pliocene to that at present forming in the gulf of St. Lawrence. Further, that this boulder clay shows in every place where I have been able to examine it, evidence of sub-aquatic accumulation, in the presence of marine shells or in the unweathered state of the rocks and minerals enclosed in it, conditions which, in my view, preclude any reference of it to glacier action, except possibly in some cases to that of glaciers stretching from the land over the margin of the sea, and forming under water a deposit equivalent in character to the 'boue glaciare' of the bottom of the Swiss glaciers. But such a deposit must have been local, and would not be easily distinguishable from the marine boulder clay. While writing these notes I have had the advantage of reading the interesting papers of Messrs. Jamieson, Bryce and Crosskey, on the boulder clay of Scotland,* which in character and relations so closely resembles that of Canada, but I confess several of the facts which they state lead me to infer that much of what they regard as of sub-aerial origin must really be marine, though whether deposited by ice-bergs or by the fronts of glaciers terminating in the sea, I do not pretend to determine. It must however be observed that the antecedent probability of a glaciated condition is much greater in the case of Scotland than in that of Canada, from the high northern latitude of the former, its more hilly character, and the circumstance that its present exemption from glaciers is due to what may be termed exceptional and acci-

---

* Journal of Geological Society for August, 1865.

dental geographical conditions; more especially to the distribution
of the waters of the Gulf stream, which might be changed by a
comparatively small subsidence in Central America.  To assume
the former existence of glaciers in a country in north latitude 56°,
and with its highest hills, under the present exceptionally favourable conditions, snow-capped during most of the year, is a very
different thing from assuming a covering of continental ice over
wide plains more than ten degrees farther south, and in which,
even under very unfavourable geographical accidents, no snow can
endure the summer sun, even in mountains several thousand feet
high.  Were the plains of North America submerged and invaded
by the cold Arctic currents, the Gulf stream being at the same
time turned into the Pacific, the temperature of the remaining
North American land would be greatly diminished ; but under
these circumstances the climate of Scotland would necessarily be
reduced to the same condition with that of South Greenland or
Northern Labrador.  As we know such a submergence of America
to have occurred in the Post-pliocene period, it does not seem
necessary to have recourse to any other cause for either side of
the Atlantic.  It would, however, be a very interesting point to·
determine, whether in the Post-pliocene period the greatest submergence of America coincided with the greatest submergence of
Europe, or otherwise.  It is quite possible that more accurate
information on this point might remove some present difficulties.
I think it much to be desired that the many able observers now
engaged on the Post-pliocene of Europe, would at least keep before
their minds the probable effects of the geographical conditions
above referred to, and enquire whether a due consideration of
these would not allow them to dispense altogher with the somewhat
extravagant theories of glaciation now agitated.*

---

* While these sheets were in the press, I have seen with much gratification,
that Jamieson has recognized in Caithness a truly marine boulder-clay, holding
those elongated and striated stones heretofore regarded as characteristic of
glacier action ; but which are frequent in the marine boulder-clays of Canada,
and in the bed of the present Arctic current.

# THE MUSK-RATS AS BUILDERS AND MINERS.*

## By J. K. Lord, F.Z.S.

The genus Fiber has hitherto been based on a solitary species, the well-known musk-rat, the *Fiber zibethicus* of zoologists, the musquash of Canadian trappers and fur traders, the ooklak of the inland Indians west of the Rocky Mountains. Strictly American mammals, musk-rats, true to their native proclivities, are habitual wanderers, regardless of even ' squatter's preemptive law,' unscrupulously seize on ' new locations ' that best befit their tastes and requirements.

A summer travelling party of musk-rats, on discovering a desirable spot for a settlement, at once appropriate it. One species sets to work and erects neat little dwellings, that are always placed in the water ; the building materials fringe the pool, fixed on as the village site. The other species, *diggers* by profession, scorn the builder's art, and *excavate* houses on the bank of some lazy stream or muddy pool.

The requisite establishments complete, the emigrants settle quietly to the ' struggle for existence,' and patiently bear as best they can, the ills that musquash, like all other flesh, is heir to.

A happy adaptability to extreme climatal changes, enables the musk-rat to endure the scorching heat of an inter-tropical sun, or the nipping cold of an Arctic winter, with trifling inconvenience either to its health or happiness. Throughout the length and breadth of Canada—tenanting the shoals of its countless lakes, the banks of its many rivers, its oozy swamps and muddy, stagnant pools—musk-rats are always to be found. Away into the trackless wastes of the Hudson Bay Company ; by the lone, still ponds scattered over the sunny prairies, or hid neath the shadows of the

---

* Fiber zibethicus, Musk-Rat.
Synonym.—*Castor zibethicus*, ' Lin. Syst. Nat.,' i., 1766.
*Mus. zibethicus*, ' Gmelin Syst. Nat.,' i., 1788.
*Myocastor zibethicus*, ' Kerr's Linnæus,' 1792.
*Fiber zibethicus*, Cuv., R. A. I., 1817, 192.
*Lemmus zibethicus*, ' Fischer Synop.,' 1829, 289.
*Ondatra zibethicus*, ' Waterhouse Mag. Nat. Hist.,' iii., 1839, 594.
Musk Beaver, ' Pennant's Arct. Zool.'
Musquash, Wach-usk of the Crees and Hurons (the animal that sits on the ice in a round form).
Nov. Sp.—*Fiber osoyoosensis* (Lord), ' Proc. Zool. Soc.,' London, 1863.

lofty pines; in dark, miry wastes, amid fungoid growths, sedge plants, and perpetual decay; along the banks of tortuous rivers, from their sources—mere mountain burns, trickling down the craggy sides of the Rocky Mountains—to their mingled exit into the Atlantic, as the great St. Lawrence;—musk-rats live, thrive, and multiply. Cross the snow-clad heights of the Rocky Mountains, and descend their western slopes, through hotter lands, to the shores of the Pacific; from the Rio-Grande to the desolate regions of Arctic America; through fertile California; grassy, flower-decked Oregon; Washington Territory, with its deserts and mountains; and the densely-timbered wilds of British Columbia, to its junction with Russian America; on rocky Vancouver Island, as well as on every island of any size in the Gulf of Georgia;— musk-rats have found their way, built and burrowed. It was once supposed, that the musk-rat had made its way to the Asiatic side of Behring's Straits, but there can be but little doubt the skins obtained from Kamschatka and Tschucktchis are traded, or bartered, from native tribes living on the American shores.

There are many structural points of similarity betwixt the musk-rats and *Arvicolas*, or ' field-mice ;' still the peculiarly formed feet, flattened tail, much larger size, and singularities of habit in the former, distinctly separates the two genera. Indeed, the musk-rat seems to fill a gap, as it were, between the field-mice (*Arvicolinæ*) and the porcupines (*Hystricidæ*). The sub-family (*Casterinæ*) which the famed beaver represents, connects the squirrels and marmots (*Sciurissæ*), on the one hand, with the gophers (*Geomyinæ*) on the other. The teeth of the musk-rat are of arvicoline type. The first and third molars are longer than the second, the second being wider than either of the other two. The grinding surface of the first molar has two indentations or reëntrant angles on each side; the second, two outside and one inside; the third, three outside and two inside. The first and third grinders have five prisms or projections on their surfaces, the second four. The loops of enamel extending across the tooth, and joining the enamel that encases the surface, completely isolates the patches of dentine; thus a mill-stone is formed by this most simple contrivance, that improves in grinding power the more it is worked, and never needs roughing with the stone-cutter's hammer.

In the lower jaw the first molar is much larger than the second and third, which are about equal in length and width. The first having five indentations inside and four outside. The other grinders have each two on either side ; the angles are alternate. The upper cutting, or incisor teeth, are broader than the lower, plane in front, but bevelled off at the outside edges, the lower being more rounded away than are the upper. Like the teeth of all the rodents, they are admirably constructed chisels, that by a simple arrangement of hard and softer material, sharpen themselves, the cutting edges becoming keener in proportion to the density of the material gnawed. The musk-rat's mouth is truly a marvellous mill, worked by machinery that needs neither steam or water-power to drive it. Its millstones—by the side of which man's best contrivance is but a bungle—never wear smooth, nor deteriorate in grinding capabilities, however hard the ' miller ' works. To supply the mill are admirable nippers that never blunt, and always remain the same length, wear and growth being so admirably balanced.

A very marked peculiarity in the skull of the musk-rat is the curious shape of the temporal bone ; so compressed is it betwixt the orbits as to narrow the skull into a mere isthmus, not at all wider than the extreme end of the muzzle. Parietals very small ; occipital foramen nearly circular.

*Fiber osoyoosensis* Lord. *Sp. char.*—In total length 3¼ inches shorter than *Fiber zibethicus* Cuv. ; in general size much smaller. General hue of back jet-black ; but, the hair being of two kinds, if viewed from tail to head, it looks grey—the under fur being fine, silky, and light grey in colour ; concealing this on the upper surface are long coarse black hairs ; the belly and sides somewhat lighter ; head broad and depressed ; neck indistinct ; ear small, upper margin rounded ; eye small and black ; whiskers long, and composed of about an equal number of white and black hairs ; incisors nearly straight, on the external surface orange-yellow. The thumb of the fore-foot is quite rudimentary ; the third claw is considerably longer than the second and fourth. The hind feet are singularly twisted, the inner edges being posterior to the outer. This simple modification of position, gives the animal immense power in swimming. The feet are then bent towards each other ; in the backward stroke, the full expanse of the flat soles pushes against the water, sending the swimmer forwards ; in the forward stroke the feet are ' feathered,' like rowers feather an oar, passing through the water edge on, offers the least possible resistance. The claws on the hind feet are small, compressed, and but slightly curved. The skin covering the under surfaces of the feet is black, wrinkled, perfectly naked, and keenly sensitive to tactile impressions. A distinct web joins the digits for about half their length ; the upper parts of the feet are clothed with short lustrous hairs, terminating at the sides in a fringe of stiff bristles, which increase the surface, and give additional force in swimming. Tail nearly as long as the body without the head, cylindrical at the base, then flattened to the point. The tail curves somewhat to a sickle shape; being readily bendable towards the belly, its point can be made to touch the inferior surface of its base ; in this position it is almost circular, like a hoop. This is a highly important arrangement, indispensable to the musk-rat. A more perfect rudder was never designed than is this flexible tail. If swimming when freighted, and a stiff breeze curls the water into miniature waves, the musk-rat drops its tail, and bending it more or less according as it needs extra steering power, guides itself straight for the desired haven. In calm weather and smooth water the rudder is carried horizontally, and a slight lateral motion close to the surface, suffices to guide the living ship. It is worth while to note, *en passant,* how differently the beaver's rudder is built, as compared with that of the musk-rat's—a difference easily accounted for when we know their respective habits. The beaver never uses its tail as a trowel, and has no more idea of ' lath and plaster ' than a hippopotamus has of a polka. This story is a myth, and the sooner the absurd fables of plastering, and " using the wondrous tail as a trowel," are sponged from out all books on natural history the better.

The beaver, with a heavy log of green timber (that would sink like a stone if free) clasped between its fore-legs, swims for its house. The counterpoise to this overweight at the bows is the downward pressure of the flat tail on the water, flattened more horizontally than the musk-rat's. Indeed, the tail of the beaver is much like an ox-tongue in shape. The musk-rat, conveying such materials through the water as are light, needs only powerful rudder-power, having no forward weight to counterbalance. The tail is covered by small hexagonal scales, with a few long, coarse hairs irregularly scattered over it. The skull differs from *Fiber zibethicus* in being much smaller, $2\frac{1}{8}$ inches in length, $1\frac{1}{8}$ inch in width, very much shorter from the anterior molar to incisors ; nasal bones much more rounded at their posterior ends, the superior outline less curved ; postorbital process not nearly so much developed ; the cranial portion of the skull in its upper outline is much less concave and smoother ; superior outline of occipital bone not so prominent or strong ; incisors shorter and much straighter ; molars much smaller, but in general outline similar.

And now I must ask my readers to accompany me, in imagination, to the Osoyoos Lakes, on the eastern side of the Cascade Mountains, where my attention was first directed to the rush-building rat, as being distinct in species from that which burrows in the mud banks. The specific name *osoyoosensis* was given in commemoration of the locality.

This magnificent piece of water is formed by the widening out of the Okanagen river as it passes through a deep valley, walled in by massive piles of rock. The Osoyoos Lake may be defined as one huge lake, or three smaller ones, with equal correctness ; as a narrowing in at particular points, gives the appearance of an actual division into separate lakes. The ' boundary-line ' runs through its centre, so that one half the lake belongs to Britain (its northern half), the southern to the United States. The shore is sandy, like a sea-beach, and strewn thickly with fresh-water shells along the ripple line, gives it quite a tidal aspect. On either side, a sandy, treeless waste stretches away to the base of the hills, so carpeted with cacti—which grow in small knobs covered with spines, like vegetable porcupines—that walking on it without being shod with the very thickest boots, is to endure indescribable torture; the prickles are so sharp and hard that they slip through ordinary leather like cobbler's awls. I had to tie up both dogs and horses, for the latter, getting the prickly knobs into their heels, kicked and plunged until exhausted. The dogs at once got three or four fast to their feet ; when impatiently seizing the vegetable pests, the prickles stuck with like pertinacity to the tongue and cheeks. I have no hesitation in saying a dog must inevitably die from starvation if he ventured to cross this waste alone ; once getting the cactus prickles in his mouth, unaided he could never free himself. A low ' divide ' separates this valley from the Similkameen. The water from the lakes eventually finds its way into the Columbia river. If there is an Eden for water-birds, Osoyoos Lakes must surely be that favoured spot. At the upper end a perfect forest of tall rushes, six feet in height, affords ducks, grebes, bitterns, and a variety of waders, admirable breeding haunts ; safe alike from the prying eyes of birds that prey on their kindred, and savages that devour anything.

The water, alive with fish of many species (permanent residents), becomes during ' the season ' crowded with lordly salmon like a fashionable watering-place ; thus affording a perpetual banquet to birds that devour fishes.   The tempting, juicy mollusks, " like turtle," seem palateable to all, be the diners scale-clad or feathered.   On one side of this lake is a swamp, in which are numerous pools, some of them deep in the middle, shoal at the sides to a few inches, all alike fringed with a tall growth of rushes. In these aquatic snuggeries, ducks, literally swarm thick as bees round thorn-blossoms.   Here, too, musk-rat houses may be likened to cities rather than villages ; the inhabitants—swimming idly about, just diving out of the way if I came too near, reappearing a short distance off—evidently deemed me an impudent intruder.

For years I have been in the habit of seeing these rush houses (which I shall presently describe), but took it for granted there existed but *one* species of musk-rat, whose winter quarters was the rush house ; its summer residence a tunnel excavated in a mud bank.   Sir John Richardson (Fa. Bo. Am.), after describing the ' winter huts,' goes on to say, " In summer the musquash burrows in the banks of the lakes, making branched canals many yards in extent, and forming its nest in a chamber at the extremity, in which the young are brought forth."   Another author writes, " They live in curiously-constructed huts, in a social state during winter ; in summer, these creatures wander about in pairs, feeding voraciously on herbs and roots."   Charlevoix adds, " They build cabins, nearly in the form of those of the Beaver, but far from being so well executed ; their place of abode is always by the waterside, so that they have no need to build causeways." Captain John Smith was in all probability the first who gave any account of the musquash, in a work published in the year 1624.   He says, " The musascus is a beast of the form and nature of a water-rat, but many of them smell exceedingly strong of musk." " We are not, however, aware that these nests are made use of by the musk-rat in spring for the purpose of rearing its young ; we believe these animals *always* for that purpose resort to holes in the sides of ponds, sluggish streams, or dykes."—Aud. and Bach.

Seated on a sandy knoll, I contemplated, measured, and began to skin my prize.   It occurred to me that there were no mud banks near, into which these rats could burrow, and according to the statement of the authorities, at this very time, they ought to have been in their summer holes.

My first proceeding was to hunt carefully round the lake to discover, if possible, some evidence of a burrow—not a trace of such could I find; next the rush houses underwent a rigid scrutiny. In each musk-rats were living, and more than this, whole families had clearly resided in the several mansions for a very long time. I now felt convinced there must be two distinct species, one a miner, the other a builder ; and further, that the two species had been classed together by observers, under the supposition that they changed quarters, in accordance with the seasons. The next thing was to prove my supposition based on correct data.

Tents were soon after struck, and the lake, with all its living treasures, abandoned to nature and the red man.

We must take up our story at Fort Colville, one of the earliest trading posts of the Hudson Bay Company, situated on a gravelly plateau, close to the Kettle Falls, on the Columbia river, about a thousand miles from the sea.

The two weary winters passed in this solitary spot were cold enough to satisfy an Esquimaux, the temperature often as low as thirty and thirty-two degrees below zero, with deep snow covering the ground for full six months of the twelve. Through the gravelly valley leading from the Fort to the hills, wound a sluggish muddy stream, with deep banks on either side, in which dwelt whole colonies of musquash. About a mile and a half from the stream, divided from it by a steep ridge of rocks, was a sedgy flat, surrounding a deep, quiet pool, so overshadowed and shut in by a brake of bulrushes, as to be hidden, until its margin, reached by wading and cutting a trail through the reedy fringe, revealed its water, and a city of musquash-houses scattered like hay-cocks over the entire surface.

In the bright, glowing sunshine of mid-summer, I carefully watched the stream and pool, fully satisfying myself that both localities were densely populated ; and, further, that ' builders ' and ' miners ' were blessed with infant workers, born, some in the rush dwellings, others in the nurseries of the mud tunnels. So far so good, nothing more could be done until winter. On carefully comparing several of the musk-rats shot in the pool, with those brought from Osoyoos Lake, I found them to be specifically alike, but differing most markedly from the rats inhabiting the Colville stream ; others procured from very distant mud banks, east and west of the Cascade Mountains, tallied exactly with these and each other, as did a series of rush-building-rats from widely separated localities. Up to this point, I had proved that both holes and lodges were occupied in July, and the rats inhabiting them differed in several distinctive characters always constant, though extended over a series of specimens, from remote and proximate districts.

The fur clothing of the two species (as I now venture to call them) seemed to my mind designedly coloured to facilitate concealment. The mud-rat's reddish, rusty-brown suit, closely resembled the furruginous tint peculiar to the gravelly soils prevailing in the north-west, and its habit is, when frightened, to dive, or if under water, to at once descend to the bottom, there to stir up the mud with all its might. In a second, the course of the fugitive is traceable only by clouds of mud rolled up into the water, like smoke into the air. Thus hid, escape is easy.

In clear water, too, small roadways are distinctly visible in every direction, threading the bottom of the stream like the lines on a map of railways, trails through which they travel to the different landings and doorways.

The *rush-rat's* black jacket is equally in keeping with the still dark water in which it swims, builds, and enjoys life ; or the sombre stalks amidst which it rambles and feeds. I know no prettier sight than that of watching a musk-rat village. As the shadows lengthen, and the mingled sounds of day die imperceptibly away, and—save the whisper of the breeze as it rattles the tall rushes, the muffled cry of the owl soaring over the marsh, the 'quack' and 'whistle' of the bald-pate (*Mareca amer*), sure herald of coming night, and the throb of invisible wings—no sounds are audible. In this quiet eventide, the entire rat population steal out to swim, flirt, quarrel, or feast, as the custom is in musquash society. So like are the swimmers to dark sticks floating on the surface, that save the tiny wake made as they paddle on, the keenest eye can hardly detect the difference. The slightest noise indicative of danger, plunging sounds over the pool as though a heap of stones hurled into the air, were falling into the water like rain-drops, warns one the revellers are gone. They soon, however, reappear, some to sit on the domes of their houses in the position of begging dogs, holding between their fore-feet a dainty on which to sup ; others to swim ashore, and forage amidst the rushes and sedge-plants, perhaps to be pounced on by the mousing-owl ; whilst the remainder seem to have no definite occupation, but swim or dive for sheer enjoyment. I can recal many long

evenings spent by some lone pool, watching these industrious little
animals ; too earnest in my vigils to note passing time, as stars
one by one gemmed the sky, and night with silence came down
upon the earth.

Winter came all too soon in October, heavy snow, and biting
blasts, sent the hybernators to their quarters, the lingering migrants
to their southern retreats, the deer to the depths of the forests,
the insects, some to their final home, others into torpidity, hid in
cleft or cranny.

If previous statements be true, no musk-rats will be found
tenanting the mud-holes, but all snugly ensconced in rush-mansions
in the pool.

On a piercing cold December morning, I waded through the
snow to the miner's quarter, my aid and guide, a red-skin, equipped
with pick, shovel and spear, to do the digging and capturing ; if
the musquashes, as I felt convinced was the case, had not abandoned their dwellings. It was no easy job breaking through the
frozen ground ; but the Indian warmed to his work, then I took
a spell, and so on, until the subterranean galleries were one after
the other laid open. No rats ; we were not far enough in. At
length we, by digging on, came plump upon a large vestibule, and
in it, coiled up semi-torpid and stupid, was a family of ' miners ;'
a goodly heap of dry grass and leaves formed an admirable bed.
The sleepers were hardly alive to danger, too drowsy to make any
attempt at escape. No food was stored, but they lay huddled
together for mutual warmth, as pigs do in straw. There were no
holes visible through the snow, but several had been dug through
the ground, to give, I imagine, admittance to the air.

This was a grand discovery. If like success attend our assault
on the builders, my theory will be proven.

The pool was frozen hard enough to have borne ten men,
enabling us to walk easily to the rush-houses, which were built in
from three to four feet water. I could discover no holes, though
quite three feet of dome in each house was clear above the ice.
On removing the snow, and tearing open the intertwined rushes,
there, rolled together in a grassy nest, as we had found the miners,
were many builders, doing their quasi-hybernation. This clearly
proved there were two kinds of musk-rats, that differed in habit,
size and colour. The skulls also showing structural variations,
left no further doubt. Two species for the future must characterize the genus Fiber, the second being *Fiber osoyoosensis*.

The number of young produced at a birth varies from four to seven, and it is by no means uncommon for a female to have three litters in a year ; and well for the musk-rats is it that nature has given them such powers of increase. Their enemies are legion. Birds of prey are ever watching for them ; indeed, it is difficult to save a trapped rat from the feathered banditti, ready on the shortest notice to tear the prisoner from the iron teeth of the trapper's snare.

The robber gang of weasels are untiring foes, hunting the rats night and day on the land and in the water. Their greatest enemy, however, is the trapper, be he red or white man. Five hundred thousand musk-rat skins were at one time annually imported from the Hudson Bay Company's territories. At the last fur sale in Fenchurch Street, in August, 1865, 93,787 skins of musquash were sold—a small proportion only of the yearly supply. The fur is used for various purposes, the bulk finding its way to foreign markets. The musk glands furnish the powerful, pungent odour from whence the animal derives its name, not to my nose the least like commercial musk. In the spring musk-rats really scent the air, and at this time the tails are taken off the trapped skins, tied in bundles, dried, and eventually sold in the bazaars at Constantinople, for ladies wherewith to perfume their cloths. The two glands are situated close to the base of the tail. Indians, white traders, trappers and settlers alike devour the muk-rat's body. To cook it *secundem artem*, after skinning, the glands should be carefully removed ; the body, split and gutted, is skewered on a long, peeled wand, and carefully grilled over a brisk camp-fire.

There are various modes of trapping musquash. If by steel trap, the trap is usually placed on a log, in the rat's water way, about four or five inches below the surface, with a bait suspended over it. In trying to reach this seductive morsel the hind feet are secured in the iron snare, which has a long string and cedar log float attached, to mark its whereabouts, as the prisoner drags it on the muddy bottom of a stream, or the deeper water of the pools. Others are caught in a kind of figure-of-4 trap, but by far the larger number are speared. The food of the musquash is of most varied character ; in the summer, grass, roots of marsh plants, the green bark from the young cotton-wood trees, and the stalks of succulent vegetation, constitutes their general fare. Though rodents, and in a measure vegetarians, they never refuse flesh if it can be obtained, and rather enjoy at times doing the cannibal. It is no infrequent occurrence for a hungry band to set upon their relative when fast by the legs in a trap, tear it to pieces, then devour the fragments as hounds are wont to rend and eat a fox. Sir J. Richardson tells us they have been known to eat one another in their houses, if unusually hungry, a statement I can quite believe, although it has never been my good or ill fortune to witness a musquash famine. I have often shot a duck that has fallen into the centre of a musk-rat pond ; waiting and wishing for a friendly breeze to waft the prize ashore, I observe it moves slowly, propelled by some unseen power, it nears a rush-house, bobs and bobs like a float as a fish tugs impatiently at the bait, then suddenly disappears. Musk-rats are the thieves that dine sumptuously at my expense. River mussels and craw-fish are also largely consumed by the musk-rats. They either crack the shells of the *Unios* with their strong teeth, or, taking them on the land, let them remain until, panting for air, the shells are opened, when the rat pounces in and devours the inmates. Not only are mussels eaten, but all fresh-water mollusks share a like fate, if discovered by prowling musk-rats.

It may be as well to say a few words, in conclusion, about their systems of building. The rush-houses are built in from three to four feet water. A solid pier, composed of sticks, rushes, grass, mud and small stones, is raised from the bottom to a height of some inches above the surface ; over this the dome-shaped roof is thrown, made of intertwined rushes with mud and sticks worked in amongst them ; the bed is placed on the centre or pier, and the entrance is invariably beneath the surface of the water. I do not believe this dome is in any degree impervious to water ; whenever I have opened a house in summer, it has invariably been wet ; and during blazing hot weather it must be a great advantage to the rush-rats, assisting to keep them cool—an advantage equally enjoyed by the ' miners,' whose houses are always wet in summer. In winter the water freezes, and hence cannot wet the insides of the domes or mud galleries. The grass and other material carried in for the winter bed must manifestly get wet in the transport, but rapidly drains and dries when the water solidifies. I do not believe in the possibility of an animal formed as the musk-rat making a waterproof fabric out of rushes and mud. One thing has always puzzled me in their engineering : how they manage to keep down the materials forming the centre or pillar, preventing

light substances from floating until the aggregated weight of
stones, mud, wetted rushes, and sodden sticks becomes, *en mass*,
specifically heavier than water, is a secret I was never able to discover.   They always work at night, hence it is impossible to
watch their operations.

The pleasure of describing the habits of these interesting
animals must be my excuse for these lengthy notes.   A new
species, like gold, usually tempts its finder to wander beyond the
limits of prudence ; if such has been my failing, I crave for-
giveness, and conclude with the sentiments of Wordsworth—

> " To the solid ground
> Of nature trust. the mind that builds for aye :
> Convinced that there, there only, she can lay
> Secure foundations."

*From the Intellectual Observer.*

---

# A CATALOGUE OF THE CARICES COLLECTED
## by John Macoun, Belleville, C. W.

The following list embraces ninety species, many of which
have not hitherto been published as Canadian, and three of which
are new.   All the species have been critically examined and
determined by Prof. Dewey, of Rochester, U. S., the eminent
caricographer ; his descriptions of the new species are cited from
Silliman's Journal for March, 1866.

### Nat. Ord. CYPERACEÆ—Genus CAREX Linn.

C. gynocrates, Wormsk. : Cedar swamps North Hastings ; Big
swamp Murray ; on a mound in a swamp near Belleville Railroad
Station.

C. polytrichoides, Muhl. : Cedar swamps ; common.

C. Backii, Boott : Rocky ground vicinity of Belleville and
Shannonville ; scarce.

C. bromoides, Schk. : Marshes and borders of ponds; scarce.

C. siccata, Dewey : Sandy plains; abundant around Castleton.

C. disticha, Hudson—var. Sartwellii, Dewey : Small marsh
west of Belleville College ; rare.

C. teretiuscula, Good. : Marshes along the Bay of Quinté ;
marshes and swamps ; abundant.

C. prairea, Dewey : Marshy border of Round Lake, Peterboro County ; big swamp, Murray.   Local ; abundant.

C. vulpinoidea, Michaux : Low meadows ; very common.

C. stipata, Muhl. : Along rivulets in wet meadows ; common.

C. sparganioides, Muhl. : Low thickets and along fences ; uncommon.

C. cephalophora, Muhl. : Woods and dry meadows, Belleville and Shannonville ; frequent.

C. Muhlenbergii, Schk. : Dry sand hill, Belmont, Peterboro County ; rare.

C. rosea, Schk. : Cedar swamps and wet woods ; common.

C. rosea—var. radiata, Dewey : Dry open woods and thickets ; frequent.

C. retroflexa, Muhl. : Wet woodlands, five miles south of Belleville ; rare.

C. tenella, Schk. : Abundant in all cedar and tamarack swamps.

C. trisperma, Dewey : Cedar swamps ; common.

C. tenuiflora, Wahl. : Cedar swamps four miles west of Belleville.   Abundant in a cedar swamp one mile beyond the Jordan, Hastings Road.

C. canescens, Linn. : Abundant in a wet meadow near Belleville.   Sphagnum swamps, North Hastings.

C. canescens, Linn.—var. vitilis, Carey : Borders of cedar swamps and low woods, Hastings County.

C. Deweyana, Schw. : Rich low woods in tufts ; abundant.

C. stellutata, Good. : Cedar and sphagnum swamps ; also low woods.

C. scirpoides, Schk. : Border of Hooper's Lake, Hastings Road ; rare.

C. sychnocephala, Carey : Border of the Millpond, Hastings Village, Madoc.   Low meadow along the Moira, Marmora.

C. scoparia, Schk. : Boggy woods and wet meadows ; common.

C. lagopodioides, Schk. : Border of water holes in meadows and fields.

C. cristata, Schw. : Low woods and meadows ; abundant.

C. festucacea, Schk. : Wet meadows and borders of woods, abundant ; all the varieties common.

C. straminea, Schk. : Low meadows near Belleville ; depressions in rocky ridges at Shannonville.

C. aperta, Boott : Border of a small lake, Hastings Road, Tudor ; rare.

C. stricta, Lam. : Wet meadows near Belleville ; meadows, Brighton.

C. aquatilis, Wahl. : Marshes along the Bay of Quinté ; wet meadows, Belleville.

C. lenticularis, Michx. : Crevices of rocks back of the old saw mill, Marmora Iron Works ; growing almost in the waters of Crow River ; abundant.

C. crinita, Lam. : Low banks of streams ; common.

C. limosa, Linn. : Peat-bog five miles north of Colborne ; rare.

C. irrigua, Smith : Big swamp Murray ; Sphagnum swamps, North Hastings ; frequent.

C. Buxbaumii, Wahl. : Border of Hooper's Lake, Hastings Road ; rare.

C. aurea, Nutt. : Low boggy meadows and sphagnous swamps ; common.

C. tetanica, Schk. : Woods east of Belleville ; very rare.

C. vaginata, Tausch : In cedar swamps near Belleville and Trenton ; abundant.

C. granularis, Muhl. : Wet meadows ; abundant.

C. conoidea, Schk. : Wet meadows east of Belleville ; scarce.

C. grisea, Wahl. : Meadow east of Belleville ; very rare.

C. formosa, Dewey : Low meadows and moist woods ; frequent.

C. gracillima, Schw. : Wet woods ; common.

C. plantaginea, Lam. : Rocky slopes in woods : Brighton and Huntingdon.

C. platyphylla, Carey : Dry rocky woodlands near Belleville ; frequent.

C. digitalis, Willd. : Hillside, North Hastings ; dry meadow, Brighton ; meadows near Belleville.

C. laxiflora, Lam. : Rich moist woods ; many varieties.

C. oligocarpa, Schk. : Gibson's Mountain, Prince Edward Co. ; hillside Port Hope ; rare.

C. Hitchcockiana, Dewey : Dry sandy field, Seymour ; very rare.

C. eburnea, Boott : Dry limestone rocks, banks of Moira and Trent.

C. pedunculata, Muhl. : Dryish cedar swamps near Belleville.

C. umbellata, Schk. : Border of the Oak-hill Pond, Sidney ; rare.

C. Novæ-Angliæ, Schw.—var. Emmonsii, Carey ; Rocky woods and banks near Belleville.

C. Pennsylvanica, Lam. : Woodlands and thickets ; common.

C. varia, Muhl. : Dry rocky ledges near Shannonville and Belleville.

C. Richardsonii, R. Brown : Dry field and thickets near Belleville and Trenton.

C. pubescens, Muhl. : Moist woods and meadows ; frequent.

C. miliacea, Muhl. : In a ravine on Simon Terrill's farm, Brighton ; scarce.

C. scabrata, Schw. : Margins of springs and woodland brooks, Brighton ; also near Port Hope.

C. arctata, Boott : Woods rear of Picton ; wet meadows near Wooler, Brighton.

C. debilis, Michx. : Woods and meadows, Brighton ; scarce.

C. flexilis, Rudge : In a cedar swamp near Trenton ; rare.

C. flava, Linn. : Abundant in old beaver meadows, North Hastings.

C. Œderi, Ehrh. : Wet sand, Presqu'ile Point, Lake Ontario ; also on Wellington Beach ; abundant.

C. filiformis, Linn. : Peat bogs and beaver meadows ; abundant.

C. lanuginosa, Michx. : Low wet meadows : common.

C. Houghtonii, Torrey : Dry rocky hills, Marmora, Tudor and Grimpsthorpe.

C. lacustris, Willd. : Marshes and swamps ; common.

C. aristata, R. Brown : Low wet ground, three miles west of Belleville ; scarce.

C. trichocarpa, Muhl. : Low marshy meadow rear of Picton ; low meadow along Crow River at Marmora works.

C. comosa, Boott : Marsh near Weller's Bay, Lake Ontario ; also Big swamp Murray ; scarce.

C. Pseudo-Cyperus, Linn. : Swamps and bogs ; common.

C. inirata, Dewey—var. minor : Border of a small pond in a meadow east of Belleville ; very rare.

C. hystricina, Willd. : Wet meadows ; common.

C. tentaculata, Muhl. : Wet meadows near Belleville ; also Presqu'ile Point.

C. intumescens, Rudge : Woods and new meadows ; common.

C. Canadensis, Dewey : Border of a small pond in a meadow, lot No. 6, 10th range, Seymour ; abundant.

C. lupulina, Muhl. : Wet meadows ; common.

C. Macounii, Dewey : Along a small stream on lot 7, 10th range of Seymour ; rare.

C. retrorsa, Schw.: Marshy meadows, and along small rivulets.

C. Schweinitzii, Dewey : In a wet swampy meadow, near Baltimore, Northumberland Co. ; abundant.

C. Hartii, Dewey : Border of a small stream in F. Macoun's farm, Seymour.

C. Bella-villa, Dewey : In a ditch about four miles north of Belleville, along the gravelled road leading to Stirling.

C. monile, Tuckerman : Low meadows along the Moira, North Hastings. Also, C. Vaseyi, Dewey—which proves to be a young state of this plant.

C. ampullacea, Good. : Ponds in meadows, also in swamps ; common.

C. cylindraca, Schw. : Swamps and wet meadows ; abundant.

C. longirostris, Torrey : Rocks, Gibson's Mountain ; ' Big Boulder ' of the Trent valley ; rocks, Marmora.

---

### DESCRIPTIONS OF THE NEW SPECIES.

---

CAREX HARTII, Dewey : Spicis staminiferis 1–3, sæpe 2, interdum 1, vel raro nulla, cylindraceis gracilibus variis erectis, suprema longiore in medio vel supra vel infra fructifera, sessilibus squamas lanceolatus acutas subfuscas ferentibus ; spicis pistilliferis 2–7, vulgo 4, cylindraceis oblongis sublaxifloris et infra præcipue subremotis plerumque erectis foliaceo-bracteatis, superioribus sessilibus sæpe ad apicem staminiferis, inferioribus exserto pedunculatis interdum supra staminiferis infimis duobus longo-exserto-pedunculatis interdum recurvis, cum bracteis culmum superantibus ; fructibus *tristigmaticis* ovatis inflatis vel conico-ellipticis longo-rostratis et teretibus bidentatis nervosis infra teretibus et stipitatis lævibus divergentibus et adultis prope retrorsis, squama lanceolata acuta margine albida latera fusca multum longioribus ; culmi foliis longis strictis modosis per-angustis margine scaberrimis et sæpe culmum lævem plusquam duplo præcedentibus.

Culm 15–25 inches high, erect, slender above, smooth except the highest part of the edges, with bracts and leaves surpassing the culm, and the leaves very narrow and long, often more than twice the length of the culm and very scabrous on the edges, knotted : spikes very variable ; the wholly staminiferous 1–3, commonly 2, nearly half 1, very rarely 3 or none, cylindric, slender, sessile ; some staminiferous have a few fruit in the middle or at the base or vertex; the terminal much the longest, and all clothed with lanceolate acute scales ; pistilliferous spikes 2–7, usually 4, the highest with stamens at the summit or in the middle or both and sessile, the next higher exsert pedunculate and erect, the lowest one or two very long-exsert pedunculate sometimes recurved, and the lowest sometimes staminate at apex, all oblong-cylindric, $\frac{1}{2}$ to $2\frac{1}{2}$ inches long, mostly erect, rather distant, loose-flowered, especially below, bracteate and the lower with long-leafy bacts surpassing the culm and rough-edged ; stigmas 3 ; fruit ovate-conic, inflated, long conic-rostrate, bidentate, nerved, tapering below, and stiped, diverging or nearly retrorse in maturity, much longer than the slender ovate lanceolate scale.

Wet grounds, Dundee, Yates Co., N. Y., discovered by Dr. S. Hart Wright. Ludlowville, Tompkins Co., H. B. Lord. Hastings Road, Canada West, J. Macoun.

The retrorse fruit brings up *C. retrorsa*, but the difference in the spikes and culm and fruit is too great, and the achenia are very dissimilar. *C. retrorsa* has achenia long and round sub-tri-quetrous ; the other has shorter triquetrous achenia tapering from the middle toward each end, and not roundish.

Var. BRADLEYI, Dewey : Staminate spikes less various ; pistillate spikes more loosely flowered ; fruit smaller ; and plant more slender.

Wet grounds, Greece, ten miles west of Rochester, Dr. S. B. Bradley. Here Dr. B. had discovered *C. mirata*, and was search-ing for its rediscovery, 1861. Also, at Belleville, Canada West, J. Macoun.

C. VAGINATA, Tausch, 1821 : Spicis distinctis ; staminifera unica oblonga culmo stricto fulta vel " sub-anthesi rectangulè refracta ;" pistilliferis sub-binis oblongis laxifloris remotis erectis linearibus exserto-pedunculatis lato-vaginatis ; fructibus *tristig-maticis* triquetro-ovatis basi attenuatis brevi-rostratis bidentatis, squama oblonga sub-obtusa longioribus ; culmo lævi foliato, foliis longis lato-linearibus margine supra scabris, bracteæ vagina vix foliaceum cuspidem abruptam ferente ; culmo perlævi.

This plant is widely spread over Germany and Scandinavia, but it is so variable that Kunze in 1840–50 gave twelve synonyms in the nineteen authors he quotes on this species, and omitted the name given by Fries, *C. sparsiflora*. In my specimens from Europe, and one of them from the hand of Fries (in my collection), there is too great a difference for identity of species; and if so, different plants may have been confounded by some authors. The one from Fries has a pair of too close-fruited spikes, scarcely sheathed, too nearly sessile, and bracts too leaf-like. The others correspond chiefly to the above description, authorized by those of Fries, Lang, Anderson, Kunze and Steudel. In Hooker's Flora Bor.-Amer., Dr. Boott gave *C. phæostachyæ*, Smith as synonymous with *C. vaginata* Tausch, as does Kunze also, and credited it to Greenland, Fort Norman on Mackenzie River, and Rocky Mts. It is doubtless the European plant. Dr. Gray informed Mr. Paine, who had found a variety in this vicinity, that *C. vaginata* had been found near Montreal by the late Mr. Macrae, and later at " Rivière-du-Loup by W. Boott." A recent examination of some of Dr. Macrae's plants by Prof. Brunet of Quebec, did not detect any plant of that name. I had hoped to ascertain whether the Montreal specimens agreed with the European or with the varities found by Mr. Paine. This differs however from the European in so many particulars that a more full account is given under the following name.

Var. ALTO-CAULIS, Dewey: Spica staminifera brevi cylindracea erecta vel infra " *rectangulè fracta ;*" pistilliferis spicis 1–3, sæpe 1, vulgo 2, per-rarò 3, cylindraceis brevibus laxifloris vel alterno-fructiferis sub-vicinis vel remotis, suprema subsessili, infirma interdum subradicali exserto-pedunculata, bracteatis vaginantibus, fructibus *tristigmaticis* ovatis ovato-conicis ellipticis interdum obovatis infra teretibus substipitatis subtriquetris lævibus nervosis brevi-rostratis bidentatis, rostro recto vel refracto, squama subacuta duplo longioribus : culmo alto-cauli infra lævi inclinato longi- et arcti-foliaceo : vagina angusta cum folio.

Culm 12–30 inches high, very slender and nearly filiform above, stiff and inclined, with culm leaves about half as long, sometimes longer ; staminate spike single, short-cylindric or oblong, often distant from upper pistillate, erect or with *stem bent rectangularly above* and near that pistillate, with scales oblong and obtuse, green on the back and reddish on the sides or wholly ; pistillate spikes 1–3, often 1, commonly 2, very seldom 3, cylindric, short, erect, loose-flowered or alternate-fruited, near or often quite remote ; lowest rarely subradical, long-pedunculate, upper sometimes nearly sessile, lower enclosed or exsertly pedunculate, bracteate with a narrow and longer foliate sheath ; stigmas three ; fruit ovate or ovate-conic-elliptic, sometimes obovate-triquetrous, tapering below, stiped, short-rostrate and the beak often turned one side or refracted, two-toothed, smooth, near twice longer ŏr rarely little longer than the ovate or oblong obtuse or sub-acute scale.

Discovered in a marsh in Bergen, twenty miles west of Rochester, by Rev. J. A. Paine ; the first known locality in the United States ; fruit mature in June, 1865. On some of the Bergen specimens, the *refraction* of the *culm* below the staminate spike and of the *beak* of the fruit, especially of the early mature plants, is striking. Both of these curious particulars are found on many of the European specimens. The former is given in the description of Kunze and Steudel as a common fact, and in some popular remarks of Fries ; and the latter is alluded to, with the other, as of no consequence, by Andersson in his Cyperaceæ Scandinaviæ ; while Lang states of the former that he had examined it on the *C. vaginata* cultivated in a botanic garden, but had never found it on one of the numerous specimens he had collected, or growing in their indigenous state. Of course Dr. Lang did not introduce the *refraction* of the stem into his description of this species.

The height of *C. vaginata* (5 to 12 inches by Steudel), the greater width of the leaves (foliis latis, Lang) ; the short cuspid-like leaf or termination of the broad sheath in Andersson, so clear on the specimens from Europe and on the figures of Kunze and Andersson, the more thick and coarse leaves and more stocky form, as well as differences in the fruit, distinguish the Bergen plant from the European.

C. MACOUNII, Dewey : Spicis variis *ordinatis* distinctis vel *inordinatis* cylindraceis erectis bracteatis ; ordinatorum stamini-feris 2, inferiore breviore longo-bracteata, terminali longiore, squamas longas graciles lanceolatas infra sparsas ferentibus ; et pistilliferis 4, suprema subsessili, cæteris remotis longo-peduncu-latis: ordinatorum terminali staminifera longa et fructifera pistillis paucis supra vel medio vel infra interpositis, vel interdum terminali apicem pistillifera et dimidio inferiore fructifera, tunc terminali pistillifera longa et in medio vel basi pauco-staminifera ; spicis pistilliferis subquinis cylindraceis erectis laxifloris, inferioribus longa exserto pedunculatis, infirma apice vel medio raro stamini-fera: fructibus *tristigmaticis* ovatis longo-conico-subinflatis lævibus nervosis brevi-furcatis substipitatis longo-gracili-rostratis divergentibus vel rectangule separatis, squamam ovato-lanceolatam ad basin aequantibus vel supra superantibus ; bracteis foliisque margine vix scabris et culmo lævi longioribus ; culmo foliis basin breviore.

Culm one to two feet high, erect, smooth ; bracts and leaves long, narrow, linear-lanceolate, the lower much surpassing the culm, smooth but slightly scabrous on the edges, nodose ; spikes six, cylindric, pedunculate ; the pistillate 1–2½ inches long, sessile above and sheathed exsert-pedunculate below, very *variable ;* as (1.) regular, staminate spikes 2, terminal, cylindric, long, the lower short with a long slender bract, both bearing long lanceolate scales very lax below, and the pistillate 4, uppermost subsessile and the others remote, long pedunculate, erect ; (2.) irregular, staminate spike terminal long, with a few scattered fruit at the vertex or in the middle or below, and pistillate 5, with some stamens at the vertex of the upper, sometimes the terminal 2–3 inches long and upper half pistillate with the lower half staminate, sometimes the terminal pistilliferous long with few stamens in the middle or at the base, sometimes the lowest pistillate with some stamens at its apex and in the middle ; all the pistillate loose-flowered, especially below ; stigmas 3 ; fruit ovate long-conic, inflated at base, rostrate with beak slender and bidentate, diverging or nearly rectangular below, smooth, nerved, generally longer than the narrow oblong acute and awned or ovate-lanceolate scale, or at the base of the lower spikes the fruit is sometimes scarcely longer than the scale ; plant straw-yellow.

At streams in Seymour, Northumberland Co., Canada West, J. Macoun, whose name the discovery honors. Though related to *C. folliculata* L., it seems quite different, and the achenia wholly unlike ; future forms may show more clearly its relations.

C. CANADENSIS, Dewey : Spicis distinctis ; staminifera unica perlongo-cylindracea erecta remota et bractea foliata e basi distante, squamas longas latas lanceolatas ferente : spicis pistilliferis 1–3, vulgo 2, sæpe 1, per-raró 3, oblongis cylindraceis erectis subensifloris, inferiore interdum brevi-ovata et sæpe per-longo-pedunculata; fructibus *tristigmaticis* ovato-conicis inflatis conico-rostratis bifurcatis subtriquetris nervosis glabris, squama ovata brevi-acuta vel aristata plus duplo longioribus ; bracteis foliisque margine supra scabris culmum lævem superantibus.

Culm 15–24 inches high, erect, rather slender, very smooth, leafy towards the base; leaves and bracts surpass the culm ; spikes distinct ; terminal staminate long-cylindric, remote from its bract and more from the pistillate, erect and slender, covered with long broad lanceolate scales ; pistillate spikes 1–3, commonly 2, often 1, very rarely 3, cylindric, oblong, erect, the lowest sometimes short and ovate and long exsert-pedunculate, bracteate and sheathed, sub-close-fruited ; stigmas 3 ; fruit ovate, inflated, conic-tapering into a 3-sided beak, which is rather deep bifurcate and sub-scabrous on the edges, nerved and smooth, more than twice longer than the ovate acute or awned scale ; plants yellowish.

Small ponds at Seymour, Northumberland Co., Canada West, J. Macoun.  I have seen nothing like it in the specimens obtained by me.  It has been referred to *C. lupulina*, but the achenia much differ, as well as the spikes and fruit.

C. BELLA-VILLA, Dewey : Spicis staminiferis 2–3, ferð 3, cylindraceis erectis vulgo approximatis sub-remotis, terminali longiore et omnibus bracteatis sessilibus longo-squamiferis ; pistilliferis vulgo 2, interdum 1, cylindraceis erectis exserti-pedunculatis brevi- et lato-vaginatis per-laxifloris suprema apice staminifera ; fructibus *tristigmaticis* longis gracilibus ovato-lanceolatis conicis basin inflatis nervosis lævibus per-divergentibus rectangule positis vel sub-retorsis rostro longi-bifurcato subtriquetro longo-stipitatis, squamam longam lanceolatam dorso viridem infra subæquantibus supra præstantibus bracteis foliisque margine scabris culmum foliatum superantibus.  Achenium est triquetrum infra teres supra brevi-rotundum triquetrum.

Culm about 1½ foot high, erect, strong, leafy toward the base, rough a little on the upper part ; bract-leaves rise from short broad sheaths, and with the leaves surpass the culm ; staminate spikes 2–3, commonly 3, cylindric, erect, near or sub-remote, the terminal often longer, all sessile and bearing long lanceolate scales, rough to the eye but soft to the touch ; pistillate spikes commonly 2 and rarely 1, cylindric, exsert-pedunculate, erect, very loose-flowered, short and broad sheathed, the highest staminate at the apex and nearly sessile, the lowest sub-remote ; stigmas 3 : fruit long, slender, ovate-lanceolate, conic, nerved, smooth, diverging and horizontal or sometimes retrorse, stipitate, with a back deeply bifid or bifurcate, quite equalling the scale at the base and exceeding the scale at the upper part of the spike.  Plant yellowish.

Near Belleville, Canada West, J. Macoun : a fine species.

## NOTES ON THE " SPECTRUM FEMORATUM."

### By Alex. S. Ritchie.

The order of Orthoptera, to which this insect belongs, is remarkable for the singularity of developement which characterizes individuals of some of its families; especially, those exotic species as the *Mantis religiosa* from the south of France, the *Phillium siccifolium*, or walking-leaf; and in the *Ectatosoma tiaratum* monstrosity reaches its acme ; the last named insect has dilated spined legs, a swollen body, and appendages also spined. I had the pleasure of seeing a specimen of this insect in a private collection in New York. The appearance of the *Spectrum femoratum* is no less wonderful, having a long cylindrical body, resembling a little broken twig and hence the popular name of Walking-stick. The only entomologist who has treated on the habits of the Phasmidæ is Stoll ; Kirby quotes him when speaking of this family of insects; with a few exceptions the order of Orthoptera has been less studied than any of the others.

There are two localities near Montreal where I have found this insect, namely, on the bass-wood trees on the north-east side of the mountain and on Logan's farm ; to one who is not in the habit of collecting insects it is very difficult to observe them, they are generally slow and quiet in their habits when undisturbed, and their resemblance in colour to the bark of the trees on which they feed makes it difficult to notice them, except to the prying eye of the entomologist ; in fact the general question asked me is—are these insects found in Canada ? and the enquirer generally says, 'tis strange I have never seen any of them before. A friend of mine told me that while he was sitting reading in the vicinity of Niagara Falls, something fell on his book, which he said resembled a dried twig, but he was more astonished when he perceived the twig (as he called it) was possessed with life, and immediately walked off.

I am not aware of any other species than the *Spectrum femoratum* being found in Canada ; they are apterous in both sexes, the male (as is generally the case among insects) being the smallest. The *Diurna chronos* of Van Dieman's Land, has wings ; there is a specimen of this insect in the University Museum ; another winged species is found in Virginia.

Having studied the habits of this insect for some time, I shall mention a few facts from actual observation, illustrating the peculiar adaptation for its comfort and place in the animal economy, and having also dissected carefully, and examined its external and internal anatomy with the microscope, I may be able to mention some new facts hitherto unobserved.

We shall first look at the habits of this little creature; the only time they are to be seen in numbers, is during the latter part of August and the month of September, when the males are in search of the females; you find them in rows on the bark of trees, their anterior legs stretched out horizontally in a level with the body; at other times they are rarely met with, as they are peculiarly solitary in their habits; they are not easily disturbed by the approach of any one, as instinct teaches them that they are not easily observed; however, when touched, the anterior legs are dropped, and they make good their escape in rather an active manner; their motion is ambulatory, or a kind of trot. They are exclusively herbivorous, living on the leaves of trees.

We shall now look at their external anatomy; the body is long and cylindrical, head and eyes small, legs long, very perfect mouth, antennæ long and setaceous, the feet are armed with two claws, and have a pulvillus or cushion, colour varies in the sexes, the length of any of the males which have come under my notice, has been from $2\frac{1}{4}$ to $2\frac{3}{4}$ inches exclusive of antennæ, the antennæ measuring about $2\frac{1}{4}$ inches; the length of the female from 3 to $3\frac{1}{2}$ inches, antennæ 2 inches; the body of the male is more slender, and the colour of the legs of a green shade. The number of joints in the antennæ of those I have examined, amount to fifty-eight in the female, and seventy-two in the male, the joints gradually shortening to the tips; the eyes are small. One thing I may observe here, that there are no ocelli or simple eyes on the head of these insects, a fact about which there has been some dispute. Latreille who has also examined them, bears me out in this, although Kirby says that three very visible ones are distinguished in the winged species.

The trophi or organs of the mouth are well developed, serving both for cutting and grinding their food. The mandibles are rounded and blunt, the maxillæ or lower jaws are obtuse, the labial palpi are four jointed, and the maxillary palpi three jointed. Another fact which I noticed, is the presence of a spur at the base of the femur, which evidently has been overlooked, as **Kirby** states that their legs are without spurs or spines. I find this spur on the second and posterior pair of legs in the male well developed, and smaller in the female ; the tarsi are five jointed, with a rudimentary or psuedo joint. The body is divided into eleven dorsal, and seven ventral segments.

The internal anatomy of these insects is typical of the class, only there are fewer convolutions of the intestinal canal, respiration is effected in the same manner as in the class insecta, by means of trachea, having an outlet by spiracles placed two on each segment. This dissection was made on a female ; the eggs are attached by a thin membrance to the back of the insect, under the dorsal vessel or heart. I examined the ovaries and saw clusters of eggs in every stage of development, from the simple cell with a nucleus, to the more advanced oval shape, with the germinal spot clearly visible, they taper from the size of a pin's head to appearance under the microscope to that of a three cent piece, in this state they are all attached by the end. I opened one of the eggs laid by the insect and saw the germinal spot more advanced.

I obtained this specimen on the morning of the 12th of September when she commenced laying, at noon on the 13th she had deposited twenty-eight perfectly formed eggs ; but on looking at her a few hours afterward she was dead, the eggs look like a miniature French bean, they have a depression on the inner side like the eye spot in that seed, and have a capsule fastened by a hinge like ligament on one side, to aid the young spectrum in making a more easy entrance into the world.

The largest egg belonging to any known insect, is the egg of *Phasma dilatatum*, one of this family it is figured in the fourth volume of the Linnean transactions ; it measures five lines in length and three lines in breadth or from a quarter to half an inch approaching the size of some of the humming birds eggs.

In this family are also some of the largest known insects ; they are natives of South America, Australia and the more southern latitudes.

The *Phasma gigas* measures about seven inches long by about seven-eighths of an inch broad. The *P. titan* of Macleay, a winged species, measures eight and one-half inches long, and three-fourths of an inch broad, longitudinal expansion of its wings, seven and one-half inches, transverse expansion, two and three-fourth inches. *P. dilatatum* is another giant in the insect world.

Very little is known of the larval state of this insect, and very little difference of appearance is observed, the metamorphoses not being complete ; size appears to be the only distinction, a succession of moults or excuviations bringing the young spectrum to the imago or perfect state.

We shall now look at the adaptation of this little creature for its place in the animal economy. First we may wonder why wings were denied it, nature answers this question ; instead of being a rover like some other insects, whose food is more precarious or uncertain, and has to be hunted, to those wings are given, but to—our humble neighbour born near its food, which, while spring time and harvest remain, trees will grow and put forth their buds, it manages to live and move and have its being.

We see also that it is gifted with a long leg to enable it to walk over the rough bark (full of hills and hollows) of the basswood on which it is generally found, where a short leg would not be so well suited, it is able to surmount those diffiulties with its long steady step, with its long body we can easily see that a short leg would not be so serviceable ; then the cushioned feet enable it to hold with greater security.

We may ask why those eyes on the crown of most insects were denied ; the dragon-fly and other insects to hunt their prey on the wing require to be pretty sharp-sighted, require to see above, around, and I may say, behind them ; but the Femoratum walks leisurely along, its food is there before it as it were, its residence is among the leaves, (except towards the close of its existence, when we find them on the bark looking for their mates,) where it manages to get at it without the quick visual organ of those insects that live by hawking.

The mouth is also well adapted ; we can see the use of the grinders in ruminating animals, as well as the incisors in carnivorous so even in the insect-world the divine mechanician has supplied the wants of the little spectrum.

---

## THE EVIDENCE OF FOSSIL PLANTS AS TO THE CLIMATE OF THE POST-PLIOCENE PERIOD IN CANADA.

By J. W. Dawson, LL.D., F.R S., F.G.S., Principal of McGill College.

---

The importance of all information bearing on the temperature of the Post-pliocene period, invests with much interest the study of the land plants preserved in deposits of this age. Unfortunately these are few in number, and often not well preserved. In Canada, though fragments of the woody parts of plants occasionally occur in the marine clays and sands, there is only one locality which has afforded any considerable quantity of remains of their more perishable parts. This is the well-known deposit of Leda clay at Green's Creek on the Ottawa, celebrated for the perfection in which the skeletons of the capelin and other fishes are preserved in the calcareous nodules imbedded in the clay. In similar nodules, contained apparently in a layer somewhat lower than that holding the ichthyolites, remains of land plants are somewhat abundant, and, from their association with shells of *Leda truncata*, seem to have been washed down from the land into deep water. The circumstances would seem to have been not dissimilar from those at present existing in the north-east arm of Gaspé Basin, where I have dredged from mud now being deposited in deep water, living specimens of *Leda limatula* mixed with remains of land plants.

In my examinations of these plants, I have been permitted to avail myself of a considerable collection in the museum of the geological survey of Canada, and also of the private collections of Mr. Billings, of Prof. Bell of Queen's College, and of Sheriff Dickson of Kingston. An imperfect list of these plants was published in my paper on the Post-pliocene of Canada in this Journal, and which was reproduced in 'Geology of Canada,' 1863. Since that time I have obtained some additional material, and have carefully re-examined all the specimens with the aid of collections of recent northern plants. I have also explored the locality in which the greater number of these remains were found. The principal points to which my attention has been directed are,— (1) The correct determination of the species of plants found; (2) The climate which they would indicate; and, (3) The portion of the Post-pliocene period to which they belong, with its probable geographical conditions.

### I. Species of Plants Found.

Under this head I shall give in detail only those species which I am able, from the fragments found, to determine with tolerable certainty.

1. *Drosera rotundifolia* Linn. In a calcareous nodule from Green's Creek, the leaf only preserved. This plant is common in

bogs in Canada, Nova Scotia and Newfoundland, and thence, according to Hooker, to the Arctic circle.  It is also European.

2. *Acer spicatum* Lamx.  (*Acer montanum* Aiton.)   Leaf in a nodule from Green's Creek.  Found in Nova Scotia and Canada, also at Lake Winnepeg, according to Richardson.

3. *Potentilla Canadensis* Linn.   In nodules from Green's Creek ; leaves only preserved.  I have had some difficulty in determining these, but believe they must be referred to the species above named or to *P. simplex* Michx., supposed by Hooker and Gray to be a variety.  It occurs in Canada and New England, but I have no information as to its range northward.

Fig. 1. Gaylussaccia resinosa.

4. *Gaylussaccia resinosa* Torrey and Gray.  Leaf in nodule at Green's Creek.  Abundant in New England and in Canada, also on Lake Huron and the Saskatchewan, according to Richardson.

Figs. 2 and 3. Populus balsamifera.

5. *Populus balsamifera* Linn. Leaves and branches in nodules at Green's Creek. This is by much the most common species, and its leaves are of small size, as if from trees growing in cold and exposed situations. The species is North American and Asiatic, and abounds in New England and Canada. It extends to the Arctic circle, and is abundant on the shores of the Great Slave Lake and on the McKenzie River, and according to Richardson constitutes much of the drift timber of the Arctic coast.

Fig. 4. Wood of Populus balsamifera.

6. *Thuja occidentalis* Linn. Trunks and branches in the Leda clay at Montreal. This tree occurs in New England and Canada, and extends northward into the Hudson Bay Territories, but I have not information as to its precise northern range. According to Lyell it occurs associated with the bones of Mastodon in New Jersey. From the great durability of its wood, it is one of the trees most likely to be preserved in aqueous deposits.

7. *Potamogeton perfoliatus* Linn. Leaves and seeds in nodules at Green's Creek. Inhabits streams of the Northern States and Canada, and according to Richardson extends to Great Slave Lake.

8. *Potamogeton pusillus.* Quantities of fragments which I refer to this species occur in nodules at Green's Creek. They may possibly belong to a variety of *P. hybridus* which, together with *P. natans*, now grows in the river Ottawa, where it flows over the beds containing these fossils.

9. *Cariceæ* and *Gramineæ.* Fragments in nodules from Green's Creek, appear to belong to plants of these groups, but I cannot venture to determine their species.

10. *Equisetum scirpoides* Michx. Fragments in nodules, Green's Creek. This is a widely distributed species, occurring in the Northern States and Canada.

11. *Fontinalis.* In nodules at Green's Creek there occurs, somewhat plentifully, branches of a moss apparently of the genus ·*Fontinalis*.

12. *Algæ.* With the plants above mentioned, both at Green's Creek and at Montreal, there occur remains of sea-weeds. They seem to belong to the genera *Fucus* and *Ulva*, but I cannot deter-

Fig. 5. Frond of Fucus.

mine the species. A thick stem in one of the nodules would seem to indicate a large *Laminaria*. With the above there are found at Green's Creek a number of fragments of leaves, stems and fruits, which I have not been able to refer to their species, principally on account of their defective state of preservation. Additional specimens may possibly in time resolve some of them.

## II. Climate indicated.

None of the plants above mentioned is properly Arctic in its distribution, and the assemblage may be characterized as a selection from the present Canadian flora of some of the more hardy species having the most northern range. Green's Creek is in the central part of Canada, near to the parallel of 46°, and an accidental selection from its present flora, though it might contain the same species found in the nodules, would certainly include with these, or instead of some of them, more southern forms. More especially the balsam poplar, though that tree occurs plentifully on the Ottawa, would not be so predominant. But such an assemblage of drift plants might be furnished by any American stream flowing in the latitude of 50° to 55° north. If a stream flowing to the north it might deposit these plants in still more northern latitudes, as the McKenzie River does now. If flowing to the south it might deposit them to the south of 50°. In the case of the Ottawa, the plants could not have been derived from a more southern locality, nor probably from one very far to the north. We may therefore safely assume that the refrigeration indicated by these plants would place the region bordering the Ottawa in nearly the same position with that of the south coast of Labrador fronting on the Gulf of St. Lawrence, at present. The absence of all the more Arctic species occurring in Labrador, should perhaps induce us to infer a somewhat more mild climate than this.

The moderate amount of refrigeration thus required, would in my opinion accord very well with the probable conditions of climate deducible from the circumstances in which the fossil plants in question occur. At the time when they were deposited the sea flowed up the Ottawa valley to a height of 200 to 400 feet above its present level, and the valley of the St. Lawrence was a wide arm of the sea, open to the Arctic current. Under these conditions the immense quantities of drift ice from the northward, and the removal of the great heating surface now presented by the low lands of Canada and New England, must have given for the Ottawa coast of that period a summer temperature very similar to that at present experienced on the Labrador coast, and with this conclusion the marine remains of the Leda clay as well as the few land mollusks whose shells have been found in the beds containing the plants, and which are species still occurring in Canada, perfectly coincide.

The climate of that portion of Canada above water at the time when these plants were imbedded, may safely be assumed to have been colder in summer than at present, to an extent equal to about 5° of latitude, and this refrigeration may be assumed to correspond with the requirements of the actual geographical changes implied. In other words, if Canada was submerged until the Ottawa valley was converted into an estuary inhabited by species of *Leda*, and frequented by capelin, the diminution of the summer heat consequent on such depression, would be precisely suitable to the plants occurring in these deposits, without assuming any other cause of change of climate.

### III. Age of the Deposits.

I have arranged elsewhere the Post-pliocene deposits of the central part of Canada, as consisting of, in ascending order ; (1) The Boulder Clay ; (2) A deep-water deposit, the Leda Clay ; and, (3) A shallow-water deposit, the Saxicava Sand. But although I have placed the boulder clay in the lowest position, it must be observed that I do not regard this as a continuous layer of equal age in all places. On the contrary, though locally, as at Montreal, under the Leda clay, it is in other places and at other levels contemporaneous with or newer than that deposit, which itself also locally contains boulders.

At Green's Creek the plant-bearing nodules occur in the lower part of the Leda clay, which contains a few boulders, and is apparently in places overlaid by large boulders, while no distinct boulder clay underlies it. The circumstances which accumulated the thick bed of boulder clay near Montreal, were probably absent in the Ottawa valley. In any case we must regard the deposits of Green's Creek as coeval with the Leda clay of Montreal, and with the period of the greatest abundance of *Leda truncata*, the most exclusively Arctic shell of these deposits. In other words I regard the plants above mentioned as probably belonging to the period of greatest refrigeration of which we have any evidence of course not including that mythical period of universal incasement in ice, of which, as I have elsewhere endeavoured to show, in so far as Canada is concerned, there is no evidence whatever.

The facts above stated in reference to Post-pliocene plants, concur with all the other evidence I have been able to obtain, in the conclusion that the refrigeration of Canada in the Post-pliocene period consisted of a diminution of the summer heat, and was of no greater amount than that fairly attributable to the great depression of the land and the different distribution of the ice-bearing Arctic current.

In connection with the plants above noticed, it is interesting to observe that at Green's Creek, at Pakenham Mills, at Montreal, and at Clarenceville on Lake Champlain, species of Canadian *Pulmonata* have been found in deposits of the same age with those containing the plants. The species which have been noticed belong to the genera *Lymnea* and *Planorbis*.*

I may also state as a curious fact, that among the nodules containing leaves, I have found some containing impressions of *feathers*, apparently of some small grallatorial bird. The substance of the feather has disappeared even more completely than in the celebrated Solenhofen specimens, but the impression is perfect, and in these hard nodular concretions might endure for any length of time. In searching for the fossil plants, I have also found an interesting addition to the fauna of these deposits in a Stickleback of the genus *Gasterosteus*.

---

## MISCELLANEOUS.

---

NEW FLUID FOR PRESERVING NATURAL HISTORY SPECIMENS ; by A. E. VERRILL.—In consequence of the high price of alcohol, a series of experiments were undertaken by me last year, with the view of finding a substitute for it in preserving the soft parts of animals. Among the various solutions and liquids tested were nearly all that have ever been recommended, besides many new ones. Chlorid of zinc, carbolic acid, glycerine, chlorid of calcium, acetate of alumina, arsenious acid, Goadby's solutions, and various combinations of these and other preparations were carefully tried, and the results made comparative by placing the same kind of objects in each, at the same time. Although each of these, under certain circumstances, have more or less preservative qualities, none of them were found satisfactory, especially when the *color* and *form* of the specimen are required to be preserved as well as its structure.

---

* Canadian Naturalist, 1860, p. 195 ; ' Geology of Canada,' 1863, p. 928.

As a test for the preservation of color, the larvæ of the tomato-worm (*Sphinx quadrimaculata*) was used. These larvæ are difficult of preservation with the natural form and color, nearly always turning dark brown and contracting badly in alcohol and most other preparations.

As a result of these experiments the following solutions were found highly satisfactory in all respects when properly used. By their use the larvæ and recent pupæ of the tomato-worm were preserved and still retain their delicate green colors, together with their natural form and translucent appearance, while the internal organs are fully preserved. Fishes, mollusks, various insects, worms, and leaves of plants have also been preserved with perfect success and far better than can be done with alcohol. In the case of mollusks, especially, the preparations are very beautiful, retaining the delicate semi-transparent appearance of the membranes nearly as in life, with but little contraction. Another great advantage is the extreme simplicity and cheapness of the solution.

To use this fluid I prepare first the following stock solution, which may be kept in wooden barrels or casks, and labeled :

## SOLUTION A 1.

| | |
|---|---|
| Rock salt | 40 oz. |
| Nitre (nitrate of potassa) | 4 oz. |
| Soft water | 1 gal. |

This is the final solution in which all invertebrate animals must be preserved. A solution with double the amount of water may be kept if desirable, and called A 2. Another with three gallons of water will be A 3.

In the preliminary treatment of specimens the following solution is *temporarily* employed, and is designed to preserve the object while becoming gradually saturated with the saline matter, for in no case should the specimen be put into the full strength of solution A 1, for it would rapidly harden and contract the external parts and thus prevent access to the interior. Even with alcohol it is far better to place the object for a time in weak spirits and then tranfer successively to stronger, and for some objects, as Medusæ, no other treatment will succeed.

### SOLUTION B 1.

| | |
|---|---|
| Soft water...................... | 1 gal. |
| Solution A 1.................... | 1 qt. |
| Arseniate of potassa........... | 1 oz. |

Another solution with double the amount of water may be made if desired, and called solution B 2.

To preserve animals with these solutions, they are, if insects or marine invertebrates, ordinarily placed first in solution B 1, but if the weather be cool it would be better in many cases to employ first B 2, and in the case of all marine animals washing first in fresh water is desirable, though not essential. If the specimens rise to the surface they should be kept under by mechanical means. After remaining for several hours, or a day, varying according to its size and the weather, in the B 1 solution it may be transferred to A 3, and then successively to A 2 and A 1, and when thus fully preserved it may be transferred to a fresh portion of the last solution, which has been filtered clear and bright, and put up in a cabinet, when no further change will be necessary if the bottle or other vessels be properly secured to prevent the escape of the fluid by crystallization around the opening. To prevent this, the stoppers, whether of cork or glass, together with the neck of the bottle or jar, may be covered with a solution of paraffine or wax in turpentine or benzole, which should be applied only when the surfaces are quite dry and clean. The length of time that any specimen should remain in each of the solutions is usually indicated by their sinking to the bottom when saturated by it. In general the more gradually this saturation with the saline matter takes place the less the tissues contract or change in appearance. In many cases, however, fewer changes than indicated above will be effectual. I have in some cases succeeded well with but two solutions below A 1. For vertebrates, except fishes, the solution A 2, will usually be found strong enough for permanent preservation, especially when the object is small or dissected. If the entire animal be preserved, when larger than two pounds in weight, it should be injected with the fluids, especially B 1 or B 2, or an incision may be made in one side of the abdomen in vertebrates, or under the carapax of crabs, &c., to admit the fluids more freely. In preserving the animals of large univalve shells an opening should be made through the shell at or near the tip of the spire. Mammals, birds and reptiles, should be placed first in solution B 2 to obtain the best results. In cases where the use of the B. fluids would be objectionable, on account of their highly poisonous nature, a fourth dilution of solution A 1, corresponding in strength with B 1, but without the arseniate of potassa, may be substituted, and in many cases will do nearly as well, if the weather be not very hot, but the specimens in this case should be carefully watched and transferred to the stronger solutions as soon as possible, so as to avoid incipient decomposition while in the first fluids.—*Silliman's Journal.*

New Haven, Feb. 12, 1866.

ILLUMINATION UNDER THE MICROSCOPE.—At the late *soirée* at University College, two forms of Mr. Smith's (of the United States) illumination for opaque objects under high microscopic powers were exhibited. One was constructed by Messrs. Smith and Beck, of Cornhill, and the other by Messrs. Powell and Lealand. The first form closely resembles the American contrivance—so closely, indeed, that it is difficult to know in what the difference between the two consists. A bass box intervenes between the end of the microscope tube and the objective. This is pierced at the side by an aperture opposite which a table lamp is placed; within the box is a small silvered mirror, which receives the light from the lamp, and throws it down through the objective upon the object. This contrivance, thought it works admirably with such a power as the one-fifth inch, is objectionable, from the fact that it cuts off half the pencil of rays proceeding to the eye of the observer. The second form—that exhibited by Messrs. Powell and Lealand—is superior to that of Smith and Beck, and differs from the American plan in having a reflector of plain glass. The result of this alteration of the original plan is that whilst sufficient light is thrown down to illuminate the object, the rays proceeding from the latter are not partially cut off. This modification applied to the one-twelfth inch gave splendid results, and the makers allege that it may be used with one-twenty-fifth or one-fiftieth inch glasses with equal advantage.—*Reader*, Dec. 23.

THE BIRDS OF NORTH AMERICA.—D. G. Elliot of New York (27, W. 23d st.) proposes to publish a work to contain all the new and unfigured birds of America, to be issued in Parts, 19 × 24 inches in size, containing each five plates colored by hand, with a concluding part of text ; price for each part, ten dollars.  Only 200 copies will be published.  Mr. Elliot is author of a Monograph of the Pittidæ or Ant Trushes, in one volume imperial folio, with 31 plates, and a Monograph of the Tetraoninæ, Grouses, one vol. royal folio, with 25 plates ; in each of which, the birds, with two exceptions only, are represented of life-size. Subscriptions are requested.—*Silliman's Journal.*

## PUBLISHER'S NOTICE.

*Owing to various unforeseen circumstances a very great delay has occurred in the issue of this number of the Canadian Naturalist.   The remaining numbers of this volume will be issued during the present year, so that Vol. 3, New Series, will be for* 1866–7.

Montreal,
          January 12, 1867.

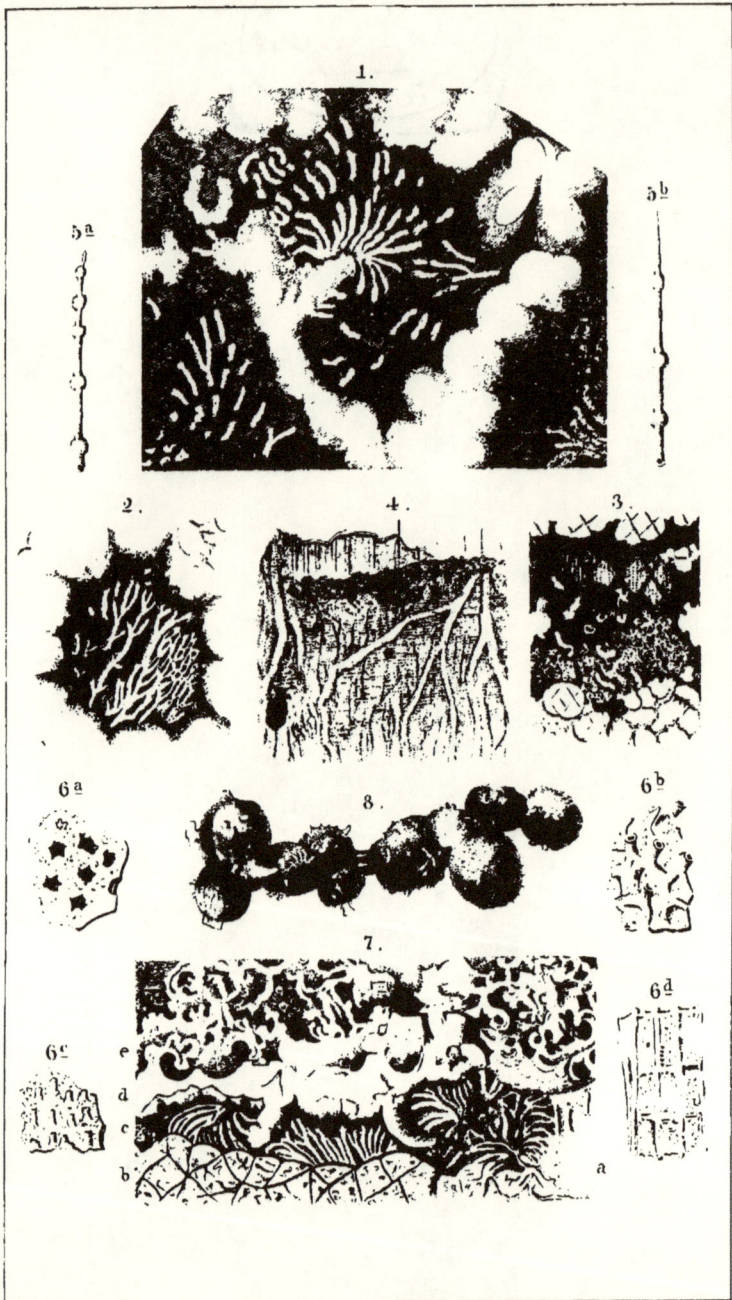

Auct. delin.

Roberts & Reinhold, Lith. Place d'Armes. Montreal.

Gümbel on Eozoön from the primitive rocks of Bavaria.

# CANADIAN NATURALIST

## SECOND SERIES.

## ON THE LAURENTIAN ROCKS OF BAVARIA.

By DR. GÜMBEL, Director of the Geological Survey of Bavaria; with a plate containing figures of two species of Eozoon.

*Translated from the Proceedings of the Royal Bavarian Academy for 1866, by Professor Markgraf.**

The discovery of organic remains in the crystalline limestones of the ancient gneiss of Canada, for which we are indebted to the researches of Sir William Logan and his colleagues, and to the careful microscopic investigations of Drs. Dawson and Carpenter, must be regarded as opening a new era in geological science.

This discovery overturns at once the notions hitherto commonly entertained with regard to the origin of the stratified primary limestones, and their accompanying gneissic and quartzose strata, included under the general name of primitive crystalline schists. It shows us that these crystalline stratified rocks, of the so-called primary system, are only a backward prolongation of the chain of fossiliferous strata ; the elements of which were deposited as oceanic sediment, like the clay-slates, limestones and sandstones of the paleozoic formations, and under similar conditions, though at a time far more remote, and more favorable to the generation of crystalline mineral compounds.

In this discovery of organic remains in the primary rocks, we hail with joy the dawn of a new epoch in the critical history of these earlier formations. Already, in its light, the primeval geologic time is seen to be everywhere animated, and peopled with new animal forms, of whose very existence we had previously no suspicion. Life, which had hitherto been supposed to have first

---

*EDITOR'S NOTE.—In revising and preparing this for the press, the original paper has been considerably abridged by the omission of portions, whose place is indicated in the text. Some explanatory notes have also been added.—T. S. H.

appeared in the primordial division of the Silurian period, is now seen to be immeasurably lengthened beyond its former limit, and to embrace in its domain the most ancient known portions of the earth's crust. It would almost seem as if organic life had been awakened simultaneously with the solidification of the earth's crust.

The great importance of this discovery cannot be clearly understood, unless we first consider the various and conflicting opinions and theories which had hitherto been maintained concerning the origin of these primary rocks. Thus some, who consider them as the first-formed crust of a previously molten globe, regard their apparent stratification as a kind of concentric parallel structure, developed in the progressive cooling of the mass from without. Others, while admitting a similar origin of these rocks, suppose their division into parallel layers to be due, like the lamination of clay-slates, to lateral pressure. If we admit such views, the igneous origin of schistose rocks becomes conceivable, and is in fact maintained by many.

On the other hand, we have the school which, while recognizing the sedimentary origin of these crystalline schists, supposes them to have metamorphosed at a later period; either by the internal heat, acting in the deeply buried strata; by the proximity of eruptive rocks; or finally, through the agency of permeating waters charged with certain mineral salts.

A few geologists only have hitherto inclined to the opinion that these crystalline schists, while possessing real stratification, and sedimentary in their origin, were formed at a period when the conditions were more favorable to the production of crystalline materials than at present. According to this view, the crystalline structure of these rocks is an original condition, and not one superinduced at a later period by metamorphism. In order however to arrange and classify these ancient crystalline rocks, it becomes necessary to establish, by superposition or by other evidence, differences in age, such as are recognized in the more recent stratified deposits. The discovery of similar organic remains, occupying a determinate position in the stratification, in different and remote portions of these primitive rocks, furnishes a powerful argument in favor of the latter view, as opposed to the notion which maintains the metamorphic origin of the various minerals and rocks of these ancient formations; so that we may regard the direct formation of these mineral elements, at least so far as these fossiliferous primary limestones are concerned, as an established fact.

So early as 1853, after investigating the primitive rocks of eastern Bavaria, which are connected with those of the Bohemian forest, I expressed the opinion that, although eruptive masses of granite and similar rocks occur in that region, the gneiss was of sedimentary origin, and divisible into several formations. I at that time endeavored to separate these crystalline schists into three great divisions, the phyllades, the mica-schists, and the gneiss formation, of which the first was the youngest and the last the oldest; all these formations having essentially the same dip and strike.

These results, obtained from very detailed geological and topographical researches, were subsequently more fully set forth in the Survey of the Geology of Eastern Bavaria, (Book IV., p. 219 *et seq.*); where I endeavored to assign local names to the subdivisions of the primitive rocks of that region. Beginning with the more recent, I distinguished the following formations:

1. Hercynian primitive clay-slate.
2. Hercynian mica-slate.
3. Hercynian gneiss. } Primary gneiss system.
4. Bojian gneiss. }

In some cases, within limited regions, I even succeeded in tracing out still smaller subdivisions. It was in this way established that definite and distinct kinds of rocks, as for example hornblende-slate and mica-slate, may replace each other and, as it were, pass into each other, in different parts of the same horizon.

After Sir Roderick Murchison had established the existence of the fundamental gneiss in Scotland, and recognized its identity with that of the Laurentian system of Canada, he turned his attention to the primitive rocks of Bavaria and Bohemia. My researches and my communications to him disclosed the important fact that these rocks belong to the same series as the oldest formations of Canada and of Scotland. On one point only was there an apparent difference of opinion between Sir Roderick and myself; which was that he was disposed to look upon the whole of the gneiss of the Hercynian mountains as constituting but a single formation, corresponding to the Laurentian gneiss of Canada and of Scotland; while I had endeavored to distinguish two divisions, the newer grey or Hercynian gneiss, and the older red or variegated, which I called the Bojian gneiss. This difference of opinion is however at once removed by the remark that I did not intend to maintain in the older gneiss the existence of a formation more ancient than the fundamental gneiss of Scotland, nor yet to assimilate the newer or grey gneiss to the more recent or so-called metamorphic series, which, according to Sir Roderick, may be clearly distinguished in Scotland from the Laurentian gneiss.

[This newer gneissic formation of the Highlands is, according to Murchison, Ramsay and others, of Lower Silurian age. Our author simply claims to have established a division in the proper Laurentian rocks of Bavaria and Bohemia. It will be seen from the recently published maps of the Laurentian region of the Ottawa, that Sir William Logan there distinguishes three great limestone formations, by which the enormous mass of Laurentian gneiss is separated into four divisions. One or two of the upper ones of these may be eventually found to correspond to the grey Hercynian gneiss of Bavaria, which is there accompanied by the Eozoon Canadense, a fossil so far as yet known characterizing the highest of the three Laurentian limestones. This grey gneiss of Bavaria appears to be lithologically distinct from the Labrador (or Upper Laurentian) series; nor do we find in the present memoir of Gumbel, any clear evidence of the occurrence either of this, or of the Huronian system, in Bavaria.—T. S. H.

After citing in this connection Sir W. E. Logan's observations on these ancient formations, which are shown, by the results of the Canadian Survey, to represent three great systems of sedimentary rocks, formed under conditions not unlike those of more modern formations, our author observes:—]

Accepting these views of the older Canadian rocks, it would naturally follow that organic life might be expected to reach back much farther than the so-called primordial fauna of Lower Silurian age, and to mark the period hitherto designated as Azoic.

Guided by these ideas, the geologists of Canada zealously sought for traces of organic life in the primitive rocks of that country. Dr. Sterry Hunt had already concluded that it must have existed in the Laurentian period, from the presence of beds of iron ore, and of metallic sulphurets, which, not less than the occurrence of graphite, were to him chemical evidences of an already existing vegetation, when at length direct evidence of life was obtained by the discovery of apparently organic forms in the great beds of crystalline limestone which occur in the Laurentian system.　Such were collected in 1858, by Mr. J. McMullen from the Grand Calumet on the Ottawa River, and were observed by Sir Wm. Logan to resemble closely similar specimens obtained by Dr. James Wilson in Burgess, a few years previously.　In 1859, Sir Wm. Logan first expressed his opinion that these masses, in which pyroxene, serpentine, and an allied mineral, alternated in thin layers, with carbonate of lime or dolomite, were of organic origin; and in 1862 he reiterated this opinion in England, without however being able to convince the English geologists, Ramsay excepted, of the correctness of his views.　Soon after this, however, the discovery of other and more perfect specimens, at Grenville, furnished decisive proofs of the organic nature of these singular fossils.

The careful and admirable investigations of Dawson and of Carpenter, to whom specimens of the rock were confided, have placed beyond doubt the organic structure of these remains, and confirmed the important fact that these ancient Laurentian limestones abound in a peculiar organic fossil, unknown in more recent formations, to which has been given the name of Eozoon.*

\*　　\*　　\*　　\*　　\*　　\*

The researches of Sterry Hunt on the mineralogical relations of the Eozoon-bearing rocks, lead him to the important conclusion that certain silicates, namely serpentine, white pyroxene, and loganite, have filled up the vacant spaces left by the disappearance of the destructible animal matter of the sarcode, the calcareous skeleton remaining more or less unchanged.　If, by the aid of acids, we remove from such specimens the carbonate of lime, (or, in certain cases, the dolomite which replaces it,) there remains a coherent skeleton, which is evidently a cast of the soft parts of the Eozoon.　The process by which the silicates have been introduced into the empty spaces corresponds evidently to that of ordinary silicification through the action of water.　It is to be noted that Hunt found serpentine and pyroxene, side by side, in adjacent chambers, and even sharing the same chamber between them; thus affording a beautiful proof of their origin through the

---

* Here follows, in the original, a lengthened analysis of the memoirs of Messrs. Logan, Dawson, Carpenter, and Hunt, published in the Quarterly Journal of the Geological Society of London, and already reprinted in the Canadian Naturalist.

infiltration of aqueous solutions, while the Eozoon was yet growing, or shortly after its death.          *          *          *

Hunt, in a very ingenious manner, compares this formation and deposition of serpentine, pyroxene, and loganite, with that of glauconite, whose formation has gone on uninterruptedly from the Silurian to the Tertiary period, and is even now taking place in the depths of the sea; it being well known that Ehrenberg and others have already shown that many of the grains of glauconite are casts of the interior of foraminiferal shells. In the light of this comparison, the notion that the serpentine, and such like minerals of the primitive limestones have been formed in a similar manner, in the chambers of Eozoic foraminifera, loses any traces of improbability which it might at first seem to possess.          *          *

My discovery of similar organic remains in the serpentine-limestone from near Passau was made in 1865, when I had returned from my geological labors of the summer, and received the recently published descriptions of Messrs. Logan, Dawson, etc. Small portions of this rock, gathered in the progress of the geological survey in 1854, and ever since preserved in my collection, having been submitted to microscopic examination, confirmed in the most brilliant manner the acute judgment of the Canadian geologists ; and furnished paleontological evidence that, notwithstanding the great distance which separates Canada from Bavaria, the equivalent primitive rocks of the two regions are characterized by similar organic remains; showing at the same time that the law governing the definite succession of organic life on the earth is maintained even in these most ancient formations. The fragments of serpentine-limestone or ophicalcite, in which I first detected the existence of Eozoon, were like those described in Canada in which the lamellar structure is wanting, and offer only what Dr. Carpenter has called an accrvuline structure. For further confirmation of my observations, I deemed it advisable, through the kindness of Sir Charles Lyell, to submit specimens of the Bavarian rock to the examination of that eminent authority, Dr. Carpenter ; who, without any hesitation, declared them to contain Eozoon.

This fact being established, I procured from the quarries near Passau as many specimens of the limestone as the advanced season of the year would permit ; and, aided by my diligent and skilful assistants Messrs. Reber and Schwager, examined them by the methods indicated by Messrs. Dawson and Carpenter. In this way I soon convinced myself of the general similarity of our organic remains with those of Canada. Our examinations were made on polished sections and in portions etched with dilute nitric acid, or, better, with warm acetic acid. The most beautiful results were however obtained by etching moderately thin sections, so that the specimens may be examined at will either by reflected or by transmitted light.

The specimens in which I first detected Eozoon came from a quarry at Steinhag, near Obernzell on the Danube, not far from Passau. The crystalline limestone here forms a mass from fifty to seventy feet thick, divided into several beds, included in the gneiss, whose general strike in this region is N.W., with a dip of 40°–60° N.E. The limestone strata of Steinhag have a dip of 45° N.E. The gneiss of this vicinity is chiefly grey, and very silicious, containing dichroite, and of the variety known as dichroite-gneiss; and I conceive it to belong, like the gneiss of Bodenmais and Arber, to that younger division of the primitive gneiss system which I have designated as the Hercynian gneiss formation; which both to the north, between Tischenreuth and Mahring, and to the south, on the south-west of the mountains of Ossa, is immediately overlaid by the mica-slate formation. Lithologically, this newer division of the gneiss is characterized by the predominance of a grey variety, rich in quartz, with black magnesian-mica and orthoclase, besides which a small quantity of oligoclase is never wanting. A farther characteristic of this Hercynian gneiss is the frequent intercalation of beds of rocks rich in hornblende, such as hornblende-schist, amphibolite, diorite, syenite, and syenitic granite, and also of serpentine and granulite. Beds of granular limestone, or of calcareous schists are also never altogether wanting; while iron pyrites, and graphite, in lenticular masses, or in local beds conformable to the great mass of the gneiss strata, are very generally present.

The Hercynian gneiss strata on the shores of the Danube near Passau are separated from the typical Hercynian gneiss districts which occur to the north, on the borders of the Fichtelgebirge and near Bodenmais and Arber, by an extensive tract, partly occupied by intrusive granites, and partly by another variety of gneiss. These Danubian gneiss strata are not seen to come in contact with any newer crystalline formation, but towards the south are concealed by the tertiary strata of the Danubian plain; while towards the N.W. they are in part cut off by granite, and in part

replaced by those belts of gneiss which accompany the quartz
ridge of the Pfahl; and belong to the red variety or Bojian
gneiss.   The grey gneiss strata of the Danube might therefore be
supposed to be older than this red gneiss, which from its relations
in the district to the N.W., between Cham and Weiden, I had
regarded as itself the more ancient formation.   But the lithological
characters of the grey Danubian gneiss are opposed to this view,
since this rock not only presents a general resemblance to the
gneiss formation of Bodenmais, which without doubt is directly
overlaid by the mica-schist of the mountains of Ossa, thus shewing
it to be the newer gneiss; but exhibits a repetition of the minor
features which characterize the gneiss district of Bodenmais.   We
find in the Danubian gneiss that same abundant dissemination of
dichroite, which gives rise to the typical dichroite-gneiss of
Bodenmais, with nearly the same mineral associations in both
cases.   On the Danube, also, interstratified beds of hornblende-
rock (at Hals near Passau), of serpentine (at Steinhag), and of
pyrites (at Kelberg, and many points along the Danube), occur, as
in the north.   On the other hand, the graphite which abounds in
the gneiss of Passau is not wanting at Bodenmais or Tischenreuth.
The interstratified syenites and syenitic granites are, in like manner,
common to all these districts; those near Passau being, however,
richer in easily decomposed minerals, such as porcelain-spar
(scapolite) and calcspar, are more subject to decomposition, and
form the parent rock of the famous porcelain clays of the region.

These resemblances lead me to refer the Danubian gneiss,
notwithstanding its apparent stratigraphical inferiority to the red
gneiss, to the newer or Hercynian formation; and to explain its
apparently abnormal relations by assuming a fault running along
the strike from N.W. to S.E., through which the older gneiss of
the Pfhal is brought up, and seems to overlie the younger.

We shall then regard the whole of the gneissic strata characterized by dichroite, which extend on the Danube from Passau to
Linz, as equivalent to the Hercynian gneiss of Bodenmais, and
designate it as the Danubian gneiss.   We may here call attention
to the abundance of graphitic beds in it, as also to the occurrence
of porcelain clay, and of beds of iron pyrites and magnetic pyrites.
If it is true (as maintained by Dr. Sterry Hunt) that all graphite
owes its origin to organic matters, we must suppose the existence
of a primordial region peculiarly rich in organic life; since
graphite occurs here in almost all the strata, and in some places in

such quantities that it is profitably extracted, and is largely used for the manufacture of the famous Passau crucibles. In all of the numerous graphite mines, the uniform interstratification of bands and lenticular masses rich in graphite with the gneiss is here distinctly marked. A similar arrangement is seen in the sulphurets of iron, which are more abundantly disseminated in the more hornblendic strata. The localities of porcelain-earth or kaolin are in like manner confined to the strike of the gneissic strata; and are generally contiguous to certain interstratified granitic and syenitic bands, rich in feldspar. Its frequent association with porcelain-spar, (probably nothing more than a chloriferous scapolite or anorthite,) indicates that this mineral has played an essential part in the production of the kaolin. The presence of chlorine in this mineral is highly significant, and suggests the agency of sea-water in its production.

Of particular interest, from their mineral associations, are three or more parallel bands of crystalline limestone of no great thickness, which occur conformably interstratified with the gneiss of the hills near Passau. They begin near Hofkirchen, and extend north and south, from along the Danube as far as the frontier, near Jochenstein, where the Danube leaves Bavaria. These separate limestone bands, although exposed by numerous quarries, cannot be followed uninterruptedly, being sometimes concealed, and sometimes of insignificant thickness.

\*     \*     \*     \*     \*     \*

The large quarry of Steinhag already described, from which I first obtained the Eozoon, is one. The enclosing rock is a grey hornblendic gneiss, which sometimes passes into a hornblende-slate. The limestone is in many places overlaid by a bed of hornblende-schist, sometimes five feet in thickness, which separates it from the normal gneiss. In many localities, a bed of serpentine, three or four feet thick, is interposed between the limestone and the hornblende-schist; and in some cases a zone, consisting chiefly of scapolite, crystalline and almost compact, with an admixture however of hornblende and chlorite. Below the serpentine band, the crystalline limestone appears divided into distinct beds, and encloses various accidental minerals, among which are reddish-white mica, chlorite, hornblende, tremolite, chondrodite, rosellan, garnet, and scapolite arranged in bands. In several places the lime is mingled with serpentine, grains or portions of which, often of the size of peas, are scattered through the limestone with

apparent irregularity, giving rise to a beautiful variety of ophicalcite or serpentine-marble. These portions, which are enclosed in
the limestone destitute of serpentine, always present a rounded
outline. In one instance there appears, in a high naked wall of
limestone without serpentine, the outline of a mass of ophicalcite,
about sixteen feet long and twenty-five feet high, which, rising
from a broad base, ends in a point, and is separated from the
enclosing limestone by an undulating but clearly defined margin, as
already well described by Wineberger. This mass of ophicalcite
recalls vividly a reef-like structure. Within this, and similar
masses of ophicalcite in the crystalline limestone, there are, so far
as my observations in 1854 extend, no continuous lines or
concentric layers of serpentine to be observed, this mineral being
always distributed in small grains and patches. The few
.apparently regular layers which may be observed are soon
interrupted, and the whole aggregation is irregular. [This is
well shown in plates II. and III. in the original memoir, which
recall the acervuline portions, that make up a large part of the
Canadian specimens of Eozoon.—EDS.]

The numerous specimens which were subsequently collected, at
the commencement of the winter, show, throughout, this irregular
structure, which seems to characterize the Bavarian specimens of
Eozoon, as is in part the case in those from Canada. It is true
that small lenticular masses or nodules, consisting chiefly of
scapolite, measuring fifty by twenty millimeters, and even much
more, are often met with, around which serpentine is arranged in
a concentric manner ; but even here the serpentine is in small
cohering masses, and not in regular layers ; nor could I, after
numerous examinations of fragments of such masses, satisfy
myself whether I had to deal with the commencing growth of an
Eozoon, or merely with a concretionary mass; since the granular
structure of the scapolite centre could never be clearly made out.
Moreover the occurrence of these nodules, arranged in a stratiform
manner, is opposed to the notion that they are nuclei of Eozoon,
although in the parts around these nodules I could sometimes
distinctly observe tubuli, canals, and even indications of a shell-like
structure.

The portions of serpentine in the ophicalcite occur of very
various sizes, from that of a millet-seed to lumps whose sections
measure fifteen by six or eight millimeters. But I think I can
detect within certain lines, (which are not, it is true, very well

defined,) chains of serpentine grains, of nearly equal size, connected
with each other. When by means of acids the lime is removed
from these aggregates, a perfectly coherent serpentine skeleton is
in all cases obtained, which may be compared to a piece of wood
perforated by ants.     *     *     *     *     *
The surface of the serpentine grains is rounded, pitted, and
irregular ; plane surfaces and straight lines are rarely to be seen.
Even when dilute nitric or acetic acid has been used to remove
the lime, a white down-like coating is frequently found on the
serpentine, which does not answer to the nummuline wall of the
calcareous skeleton. In many cases, where the lime is very
crystalline, and the more delicate organic structure obliterated,
small tufts of radiated crystals, apparently hornblende or tremolite,
are seen resting upon the serpentine. These crystals, when seen
in thin sections, by transmitted light, may easily give rise to
errors ; their formation seems to have been possible only where the
calcareous skeleton had been destroyed, and crystalline carbonate
of lime deposited in its stead ; during which time free space was
given for the formation of these crystalline groups. In very
many cases there are seen, by a moderate magnifying power, (in
the residue from acids) deposits of small detached cylindrical
stems, with some larger ones, consisting of a white matter insoluble
in acids. These appear to be the casts of the tubuli which
penetrated the calcareous skeleton, and of the less frequent stolons,
as will be described.

The serpentine in these sections never appears quite homogeneous, but exhibits, on the contrary, irregular groups of small
dark-colored globules disseminated through the mass, without
however any definite indications of organic form. Still more
frequently, the serpentine is penetrated by irregularly reticulated
dark colored veins, giving to the mass a cellular aspect.

In certain parts of the serpentine, however, parallel lines, groups
of curved tube-like forms, and oval openings, clearly indicate an
organic structure like that of the Canadian Eozoon. The finely
tubulated nummuline wall of the chambers, which was discovered
by Carpenter, and the casts of whose tubuli appear in the
decalcified specimens from Canada as a soft white velvet-like
covering, could only be found in a few isolated cases in the
Bavarian specimens, but was clearly made out in a few fragments.
(Pl. I., 4.)     The somewhat oblique section shows the openings
of the minute tubuli.

It should be remarked that the serpentine at Steinhag occurs, not only replacing the sarcode in the carbonate of lime of the Eozoon, but also forming layers over the limestone strata, and moreover filling up large and small crevices and fissures, which have nothing at all to do with the organic structure. Especially worthy of notice are the plates of fibrous serpentine, or chrysotile, often from five to ten millimeters in diameter, which are found extending in unbroken lines through the compact serpentine.

The color of the serpentine presents all possible shades, from blackish green, to the palest yellowish green tint. Where it has been exposed to the weather, the serpentine has become of a pale brownish green, and appears changed into gymnite. The different tints are arranged in zones, and seem to mark different periods of growth. The carbonate of lime which is interposed among the grains of serpentine in the specimens from Steinhag, is either distinctly crystalline, or apparently compact. In the first case, no organic structure can be perceived; thin sections of the crystalline portions show only intersecting parallel lines; and in etched or entirely decalcified specimens, no clear evidence of the fine canal-system of the skeleton can be observed. These crystalline portions often alternate with others which are compact and but feebly translucent. In thin sections of these compact parts, the rounded forms of the delicate tubuli are very clearly discerned, provided the section is at right angles to them. In etched specimens, viewed by reflected light, these tubuli are seen to branch out in the form of tufts, exactly as described and figured by Drs. Dawson and Carpenter.

These branching and ramified tubuli rest upon the serpentine granules, and seem by anastomosis to be connected with adjacent groups. The diameter of these tubuli is from $\frac{15}{1000}$ to $\frac{20}{1000}$ millimeters. They are easily distinguishable from the delicate groups of crystals, which are also sometimes found implanted in the serpentine, by the nearly uniform thickness throughout their whole length; by their extremities, which are always somewhat crooked; and by their pipe-like form. The latter are never ramified; have a fibrous aspect; and are always straight, and terminate in a point. (Pl. I., figs. 1, 2, 3.)

Here and there are observed larger tubuli, which, so far as my observations extend, are always isolated, and nearly or quite parallel. (Pl. I., fig. 1.) Their diameter is about $\frac{7}{100}$ millimeters, and they not improbably represent those stolons or connecting channels with which Carpenter has made us acquainted.

In the decalcified specimens, delicate very slender string-like leaflets were very frequently observed, stretched between the serpentine granules; but they presented no discernible organic structure, and are perhaps only the casts of small crevices. More remarkable are the numerous canals filled with carbonate of lime, which traverse the serpentine granules, and at the surface of these are expanded into funnel shapes. They appear to represent cross connections between the calcareous skeleton.

As my object at present is merely to shew the presence, in the primitive limestones of Bavaria, of forms corresponding to the Canadian Eozoon, I will not dwell longer on these various appearances met-with in the microscopical examinations, nor on the peculiar cellular structures observed in the carbonate of lime. I will, for the same reason, only mention a specimen which exhibits, by the side of a curved main tube, a number of secondary tubuli, and farther on a parallel layer of fibres; and also another radiated form which resembles a section of a Bryozoon. It is sufficient to draw attention to the fact that, in addition to Eozoon, there are other organic remains in these crystalline limestones. There remains however to be noticed a phenomenon of some importance.

When the lime is removed by nitric or acetic acid from the interstices of the serpentine granules, there may be observed, on gently moving the liquid, extremely delicate membranes, that separate themselves from the serpentine grains, (which they covered thickly, as with a fine white down,) and now remain swimming in the liquid, so that they can readily be separated, by decantation, from a multitude of heavier particles, which, having also detached themselves from the serpentine mass, accumulate at the bottom of the vessel. These consist in great part of indistinct mineral fragments, and of small crystalline needles, together with distinct cylindrical portions, which are the broken tubuli of the Eozoon. Besides these are, here and there, distinctly knotted stems or tubules, (Pl. I., figs. 5, *a* and *b*,) which I dare not suppose to belong to Eozoon. Various other fragments of tubuli are also associated with these.

The delicate flakes, which can be obtained by evaporating the liquid in which they are suspended, shew, under a magnifying power of 400 diameters, a membranous character, and peculiar structures, which seem to be undoubtedly of organic origin.

Their forms are best understood by the figures 6, *a*, *b*, *c* and *d*. The examination of the fine slimy residues from the solution of various primary crystalline limestones, in which, from the absence of well marked foreign minerals, it may be difficult to prove the presence of distinct organic forms, will, I think, afford the quickest and readiest mode of establishing the existence of organisms.

The presence of the Eozoon in the primary limestone of Steinhag being thus established, I proceeded to examine such specimens as were at my disposal from other localities of similar limestones in the vicinity of Passau. I must here remark that these specimens, collected during my geological examinations twelve years since, were chosen as containing intermixtures of serpentine and hornblende, and not with reference to the possibility of their holding organic remains. I succeeded however in detecting at least traces of Eozoon in specimens of the limestone from Untersalzbach, (fig. 2,) from Hausbach, Babing, (fig. 3,) and from Kading and Stetting. Moreover a specimen of ophicalcite from a quarry near Srin, in the region between Krumau and Goldenkron, among the primitive hills of Bohemia, afforded unequivocal evidences of Eozoon. Von Hochstetter moreover has received specimens of crystalline limestone from the same strata at Krumau, in which Dr. Carpenter has shown the presence of Eozoon. To the same formation belong the calcareous rocks near Schwarzbach, in the vicinity of which, as near Passau, great masses of graphite are intercalated in the gneiss hills. These limestones of Schwarzbach connect those of Krumau with the similar strata near Passau, from which they are only separated by the great granite mass of the Plockenstein hills. We thus obtain a still farther proof of the similarity of structure throughout the whole range of primitive rocks of Bavaria and Bohemia; and of the parallelism of their lowest portion with the Laurentian gneiss system of Canada. I think therefore that we may, without hesitation, *place the Hercynian gneiss formation of the mountains forming the Bavarian and Bohemian frontier, on the same geological horizon with the Laurentian system.*

Farther northward, in similar gneiss hills, occupying a limited area, a crystalline limestone occurs near Burggrub, not far from Erbendorf, from which a few specimens were at hand. They were however a reddish, very ferruginous dolomite, penetrated by fibres of hornblende and epidote, and gave me no trace of organic remains.

Besides these limestones of the Hercynian gneiss, there is found in Bavaria another remarkable deposit of crystalline limestone, included in the Hercynian primitive clay-slate series on the south and south-east border of the Fichtelgebirge, in the vicinity of Wunseidel. This clay-slate formation, as we have already shewn, overlies the Hercynian gneiss and mica-slate series, and is immediately beneath the primordial zone of the Lower Silurian strata met with in the Fichtelgebirge. It would thus seem to correspond with the Cambrian rocks of Wales, and with the Huronian system of Canada, as Sir Roderick Murchison has already suggested. This view is confirmed by Fritzsch's discovery of traces of annelids in the grauwacke of Przibram, and by the occurrence of crinoidal stems and foraminiferal forms, according to Reuss, in the limestone of the primitive clay-slates of Paukratz, near Reichenstein. Thus our Hercynian mica-slate, with certain hornblendic strata and chloritic schists belonging to the same horizon, would occupy a stratigraphical position similar to the Labrador series, or Upper Laurentian, of Canada.

The crystalline limestone of the Fichtelgebirge forms in the primitive clay-slate two nearly parallel bands, which I conceive to be the outcrops of one and the same stratum, on the opposite sides of a trough. It presents several parallel beds separated by intervening beds of the conformable clay-slate.

The limestone strata near Wunseidel dip from 50° to 75° S.E., and sometimes attain a thickness of 350 feet. They are in many places dolomitic.    *    *    *    *    Spathic iron, in nests and disseminated, characterizes this rock, and by its decomposition gives rise to the valuable deposits of brown hematite, which are worked along the outcrop of the limestone band. Among the other minerals may be mentioned graphite, in crystalline plates, and also in small round grains and rounded compact masses in the limestone; besides which it frequently enters into the composition of the adjacent clay-slate, giving rise to a plumbaginous slate. Fluor-spar, chondrodite, tremolite, common hornblende, serpentine, cubic and magnetic pyrites, are among the minerals of the limestone. Quartz secretions are also met with, but are evidently of secondary origin. The hornblende forms rounded patches, remarkable twisted stripes, and banded parallel layers, often of considerable dimensions, as in the specimens from Wunseidel, which exhibit sheets of hornblende of from five to fifteen millimeters, separated by limestone layers of from fifteen to twenty millimeters in thickness. My examinations of the specimens of this nature, in my collection, have not enabled me to connect these hornblende layers with organic structure, nor to discover any traces of Eozoon in the highly crystalline limestone.

The result of my examinations of specimens of the limestone containing serpentine from the quarries near Wunseidel, from Thiersheim, and from between Hohenberg and the Steinberg, were however more successful. Fragments of the rock from near Hohenberg show irregular greenish stripes, which are made up of parallel undulating laminæ, or of elongated grains. This banded aggregate is a granular mixture of carbonate of lime, serpentine, and a white mineral, insoluble in acids, which appears to be a variety of hornblende. The grains of this aggregate have generally a diameter of $\frac{1}{10}$ millimeter.

When examined in thin sections, the calcareous portions appear for the most part sparry, and traversed by straight intersecting lines, (Pl. 1, fig. 7 $a$,) or divided into cellular spaces by small irregular bands, which, after the surface is etched, are seen in slight relief. The portions between these bands are granulated. (fig. 7 $b$.) More compact calcareous portions are however met with, and these are penetrated by delicate tufts of tubuli like those of Eozoon, (fig. 7 $c$,) and are adherent to the serpentine portions, which have nearly the same form as in the Eozoon of Steinhag, but are far smaller. (fig. 7 $d$.) In decalcified specimens, they are found to possess the same arched walls as the Eozoon. Their breadth in the cross section is generally about one tenth, and the diameter of the casts of the tubuli only about one hundredth of a millimeter. These broader serpentine portions are generally connected with an adjacent portion of lamellæ, (also composed of serpentine, or of a whitish mineral,) which are not more than one-half their size, curiously curved, and presenting highly arched and deeply incurved outlines, as may be seen in decalcified specimens, (fig. 7 $e$.) The study of these structures leaves no doubt that they are due to an organism belonging to the same group as the Eozoon. In order however to distinguish this distinctly smaller form of the primitive clay-slate series, with its minute contorted chambers filled with serpentine, from the typical *Eozoon Canadense* of the more ancient Laurentian system, it may be designated as *Eozoon Bavaricum*.

I have moreover subjected to microscopic examination a series of specimens from the same limestone horizon in the Fichtelgebirge, which, unlike those just described, showed no distinct foreign minerals, although presenting certain dense portions which seemed to indicate the presence of some foreign matter. These portions however showed only a cellular structure, like that in the specimen from Hohenberg, without any tubuli ; nor did etching succeed in developing any structure in these wholly calcareous specimens. When therefore carbonate of lime both constitutes the skeleton, and replaces the sarcode, there is evidently little hope of recognizing these organic forms. If however the flaky pellicles which remain suspended in the acid after the solution of the lime, in these almost wholly calcareous specimens, are examined, they present a very great resemblance to the similar pellicles from the Eozoon limestone of Steinhag, already figured, which have such a striking resemblance to organic forms. The careful examination of the limestone from many other parts in the Fichtelgebirge, affords evidence of organic life similar to those of Hohenberg ; thus tending more and more to fill up the interval between the Laurentian gneiss, and the primordial zone of the Lower Silurian fauna. We may therefore reasonably hope that in the study of these more ancient rock-systems, which geologists have only recently ventured to distinguish, paleontological evidence will be found no less available than in the more recent sedimentary formations. The inferences which we are permitted to draw from the discovery of organic remains in these ancient rocks, confirm the conclusion to which I had previously arrived from the study of the stratigraphical relations, and the general character of these ancient rock-systems ; viz., *that there exists, in these ancient crystalline stratified rocks, a regular order of progress determined by the same laws which have already been established for the formations hitherto known as fossiliferous.*

I cannot conclude this notice of the preliminary results obtained in the investigation of the ancient Eozoon limestones of Bavaria, without adding a few observations upon some foreign crystalline limestones. It is well known that the crystalline minerals, which in numerous localities are found in these limestones, often present rounded surfaces, as if they had at one time been in a liquid state. As examples of these, Naumann mentions apatite, chondrotite, hornblende, pyroxene, and garnet. The edges and angles of these are often rounded ; the planes curved or peculiarly wrinkled, and only rarely presenting crystalline faces ; having in short a half-fused aspect, and offering a condition of things hitherto unexplained. One of the best known instances of this is found in the green hornblende (pargasite) from Pargas in Finland. This mineral there occurs in a crystalline limestone with fluor, apatite, chondrotite, pyroxene, pyrallolite, mica and graphite; associations very similar to those of the serpentine of Steinhag. The grains of pargasite, although completely crystalline within, and having a perfect cleavage, are rounded on the exterior, curved inward and outward, and also approximatively cylindrical in form; so that they may be best compared with certain vegetable tubercles. If the crystalline carbonate of lime which accompanies the pargasite is removed by an acid, there remains a mass of pargasite grains, generally cohering, and presenting a striking resemblance to the skeleton obtained by submitting the Eozoon serpentine-limestone to a similar treatment. The tubercles of pargasite are then seen to be joined together by short cylindrical projections, which are however readily broken by pressure, causing the mass to separate into detached grains. The highly crystalline and ferruginous carbonate of lime which is mingled with the pargasite, shews no organic structure either when etched or examined in thin sections; although the pargasite presents forms similar to those observed in the serpentine of Steinhag. The surfaces of the curved cylindrical and tuberculated grains of pargasite are in part naked, and in part protected by a thin white covering. In some parts fine cylindrical growths are observed, and in others cylindrical perforations passing through the grains of pargasite. By a careful microscopical examination of the surface of these grains (Pl. I., fig. 8), numerous small tubuli, sometimes two millimeters in length, are clearly seen, and by their exactly cylindrical form may be readily distinguished from other pulverulent, fibrous and acicular crystalline mineral matters. These cylinders consist of a white substance, which contrasts with the dark green pargasite, and have the diameter of the tubuli of Eozoon, or from $\frac{3}{1000}$ to $\frac{2}{1000}$ millimeters. A single large cylinder was also observed lying obliquely across between two of the pargasite tubercles. (Pl. I., fig. 8 $a$.) In the decalcified specimens, a white mineral, probably scapolite, was observed side by side with the green pargasite; sometimes forming groups of tubercles like the latter; while in other cases a single tubercle was found to be made in part of the green and partly of the white mineral. From these observations there can scarcely remain a doubt that these curiously rounded grains of pargasite imbedded in the crystalline limestone of Pargas represent the casts of sarcode-chambers, as in the Eozoon; and that they are consequently of organic origin. From the great similarity between the forms of the pargasite grains and the Eozoon-serpentine, we may fairly be permitted to assume the presence of Eozoon in the crystalline limestones of Finland.*

Similar relations are doubtless to be met with throughout the crystalline limestones of Scandinavia, wherever such mineral species occur in rounded grains or in tuberculated forms. The notion that these forms are of organic origin, and have been moulded in the spaces left in a calcareous skeleton by the decay of animal matter, receives a strong support from the observations of Nordenskiold and Bischof. The former found in a tuberculated pyrallolite, 6·38 per cent. of bituminous matter, besides 3·58 per cent. of water ; while Bischof states that the same mineral becomes black when ignited, and when calcined in a glass tube, gives off a clear water with a very offensive empyreumatic odor.

There may also be mentioned in this connection a phenomenon which is probably related to those just described. Upon the pyritous layers which occur in the Hercynian gneiss near Boden, are found great quantities of grains of quartz, almost transparent, and with a fatty lustre, which have in all cases rounded undulating forms, precisely resembling the pargasite tubercles from Finland. Dichroite also sometimes occurs in this region in similar shapes, although it also, in many cases, forms perfect crystals. The evidence of organic forms may perhaps be found in these masses of quartz and dichroite, though their treatment will necessarily present difficulties.

A specimen of crystalline limestone, with rounded pyroxene (coccolite) grains from New York, showed, after etching by means of acids, no traces of tubuli ; but the grains of coccolite, remaining after the entire removal of the carbonate of lime, were found to be connected with each other by numerous fine cylindrical tubuli and skin-like laminæ. The surface of the rounded coccolite grains was much wrinkled, and studded with small cylindrical processes of a white mineral, sometimes ramifying, and apparently representing the remnants of a system of tubuli which had been destroyed by the crystallization of the carbonate of lime. The flaky residue from the solvent action of the acid exhibits, under the microscope, laminæ, needles, and strings of

---

* These belong to the primitive gneiss formation of Scandinavia, which the geologists of Canada, so long ago as 1855, referred to the Laurentian system.—T. S. H.

globules similar to those described in the residue from the
Eozoon ophicalcite of Steinhag, with which, and with the
hornblendic limestone of Pargas, this coccolite-bearing limestone
of New York seems to be closely related.

A fragment of ophicalcite from Tunaberg in Sweden bears a
striking resemblance to the coarser marked varieties of this rock
from near Passau. The carbonate of lime between the tubuli is
very sparry; and after its removal, a perfectly coherent serpentine
skeleton is obtained, as in the Passau specimens. The surface of
the serpentine tubercles is abundantly covered with acicular
crystalline needles of various lengths, whose inorganic nature is
unmistakeable. The sediment from the acid solution also contains
a prodigious quantity of these same small crystalline needles. On
etching a specimen of this rock with dilute acid, the same needles
were found in most places; but here and there, in isolated, less
crystalline and more solid portions of the carbonate of lime, there
were seen curved and ramified tubuli, undoubtedly corresponding
with the tubuli of Eozoon, and having the same size and manner
of grouping as in the Eozoon of Passau. The ophicalcite of
Tunaberg is therefore to be classed with the Eozoon-bearing
limestones.

A specimen of crystalline limestone from Boden in Saxony,
holding rounded grains of chondrodite, hornblende and garnet, and
furnished me by Prof. Sandberger, showed, after etching, tubuli of
surprising beauty, both singly and in groups, but only in small
isolated compact portions of the carbonate of lime. The sparry
crystallization of this mineral seems to have frequently destroyed
the cohesion of the very delicate tubuli, the fragments of which
may be observed in very large quantity in the flaky residue from
the solution.

A blackish serpentine limestone from Hodrisch in Hungary,
showed by etching no traces of tubuli. The granular residue from
its solution in acids showed under the microscope large quantities
of cell-like granules, with a central nucleus, and generally joined
in pairs, like the spores of certain lichens. More rarely however
three or four of such grains were joined together. By far the
greater part of them were of one and the same size, although
occasionally others of double size were met with. Their regularity
of form is much in favor of their origin from organic structure.

A fragment of ophicalcite from Reichenbach in Silesia, which
Prof. Beyrich kindly furnished me, showed distinct parallel bands

of serpentine with curved and undulating outlines, resembling the Eozoon ophicalcite of Canada. The etched portions show, in the carbonate of lime between the serpentine, or in the interspaces of the serpentine, the same relations as the limestone of Hohenberg from the primitive clay-slate formation. The tubuli, which have a certain resemblance with those of Hohenberg, are stuck together, as if covered by an incrustation. Further examinations of this limestone are required to determine more definitely the organic nature of its enclosures.

A fragment of similar limestone without serpentine, from Raspenau, shows not the remotest trace of any organic structure whatever. The same negative results were obtained with a specimen of granular limestone from Timpobepa in Brazil; and with a very coarsely crystalline carbonate of lime, holding chondrodite, from Amity, New Jersey. These negative results show that organic remains are sometimes wanting in the primitive crystalline limestones, as well as in those of more recent formations. The occasional absence from the primary limestones of these regular structures is therefore an indirect argument for their organic origin.

### Explanation of the Plate.

Figure 1. Section of *Eozoon Canadense*, with its serpentine replacement, showing the fine tubuli and the canal-system, from the limestone of the Hercynian gneiss formation at Steinhag; seen by reflected light, and magnified 25 diameters.

2. Section of Eozoon from the limestone of Untersalzbach; 25 diameters.

3. Section of Eozoon from the limestone of Babing.

4. Section of Eozoon from the limestone of Steinhag; 120 diameters.

5, a and b. Knotted tubuli from the insoluble residue of the Steinhag limestone; 300 diameters.

6, a, b, c, and d. Flocculi from the same residue; 400 diameters.

7. Section of *Eozoon Bavaricum*, with serpentine, from the crystalline limestone of the Hercynian primitive clay-state formation at Hohenberg; 25 diameters.

   a. Sparry carbonate of lime.

   b. Cellular carbonate of lime.

   c. System of tubuli.

   d. Serpentine replacing the coarser ordinary variety.

   e. Serpentine, and hornblende, replacing the finer variety, in the very much contorted portions

8. Aggregated grains of pargasite, remaining after the solution of the carbonate of lime, from the granular limestone rock of Pargas.

# ON THE CANADIAN SPECIES OF THE GENUS PICEA.

By the ABBÉ O. BRUNET, of Laval University

Botanists have always recognized the existence in North America of two trees which may be referred to the genus *Picea*, established by Link. They are the *Abies alba* of Michaux, and the *Abies nigra* of Poiret, (*A. denticulata*, Michaux). These two species have been imperfectly described, and are almost always confounded; some authors, moreover, have regarded them as nothing more than varieties of one and the same species. These considerations have led me to study these interesting trees in detail, and to complete, as far as possible, their history.

## Genus PICEA, LINK.

Leaves persistent, solitary, scattered, and surrounding the branches, tetragonal, stiff, marked on both sides with white lines of numerous stomata; male flowers clustered towards the ends of the branches; cones pendulous, persistent, terminal or axillary; seeds without resiniferous ducts, separating after a time from the base of the wing. Wood, almost white, with resiniferous ducts, having no distinction of alburnum or duramen; cells of the medullary rays without large pits; groups of cubic lignified cells in the older bark.

## PICEA ALBA.

The *Picea alba* is one of the most abundant trees in Canada, extending throughout the province. To the northward, following the line of the Saguenay, it is found, diminished in size, along the Mistassini, but disappears altogether about the cascades of that river (Michaux MS.) to reappear in the Hudson Bay territory; where, according to Dr. Richardson, it grows to a large size, and is the most important forest tree of those northern regions.

The *Picea alba* in favourable situations generally attains a height of from seventy to eighty feet, with a diameter of ten feet at the base; in the Saguenay district however, trees of this species are said to have been found, from 130 to 140 feet in height. These large trunks taper gradually and regularly towards the top; they are very straight, and the branches extend horizontally, and are arranged so as to form a regular pyramid, the summit of which is long and slender, giving to the tree a very

characteristic aspect.    In places exposed to the force of the tem-
pests it becomes stunted in growth, creeping as it were, along the
soil.    This is well shown in Anticosti, where, on the cliffs and at
the point of the island, these trees are seen extending from ten to
twenty feet in length, though scarcely five feet in height, and
forming a sort of hedge, which is almost insurmountable.    In the
interior of the island, however, the tree assumes its ordinary
aspect.

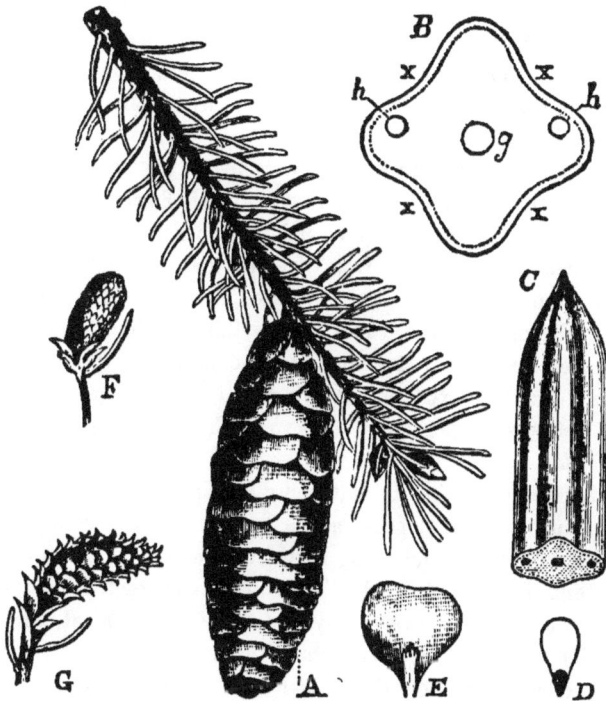

PICEA ALBA, Link.

A. Branch with cone, gathered in winter.
B. Transverse section of leaf; *g*. vascular bundles ; *h*. resiniferous
    canals ; *x*. parts of leaf where the stomata occur; ✕ 50 diameters.
C. Point of leaf, enlarged ten diameters.
D. Ripe seed with its wing.
E. Seminal scale, dorsal view.
F. End of a branch with a male flower.    (May 27, 1863.)
G. End of a branch with a female flower.    (Ditto.)

The bark of this tree is whitish upon the branches, but on the old trunks it appears as a corky tissue, ferruginous-brown in color, with a scaly rhytidoma, cracked in all directions, and separating in whitish-gray plates. Some have supposed that both the specific and vulgar names of this tree are derived from the whitish color of its bark.

The leaves are from six to ten lines in length, and about three fourths of a line in breadth, ordinarily curved, presenting few stomata on both surfaces, summit acute, but much less so than is the leaf of *Abies* (*Picea*) *Menziesii;* section of the leaf quadrangular, presenting two resiniferous ducts larger than those of *P. nigra.* The leaves of *P. alba* are much more robust than those of *P. nigra*, but their size varies very much, even upon the same individual ; the same is true of the form, which is also very variable.

The male catkins are ovate, not pedicellated, about six lines long ; length of the anthers one line. Female flowers in cylindrical catkins, violet-red in color, and ten lines in length. Cones cylindrical, reddish-brown, from one to two and a half inches in length, numerously disseminated at the extremity of the branches, and in the axils of the leaves; scales thin, six lines long, rhomboidal, entire, slightly indented at the summit. Seeds small, brown, a line long, with an oval wing of a very pale yellow color, three times that length ; embryo with from six to eight cotyledons.

This tree in the vicinity of Quebec blossoms about the end of May, and its fruit ripens in the autumn of the same year. The warmth of the following spring-time opens the scales of the cones, and liberates the seeds. These require for their germination about twenty days ; twelve days later the young plant escapes from its envelopes, and appears with its numerous cotyledons, which resemble precisely the other leaves. The plumula of the young plant is not apparent before two or three months.

The wood of the white spruce is very white, compact, and harder than that of the white pine (*Pinus strobus*). The annual rings are sometimes three lines in breadth, and are for the most part strongly marked, the autumnal wood being dark colored. The medullary rays are composed of a layer of uniform cells (figures A. and B, p. 109). The resiniferous canals (figure c.) which are distinguishable by the aid of a magnifying glass, furnish an excellent characteristic, and a ready means o distinguishing the wood of the species of *Picea* from that of any other conifers.

This wood is more subject to cracking than that of the white pine, and is liable to shrink when not perfectly dried. It is, however, much employed for flooring, on account of its greater hardness, and is largely exported from Quebec in the form of planks. It is also esteemed for its lightness and elasticity, for which quality it is employed for the ship-yards. All the houses which, in the country parts of Canada are made of hewn logs, and are known as log-houses. are constructed of white spruce, which is also employed for the frame-work of steeples, of bridges, etc. The bark of the tree furnishes curved timbers, or knees, as they are called, which are used for ship-building, although inferior to those furnished by the tamarack (*Larix Americana*). The aborigines make use of the tough rootlets, previously macerated in water, to sew the seams of their bark canoes.

The pyramidal form of this tree, the regularity and number of its branches, and its abundant foliage, make the white spruce one of the best of ornamental evergreen trees. It moreover adapts itself to almost any soil, not too solid and compact, so that it is one of the Canadian trees best fitted for plantations. The readiness with which the white spruce throws out auxilliary buds renders it fit for pruning, and enables us to make of it excellent hedges, which may advantageously replace those of hawthorn.

This sketch of the white spruce would be incomplete if we did not mention a parasitic insect, which frequents it, and causes the small galls which are often seen upon this tree. They may be observed in the spring-time at the ends of the young branches, where they are dark red in color, and resemble in miniature the fruit cones. We met them for the first time at the end of May, 1863, on the island of Orleans, and again some time later near the Château Bigot, in the rear of Quebec. Baron Osten-Sacken, after having examined the specimens which we sent him, informs us that these galls are produced by a species of *Aphis*, hitherto unknown to science.

## PICEA NIGRA.

The *Picea nigra* is even more widely spread in the north of America, than the preceding species, for it is found farther to the northward, and beyond the Saguenay, in elevated localities, where, as already remarked, the *P. alba* disappears. Michaux the elder, in his manuscript journal, informs us that the black spruce is met with, in a stunted form, upon the hills bordering on Swan Lake, and that it is only on the height of land, or water-shed between the St. Lawrence and Hudson Bay that it entirely disappears, giving place to the *Pinus rupestris* which reigns alone in those boreal regions.

The *Picea nigra* in certain localities may reach a height of seventy feet, and a diameter of from fifteen to eighteen inches, but is generally smaller, and seems to diminish in size as we go

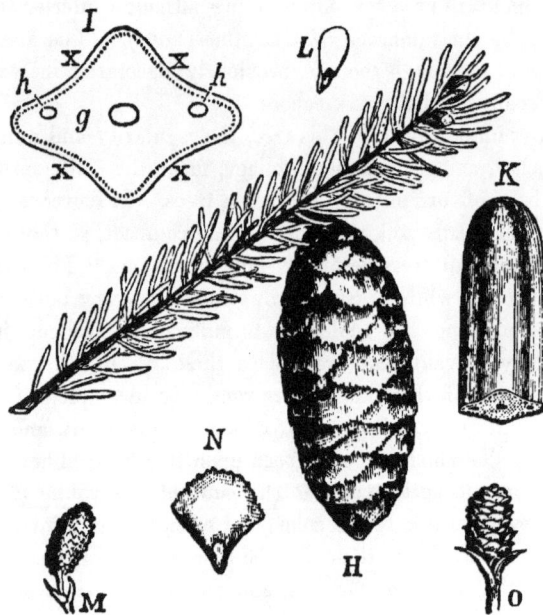

PICEA NIGRA, Link.

H. Branch with a cone, gathered in January, 1865.
I.  Transverse section of the leaf; *g.* vascular bundles; *h.* resiniferous canals; *x.* parts of the leaf having stomata; × 50 diameters.
K. Point of a leaf, enlarged ten diameters.
L. Ripe seed with its wing.
N. Seminal scale, dorsal view.
M. End of a branch with a male flower. (June 5, 1865.)
O. End of a branch with a female flower. (Ditto.)

northward. In the vicinity of Quebec its height is not above seventy feet, and in the valley of the Saguenay, it does not exceed forty or fifty feet, with a diameter of eight or ten inches. It prefers a deep, black, and moist soil, thickly covered with moss, but in places which are constantly wet or covered with water, as in peat bogs, it grows but indifferently, and rises to no great height.

The bark of the *P. nigra* is yellowish on the young branches ; the older trunks are covered with a reddish corky rhytidoma, the cracks in which are chiefly vertical, and which exfoliates at last in little plates, more or less rectangular in shape.

The leaves are from five to seven lines in length, and about three fourths of a line in breadth, flattened, and with the apex obtuse. They are of a sombre green color, and are supported on sterigmata twice as prominent as those of the preceding species. The leaves of the *P. nigra* are shorter, more closely appressed to the branches, and more flattened than those of the *P. alba.* They also present more numerous rows of stomata, amounting sometimes to not less than five or six rows on each side of the median vein, and the diameter of their resiniferous ducts is smaller.

The male catkins are ovoid, slightly pedunculate, and three or four lines in length. The female flowers are also in ovoid catkins, violet-red in color, six or eight lines in length, which are at first upright, but after impregnation are bent sharply downwards. The cones are ovoid, reddish-brown, from one inch to one and a half inches in length, slightly pedunculate ; scales thin, about six lines in length, with undulated and denticulated edges. The seeds are black, with an oval wing, smaller than that of *P. alba.* The embryo has ordinarily four cotyledons, rarely more. This tree flowers in the month of June, about a week later than the preceding species, and ripens its seeds the same year. The seeds germinate in three or four weeks, and demand a great deal of moisture. After the fall of the perisperm, the young plant generally presents four seed-leaves, which have the form of the ordinary leaves, and already present the sombre green color which characterizes the foliage of the *P. nigra.*

In the localities most favourable to the development of this species, and in places where the white pine has become rare, the black spruce is cut by the lumberers. It is manufactured into planks and boards, and the wood is employed for the same uses as that of the white spruce. The woods of these two species of *Picea* offer no perceptible differences in structure, color, lightness, or other qualities. They are equal in value, and command the same price in the Quebec market.

*Picea nigra,* var. *grisea;* gray spruce.

This spruce does not appear to differ essentially from the black spruce in its organs of fructification. Its leaves are however of a more or less dingy and grayish green, and its bark has a lighter red color than the typical black spruce. The gray spruce is found principally in poor soils. This variety often attains a very large size. We measured one of these trees in the eastern section of Rimouski, and found it to be 160 feet high, with a diameter of four feet.

In certain parts of Canada an infusion of the leaves of the *Picea nigra* is used as a common drink. The Abbé Ferland in his *Voyage au Labrador* speaks of " the little black spruce which creeps over the rocks, and whose leaves infused in hot water furnish a beverage which by the peasants is preferred to tea." It is with this plant also that is made the fermented liquor known as spruce beer. As it may not be without interest, we copy a description of the mode of preparing this beverage a century since, copied from Duhamel, (*Traité des arbres et arbustes,* Paris, 1755.)

" The white spruce \* (*épinette blanche*) which is a species of *Epicea,* having smaller leaves and cones than that cultivated in France, serves in Canada to make a wholesome beverage, which is not agreeable when tasted for the first time, but becomes so by use. As a similar drink might be made very cheaply from our own *Epicia,* I give the receipe. . . . . . . . . . .

---

\* This is evidently an error of the author, since the black spruce has always been employed for making this kind of small beer.

The French of Lower Canada apply the name of *Epinette* to several trees ; the *Larix Americana* is by them called *épinette rouge,* and the white and black spruce are respectively *épinette blanche,* and *épinette noire,* while the name of *épinette grise* is given to what we regard as a variety of the latter, *P. nigra* var. *grisea.* The origin of this word, which is not applied to any tree in France, is by no means clear. It has, however been used from an early date in the history of the colony, as will appear from the following citation from the *Histoire Naturelle du Canada,* of Pierre Boucher, 1663. " Il y a une autre espéce d'arbre qu'on nomme épinette ; c'est quasi comme du Sapin, si non qu'il est plus propre a faire des masts de petits vaisseaux, comme des chaloupes et des barques, estants plus fort que le Sapin."

" For a barrel, a boiler holding at least a quarter more is required. This being filled with water, and heated, a bundle of spruce branches, broken small, and about twenty-one inches in girth, is added, and the water is kept boiling until the bark readily peels off from the whole length of the branches.    Then a bushel of oats is r)asted by portions. in a great iron pan, about fifteen sea-biscuit,

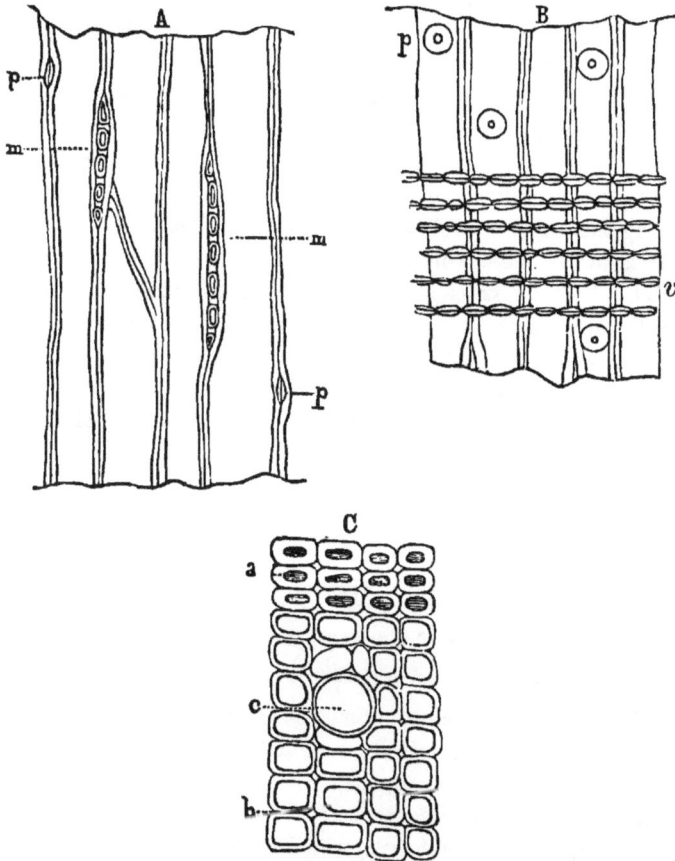

Figure A.  Longitudinal tangential section of the wood of *P. alba*.
   *c.* ligneous cellules ; *m.* medullary rays; *p.* discs ; (500 diameters.)
Figure B.  Longitudinal section, parallel to one of the medullary rays;
   *v.* medullary rays ; *p.* discs; (500 diameters.)
Figure C.  Transverse section of the same wood ; *a.* fibres of the autumnal wood ; *b.* fibres of the spring wood ; *c.* resiniferous ducts ; (300 diameters.)

   These figures were drawn by the author and engraved by Mr. G. J. Bowles.

or in place of them, twelve or fifteen pounds of bread, cut in slices, are also roasted, and with the oats, added to the boiling kettle, where they remain till the spruce branches are well cooked. These branches are now taken out, and the fire extinguished. The bread and oats then settle to the bottom, and the spruce leaves are removed by a skimmer; after which are added six quarts of molasses or syrup, or in place thereof twelve or fifteen pounds of coarse sugar. The liquid is then put at once into a fresh red-wine cask; and if it is wished to give more color to the liquor, the lees, and five or six quarts of the wine are left therein. When the liquid is only lukewarm, a pint of beer-yeast is added, the whole well stirred. to mix it, and the cask then filled to the bung-hole, which is left open. Fermentation soon begins, and much scum is thrown off; during this time the cask must be filled from time to time with a portion of the liquid which has been kept apart in some wooden vessel. If the cask is bunged at the end of twenty-four hours, the liquor is sharp and lively as cider, but if it is wished to have it milder, the cask should be filled twice a day, and not bunged till fermentation is over. This liquor is very refreshing and wholesome, and those accustomed to it drink it with pleasure, especially in summer."

---

## ON THE OBJECTS AND METHOD OF MINERALOGY.

### BY Dr. T. STERRY HUNT, F.R.S.

#### (Read before the American Academy of Sciences, Jan. 8, 1867.)

Mineralogy, as popularly understood, holds an anomalous position among the natural sciences, and is by many regarded as having no claims to be regarded as a distinct science, but as constituting a branch of chemistry. This secondary place is disputed by some mineralogists, who have endeavored to base a natural-history classification upon such characters as the crystalline form, hardness, and specific gravity of minerals. In systems of this kind, however, like those of Möhs and his followers, only such species as occur ready formed in nature, are comprehended, and the great number of artificial species, often closely related to native minerals, are excluded. It may moreover be said in objection to these naturalists, that, in its wider sense, the chemical history of bodies takes into consideration all those characters upon which the so-called natural systems of classification are based. In order to understand clearly the question before us, we must first consider what are the real objects, and what the provinces, respectively, of mineralogy, and of chemistry.

Of the three great divisions, or kingdoms of nature, the classification of the vegetable gives rise to systematic botany, that of the animal to zoology, and that of the mineral to mineralogy, which has for its subject the natural history of all the forms of unorganized matter. The relations of these to gravity, cohesion, light, electricity, and magnetism, belong to the domain of physics; while chemistry treats of their relations to each other, and of their transformations under the influences of heat, light, and electricity. Chemistry is thus to mineralogy what biology is to organography; and the abstract sciences, physics and chemistry, must precede, and form the basis of the concrete science, mineralogy. Many species are chiefly distinguished by their chemical activities, and hence chemical characters must be greatly depended upon in mineralogical classification.

Chemical change implies disorganization, and all so-called chemical species are inorganic, that is to say unorganized, and hence really belong to the mineral kingdom. In this extended sense, mineralogy takes in not only the few metals, oxyds, sulphids, silicates, and other salts, which are found in nature, but also all those which are the products of the chemist's skill. It embraces not only the few native resins and hydrocarbons, but all the bodies of the carbon series made known by the researches of modern chemistry.

The primary object of a natural classification, it must be remembered, is not like that of an artificial system, to serve the purpose of determining species, or the convenience of the student, but so to arrange bodies in orders, genera, and species as to satisfy most thoroughly natural affinities. Such a classification in mineralogy will be based upon a consideration of all the physical and chemical relations of bodies, and will enable us to see that the various properties of a species are not so many arbitrary signs, but the necessary results of its constitution. It will give for the mineral kingdom what the labors of great naturalists have already nearly attained for the vegetable and animal kingdoms.

Oken saw the necessity of thus enlarging the bounds of mineralogy, and in his Physiophilosophy, attempted a mineralogical classification; but it is based on fanciful and false analogies, with but little reference either to physical or chemical characters, and in the present state of our knowledge is valueless, except as an effort in the right direction, and an attempt to give to mineralogy a natural system. With similar views as to the scope of the science, and with far higher and juster conceptions of its method, Stallo, in his Philosophy of Nature, has touched the questions before us, and has attempted to show the significance of the relations of the metals to cohesion, gravity, light, and electricity, but has gone no farther.

In approaching this great problem of classification, we have to examine—first, the physical condition and relations of each species, considered with relation to gravity, cohesion, light, electricity, and magnetism; secondly, the chemical history of the species; in which are to be considered its nature, as elemental or compound, its chemical relations to other species, and these relations as modified by physical conditions and forces. The quantitative relation of one mineral (chemical) species to another, is its equivalent weight, and the chemical species, until it attains to individuality in the crystal, is essentially quantitative.

It is from all the above data, which would include the whole physical and chemical history of inorganic bodies, that a natural system of mineralogical classification is to be built up. Their application may be illustrated by a few points drawn from the history of certain natural families.

The variable relations to space of the empirical equivalents of non-gaseous species, or, in other words, the varying equivalent volume (obtained by dividing their empirical equivalent weights by the specific gravity), shows that there exist, in different species, very unlike degrees of condensation. At the same time, we are led to the conclusion that the molecular constitution of gems, spars and ores, is such that those bodies must be represented by formulas not less complex, and with equivalent weights far more elevated than those usually assigned to the polycyanids, the alkaloids, and the proximate principles of plants. To similar conclusions, conduce also the researches on the specific heat of compounds.

There probably exists between the true equivalent weights of non-gaseous species and their densities, a relation as simple as that between the equivalent weights of gaseous species and their specific gravities. The gas, or vapor of a volatile body, constitutes a species distinct from the same body in its liquid or solid

state, the chemical formula of the latter being some multiple of
the first, and the liquid and solid species themselves often
constituting two distinct species of different equivalent weights.
In the case of analogous volatile compounds, as the hydrocarbons
and their derivatives, the equivalent weights of the liquid or
solid species approximate to a constant quantity, so that the
densities of these species, in the case of homologous or related
alcohols, acids, ethers and glycerids, are subject to no great variation. These non-gaseous species are generated by the chemical
union, or identification, of a number of volumes or equivalents
of the gaseous species, which varies inversely with the density
of these species. It follows from this, that the equivalent weights
of the liquid and solid alcohols and fats must be so high as to
be a common measure of the vapor-equivalents of all the bodies
belonging to these series. The empirical formula, $C_{114}H_{110}O_{12}$,
which is the lowest one representing the tristearic glycerid, ordinary stearine, is probably far from representing the true equivalent weight of this fat in the liquid or solid state; and if it should
hereafter be found that its density corresponds to six times the
above formula, it would follow that liquid acetic acid, whose
density differs but slightly from that of fused stearine, must have
a formula, and an equivalent weight about one hundred times
that which we deduce from the density of acetic acid vapor,
$C_4H_4O_4$.

Starting from these high equivalent weights of liquid and
solid hydrocarbonaceous species, and their correspondingly complex formulas, we become prepared to admit that other orders
of mineral species, such as oxyds, silicates, carbonates, and sulphids, have formulas and equivalent weights corresponding to
their still higher densities; and we proceed to apply to these bodies the laws of substitution, homology, and polymerism, which
have so long been recognized in the chemical study of the members of the hydrocarbon series. The formulas thus deduced
for the native silicates and carbon-spars, show that these poly-
basic salts may contain many atoms of different bases, and their
frequently complex and varying constitution is thus rendered
intelligible. In the application of the principle of chemical homology, we find ready and natural explanations of those variations, within certain limits, occasionally met with in the composition of certain crystalline silicates, sulphids, etc., from which
some have conjectured the existence of a deviation from the law

of definite proportions, in what is only an expression of that law in a higher form.

The principle of polymerism is exemplified in related mineral species, such as meionite and zoisite, dipyre and jadeite, horn-blende and pyroxene, calcite and aragonite, opal and quartz, in the zircons of different densities, and in the various forms of titanic acid and of carbon, whose relations become at once intelligible if we adopt for these species high equivalent weights and complex molecules. The hardness of these isomeric or allotropic species, and their indifference to chemical reagents, increases with their condensation, or, in other words, varies inversely as their empirical equivalent volumes ; so that we here find a direct relation between chemical and physical properties.

It is in these high chemical equivalents of the species, and in certain ingenious, but arbitrary assumptions of numbers, that is to be found an explanation of the results obtained by Playfair and Joule in comparing the volumes of various solid species with that of ice; whose constitution they assume to be represented by HO, instead of a high multiple of this formula. The recent ingenious but fallacious speculations of Dr. Macvicar, who has arbitrarily assumed comparatively high equivalent weights for mineral species, and has then endeavoured, by conjectures as to the architecture of crystalline molecules, to establish relations between his complex formulas and the regular solids of geometry, are curious but unsuccessful attempts to solve some of the problems whose significance I have endeavoured to set forth. I am convinced that no geometrical groupings of atoms, such as are imagined by Macvicar, and by Gaudin, can ever give us an insight into the way in which nature builds up her units, by interpenetration and identification, and not juxtaposition of the chemical elements.

None of the above points are presented as new, though they are all, I believe, original with myself, and have been, from time to time brought forward, and maintained, with numerous illustrations, chiefly in the American Journal of Science, since March, 1853, when my paper on the Theory of Chemical Changes and Equivalent Volumes, was there published. I have however thought it well to present these views in a connected form, as exemplifying my notion of some of the principles which must form the basis of a true mineralogical classification.

# THE AMERICAN ASSOCIATION AT BUFFALO, AUGUST, 1866.

## ON A NEW NOMENCLATURE.

### BY PROF. S. D. TILLMAN OF NEW YORK.

The author, in this paper, gave a brief account of the amendments and alterations made in our present nomenclature, which originated with DeMorveau, Lavoisier, Bertholet and Fourcroy in France, in the year 1787. He showed furthermore, that it cannot be adapted to the new views of chemical combinations, according to the atomic system, without producing serious confusion, and rendering all our present works on chemistry comparatively worthless. He therefore proposed to let the old nomenclature remain as the exponent of the system of combining proportions, or so called " equivalents," and to give new names to atomic combinations, which would express both the views of Berzelius and Gerhardt. The method was devised by him many years ago, but until there was a general agreement among advanced chemists with regard to the numbers expressing atomic weights, it would have been useless. Under the lead of Gibbs, in this country, and Canizzaro in Europe, those of the unitary school who double the numbers represented by the symbols O, C, and S, now also double the numbers of at least fifty other symbols, and thus all objections have been removed in regard to using a system of names based upon atomic weights. The nomenclature now proposed is also adapted to the typical classification, first proposed by a distinguished member of this Association, Dr. T. Sterry Hunt, which, with a few modifications, has been very generally adopted by European chemists. Prof. Tillman's method of construction may be briefly explained in the following heads :

1. The system is based on abbreviations of the universally received names of the metals, and on the chemical symbols of the metalloids, or non-metallic elements, with such modifications as were imperatively required.

2. The name of each chemical element relates not to its mass, but only to a minimum combining proportion, termed an atom, or to some multiple of it. The atom is therefore the unit of measurement, and the starting point of the scale in each series of compounds.

3. The atomic name of each of the 50 metals now well-known, consists of two syllables, and ends with the consonant $m$.

4. The name of each of the 13 metalloids terminates with a different consonant ; arsenic and tellurium, classed by some chemists among the metalloids, are by this arrangement included among the metals.

5. The number of atoms of any element is designated by the vowel immediately preceding the terminal consonant. The numerical power of the vowels advances with the order in which they are placed in the alphabet, thus 1, 2, 3, 4 and 5 are represented by a, e, i, o and u, each having a short or stopped sound, and the same vowels, each preceded by e, and having the long or full sound, represent 6, 7, 8, 9 and 10. Other letters represent higher numbers, so that any number to 1000 is readily denoted.

6. The following metalloids are represented by their symbolic letters :  One atom of Fluorine is *af*, one atom of Bromine *ab*, one of Nitrogen *an*, one of Carbon *ac*, one of Sulphur *as*, one of Phosphorus, *ap*.  For reasons which] need not here be stated, an atom of Hydrogen is *al*, of Oxygen *at*, of Chlorine *ad*, of Iodine *av*, etc.

7. The manner of uniting these syllables may be thus illustrated : The protoxide of iron is *Ferramat ;* the sesqui-oxide of iron, *Ferremit;* the black or magnetic oxide, *Ferrimot ;* sulphate of protoxide of iron, *Ferrmasot ;* sulphate of sesqui-oxide of iron, *Ferremisoit.*

The combinations containing carbon and hydrogen are so numerous that it was found essential to use another letter, *r*, to designate carbon—*ar* and *ac* each denote an atom of carbon. Two atoms of hydrogen are designated by *h*, thus *ach* is equal to $C_2 H_2$ in the old notation. This is the important increment in several series of organic radicals. The first of the alcohol-forming radicals is *achal*, methyl ; the second, *echal*, ethyl ; the third, *ichal*, propyl ; the fourth, *ochal*, butyl ; the fifth, *uchal*, amyl, etc. These radicals play the part of monatomic metals.

The author gave specimens of the new names for several thousand compounds; and showed their application in cases of isomerism, where, for instance, ten bodies, having the same ultimate components, are distinguished by ten different names. The doctrine of substitutions was also very clearly set forth ; and derivatives were so classified and simplified as to be readily comprehended.

The author then proceeded to show the manner in which names were provided for salts containing water of crystallization, and for solutions containing either an indefinite or definite quantity of water.

In future chemical investigation, the speaker thought increasing significance must be given to the state of dilatation in which the body under consideration exists; he therefore proposed to designate every gas, and every volatile body after it is formed into vapor, by prefixing to the new name the letter $g$. For instance, carbonic oxide is *gart*, $CO$; carbonic anhydride (commonly called carbonic acid gas), *garet*, $CO_2$; sulphuretted hydrogen, *gelas*; olefiant gas, *gerlel* ; carburetted hydrogen gas, *gurol;* oxychloride of carbon gas, *garted;* etc. So of volatiles heated to the boiling point; for instance, bisulphide of carbon, *ares*, when heated to 49 ° Centigrade, is a vapor, denoted by *gares ;* water, *elat*, heated to 100 ° Cent. or steam, is *gelat.*

In conclusion the speaker proposed that the new names, if approved, should be used at first side by side with the old names, and in lieu of the notation. Chemical writers, who study brevity of expression will fully appreciate the saving of pen and type work, as seen in the following statement of a recent discovery in the old and new manners. Lossen has succeeded in replacing an atom of hydrogen in ammonia by an atom of hydrogen and oxygen, or hydroxyl, thus forming hydroxalamine, which may be thus stated: 'Lossen has succeeded in replacing *al* in *ilan* by *alt*, thus forming *altelan.'*

The speaker thus, in one paper, attempted to present to his hearers the whole chemical field ; yet, as he passed from one division to another, he only cited such examples as seemed essential to prove the copiousness and capacity of the new nomenclature. A more complete elucidation and application of it was reserved for succeeding papers.

---

## ON THE PRIMEVAL ATMOSPHERE.

Dr. Hunt adverted, in commencing, to a theory first put forward by him to explain the chemical conditions of our globe. Starting from the notion of an igneous origin, he had contended that the mass probably commenced cooling at the centre, and thus gave rise to an anhydrous solid nucleus, having a crust of silicates, with an irregular surface, while the chlorine, carbon and sulphur, together with all the hydrogen, and an excess of oxygen, formed the atmosphere. As cooling from radiation went on, the first precipitate from this dense atmosphere must have been an intensely

acid liquid, which, attacking the crust of the silicates, separated
vast amounts of silica, and became saturated with earths and
alkalies, forming the primeval sea.  This condition of things, he
claimed, was in strict accordance with the known chemical laws,
and flowed logically from the hypothesis of the origin of our
planet.  The early ocean should thus have abounded in salts of lime
and magnesia; and this is confirmed by the saline waters from the
Paleozoic rocks, which represent fossil sea-water of that ancient
period.  Dr. Hunt here referred to his extended chemical and
physical investigations of the older rocks, and their mineral
springs, in support of this view.

The stronger acids of chlorine and sulphur having been
separated from the atmosphere, a decomposition of the silicates of
the exposed portion of the earth's crust, under the influence of
carbonic acid, moisture, and heat, went on, resulting like the
modern process of kaolinization, in the production of a silicate of
alumina or clay, and carbonates of the protoxyd bases.  In this
way great quantities of carbonate of soda were formed, which,
decomposing the lime and magnesia salts of the sea, gave rise
to the first limestones, and to chlorid of sodium.  Hence the clays,
the limestones, and the sea-salt were the joint results of a process
which was slowly removing from the earth its carbonic acid,
and fitting it for the support of higher forms of life.  These
views of Dr. Hunt, first put forward in 1858 and 1859, are
gradually being received and appropriated by writers, who do not
always acknowledge the source of them.  They are here insisted
upon as preliminary to some considerations on the atmosphere of
early times, when it must have contained, in the form of carbonic
acid, the whole, or the greater part of the carbon now present in
the strata of the earth, and in bodies of fossil coal.

Simple calculations show that the carbonic acid contained in a
layer of pure carbonate of lime extending over the earth, with a
thickness of 8·61 meters, would, if set free, double the weight of
our atmosphere; and that from 13·65 meters, (about forty-four
feet), would double its volume.  It moreover appears that a
similar layer of ordinary coal, one meter in thickness, would suffice
to convert into carbonic acid the whole of the oxygen of the
atmosphere: so that if, as is probable, the whole amount of coal
and carbonaceous matters on the earth exceeds this quantity,
there must have been an absorption of the oxygen, set free during
the conversion of carbonic acid into coal, this oxygen being

probably retained by peroxyd of iron.  Disregarding this, however, and admitting that the carbonic acid, corresponding to a layer 8·61 meters of limestone [about twenty-eight feet] were present in our atmosphere, tho effect would be most remarkable.  The height of the barometric column would be doubled; the boiling point of water, raised to 121° Centigrade [250° Fahr.] ; and, as the absorptive power of an atmosphere of carbonic acid is, according to Tyndal, ninety times that of dry air, the temperature of the lower regions of the atmosphere would be greatly elevated, and the whole climatic conditions of the earth modified.  Yet, as the amount of carbonic acid required to produce these results is probably but a small proportion of that now fixed in the limestones of the earth's crust, we should find this condition of things at a period, geologically, not very remote, and in still earlier times the earth must have had a far denser and more highly carbonated atmosphere than that just supposed.  The relations of such a condition of things to the animal and vegetable world furnish fruitful themes for conjecture and experiment ; and its influence on chemical processes is not less worthy of consideration, as a single instance will show.  Some years since, I pointed out that the explanation of the almost constant association of gypsum and magnesian limestone in nature, was to be found in the fact that solutions of bicarbonate of lime and sulphate of magnesia decompose each other, with production of solutions of sulphate of lime and bicarbonate of magnesia.  By spontaneous evaporation, the former may be in part separated as gypsum ; but as in this process the bicarbonate is changed into mono-carbonate of magnesia, this partially decomposes the gypsum, regenerating carbonate of lime, and the results of the experiment in an ordinary atmosphere are imperfect.  I find, however, that by infusing into the drying atmosphere a large proportion of carbonic acid, the separation by evaporation goes on regularly, and the gypsum is deposited in a pure state, enabling us thus to realize the conditions of earlier geologic periods, when vast beds of gypsum, with their accompanying magnesian limestone, were deposited in evaporating basins at the earth's surface, beneath an atmosphere charged with carbonic acid.

Ebelman has speculated on the probable existence of a much larger proportion of carbonic acid in the atmosphere of earlier geologic times; and Dana, Tyndal, and anterior to them, the late Major E. B. Hunt, have considered its meteorological relations ;

but the chemical history of this carbonic acid, considered with reference to its origin, its fixation in the form of limestones, and and its influence on chemical processes at the earth's surface, are points for the most part peculiar to the author, and, in part, now brought forward for the first time.

---

## ON THE GEOLOGICAL STRUCTURE OF THE SOUTHERN PART OF MINNESOTA.

### BY PROF. JAMES HALL, OF ALBANY.

The object of this paper is mainly to show a clear and depicted geological structure of formations of different age, over a large part of Minnesota, heretofore regarded as deeply covered by drift deposits.

In going west from the Mississippi River at St. Paul, we pass over the older SILURIAN formations of Trenton limestone, Magnesian limestone, and Potsdam sandstone, which extend as far as the lower bend of the Minnesota, at Mankato. Beyond this, in ascending the Minnesota River, for more than one hundred miles, no palæozoic formations are at present known. Approaching the Minnesota, at New Ulm, over the high prairie from the East, we find frequent exposures of a metamorphic rock, having on its weathered surface a syenitic aspect, which is in reality a quartzite, of gray, variegated or reddish color. On the Minnesota River, at Redstone ferry, these quartzites are found to have a decided dip to the eastward or south-eastward, and we have an exposure of one hundred and fifty or two hundred feet of thickness.

TRIASSIC.—Abutting against the upturned edges of these quartzites of Huronian age, there is a series of horizontal strata, consisting of red marls, reddish and variegated, and red and gray limestones, which are referred to the Triassic system.

CRETACEOUS.—Lying upon the latter formation, and likewise horizontally stratified, is a series of marls, clays, sandstones, and beds of earthy coal, having altogether a thickness of perhaps two or three hundred feet. The sandstones contain fragments of plants or trees, and leaves of the willow, poplar, liriodendron, and magnolia, all of which are referred to the age of the Creta-ceous formations.

PRAIRIE FORMATION.—Covering all these, except in the river banks, and at intervals in the prairie, is the deposit of drift and lighter soil, constituting the Prairie formation.

From the Minnesota at Redstone ferry west-ward, the Cretaceous formation extends for forty miles unbroken, when we come again to the red quartzites, which dip in the opposite direction, or to the westward; and continues for seventy miles, coming out again at the Pipestone locality, on the Sioux valley. At some point higher up the Minnesota valley, the Cretaceous formation occupies large areas resting on Laurentian rocks.

The result of these investigations shows a portion of the outcrop of a synclinal axis on the east of the Minnesota, with a valley of forty miles in width, which has been eroded in the line of a great anticlinal axis; while beyond this is a synclinal axis; of quartzites, of similar character, which forms the foundation of the great Coteau-des-Prairies, which extends for more than four hundred miles to the northwest, rising seven or eight hundred feet above the lower prairie.

We have the evidence that the synclinal axis referred to is the highest portion of the country, while the anticlinal axis had been eroded prior to the age of the Triassic formation.

The chains of lakes of this part of the country, lie in the plateau of the synclinal axis, while the line of the anticlinal is free from this feature; and the same conditions, essentially, prevail in a portion of a more eastern synclinal, which lies to the east of the Minnesota River.

---

### ON PETROLEUM.

At the opening of the session, Dr. T. Sterry Hunt read an interesting paper on Petroleum, of which the following is a brief synopsis.

He had shown in 1861, that the mineral oil of Western Canada was indigenous in the Corniferous limestone; wells sunk in the outcrop of which have yielded, and still yield, oil in that region, and also in Kentucky, according to Lesley. At that time (1861) he called attention to the existence of petroleum in the limestones of the Trenton group, and had, since then, in the Geology of Canada, in 1863, insisted upon these Lower Silurian oils as likely to prove, in some regions, of economic importance—a prediction verified by the recent developments in the Lower Silurian strata of the Cumberland, in Kentucky, and the oil wells of the Manitoulin Islands, which latter are sunk through the Utica into the Trenton formation. Another important point, on which he had been the first to insist, was that the accumulation giving rise to productive wells, occurs along the lines of anticlinal folds, where the oil would naturally accumulate in fissures, or in porous strata, in obedience to well-known hydrostatic laws. This view, first insisted upon in a lecture published in the *Montreal Gazette* for March, 1861, was further developed in a paper on Petroleum in the *Canadian Naturalist* for July, 1861, and simultaneously by Professor E. B. Andrews in *Silliman's Journal.* Since then, this view, though frequently opposed, is gaining ground; and, according to Prof. Andrews and Dr. Newberry, is sustained by all experience in the oil fields of the United States, as it also is in Canada. This remark applies to large accumulations, and to flowing wells, but oil may doubtless flow slowly from horizontal strata containing it.

As to the origin of the petroleum, Dr. Hunt supposes that it is indigenous in the two limestone formations already mentioned, and that it may have thence risen and accumulated in overlying pervious strata, or in fissures capped or sealed by impervious beds, such as the Pennsylvania sand-rock, or quarternary gravel beds.

He is inclined to think, however, that petroleum may also be indigenous in certain sandstones of Devonian or Carboniferous age, and referred to Lesley's observations to this effect, closely agreeing with those of Wall and Cruger in Trinidad, where fossil plants are sometimes found partly converted into petroleum, and partly into lignite.

Dr. Hunt regards the process by which animal and vegetable hydrocarbonaceous tissues have been converted into solid or liquid bitumen, as a decay or fermentation, under conditions in which atmospheric oxygenation is excluded, so that the maximum amount of hydrogen is retained by the carbon; and as representing one extreme of a process, the other of which is found in anthracite and mineral charcoal, the two conditions being antagonistic, and excluding each other, and the production of petroleum implying, when complete, the disappearance of the organic tissue. Hence pyroschists, the so-called bituminous shales, and coal, are not found together with petroleum, but in separate formations, and it is to be borne in mind that the epithet bituminous applied to the former bodies is a mistaken one, since they seldom or never contain any bitumen, although, like all fixed organic bodies, they yield hydrocarbons by destructive distillation. The fallacy of the notion which ascribes petroleum to the action of subterranean heat on strata holding coal and pyroschists was exposed; and it was remarked, among arguments founded upon the impermeability of many of the petroleum-bearing strata, that the oil of the Trenton limestone occurs below the horizon of any pyroschists, or other hydrocarboneous rocks.

A discussion on the subject of Petroleum followed, in which Dr. Andrews, Prof. Hall and Prof. Newberry took part.

---

## ON THE LAURENTIAN LIMESTONES AND THEIR MINERALOGY.

### BY DR. T. STERRY HUNT, F.R.S.

The author alluded to the existence in the Lower Laurentian system of three limestone bands or formations, of great but variable thickness, which might fairly be compared with the great limestone groups of the North American paleozoic system. In addition to these, there is probably a fourth and newer limestone formation belonging to the lower or true Laurentian, besides one or more in the unconformable overlying Labrador series or Upper Laurentian. The three limestone formations first named are separated by great masses of gneissic and quartzose strata, and are intimately associated with beds in which silicates of lime and magnesia prevail, together with graphite, and various metallic ores. The minerals associated with these limestones, and their accompanying strata, were next considered, and it was shown that they occur, both disseminated in the beds, and filling fissures or veins which traverse the strata. The importance in a geological point of view of these veinstones, which from their mode of formation might be named *endogenous rocks*, was insisted upon. They may attain very great dimensions, and may include any or all of the mineral species belonging to the adjacent stratification, variously grouped, and sometimes having a banded arrangement parallel to the walls of the vein. Among the characteristic minerals of these veins are calcite, apatite, pyroxene, hornblende, serpentine, chondrodite, orthoclase, scapolite, phlogopite, quartz, garnet, idocrase, epidote, spinel, corundum, sphene, zircon, magnetite, and graphite. Some of these occasionally occur in a nearly pure state, filling the veins, as graphite, pyroxene and apatite. Veins of crystalline carbonate of lime, generally including some one or more of the preceding minerals, are often met with, and it is these which have given rise to the notion maintained in this country by Emmons, and in Europe by Leonhard and others, that crystalline limestone is either partially or entirely of eruptive origin, these calcareous veinstones having been confounded with intrusive dykes. From such veinstones a transition may be traced to those in which orthoclase and quartz prevail, often to the exclusion of lime and magnesia compounds. We have then true granite veinstones, in which tourmaline, beryl, muscovite, cassiterite, and columbite are sometimes met with. These *endogenous rocks*, in which are often concentrated the rarer chemical elements of the rocks, are to be carefully distinguished from intrusive dykes which are *exotic rocks*. Such veins are not peculiar to the Laurentian system, but are found in crystalline strata at various ages. The crystalline limestones of Scandinavia, which offer so many remarkable resemblances to those of New York, New Jersey and Canada, are however of Laurentian age, and the nature of their veins has been well understood by Scheerer.

The rounded angles of crystals of certain minerals from the calcareous veins of the Laurentian system, especially of the crystals of apatite and quartz, which Emmons had supposed to be due to a commencement of fusion, is to be regarded as the result of a partial resolution of the previously deposited crystals, and as marking a stage in the progressive filling of the veins. Crystals of orthoclase, pyroxene, sphene and zircon, though accompanying these rounded crystals, retain the sharpness of their angles, because of their permanence in the heated alkaline solutions which circulated through these yet partially filled veins. The various minerals of these veinstones have been deposited from aqueous and saline solutions, at elevated temperatures, and the experiments of Daubree and of De Senarmont, and the microscopic observations of Sorby, support this view. Plutonists begin to understand that water cannot be excluded from rocky strata, but is all-pervading, and that at greater depths, kept by pressure in a liquid state, at an elevated temperature, and having its solvent powers augmented by alkaline salts, it plays a most important part in metamorphosis, and in the formation of veinstones. The author supposed, with Mr. Hopkins, that in earlier geological periods the increase of temperature in buried strata was far more rapid than at present, so that great heats prevailed at comparatively small depths from the surface, and produced important chemical and molecular changes. The temperature at which the various silicated and other minerals, including graphite, were dissolved from the strata and crystallized in the veins, he supposed to have been, judging from various analogies, between the melting point of tin and low redness.

The distinction between the apatite, graphite and magnetite disseminated in the beds, and the same minerals in the veins, was particularly insisted upon.   As to the origin of the principal silicious minerals of the limestones, such as serpentine, chondrodite, pyroxene, rensellaerite and loganite, Dr. Hunt regards these as having been directly deposited as chemical precipitates from the seas of the time; and cites the example of the *Eozoon Canadense*, an abundant fossil of the age, found imbedded in these silicates, which enclose it, and fill the minute pores of its calcareous skeleton.   To a similar chemical precipitation he attributes the serpentines, talcs, chlorites and epidotes which occur in more recent rocks, and may be found in their incipient state before the metamorphosis of these rocks, which has for the most part only crystallized and re-arranged the already-formed amorphous silicates.   The chemical agencies which gave rise to these silicates of lime, magnesia, iron and alumina were briefly discussed, and declared to be still active, although probably to a less degree than formerly.

(*Corrected from the Newspaper Reports.*)

---

## ADDRESS TO THE MEMBERS OF THE MONTREAL NATURAL HISTORY SOCIETY,

### DELIVERED MAY 18TH, 1866.

BY CHARLES SMALLWOOD, M.D., LL.D., D.C.L., &c., President of the Society.

MY LORD AND GENTLEMEN,—The rolling wheels of time have again brought us to this our annual re-union.   Thirty-nine years have passed away since this Society was founded; and it now devolves upon me, as your President, (a position which I owe to your individual kindness,) to resign into your hands the charge you have placed in my keeping.   I felt at the outset my utter inability to fulfil those duties which my predecessors have so well and so efficiently discharged; but I relied upon your help and assistance, and was assured that what was wanting in my own personal exertions, would be supplied by your advice and help. In this, gentlemen, I have not been disappointed; and permit me now to tender to each of you individually my best and warmest thanks for the forbearance and kindness you have at all times shewn to me in those shortcomings which have occurred during my tenure of office. And while it is with feelings of gratitude that I tender to you my resignation, they are mingled, nevertheless, with feelings of pride for the honor you have conferred upon me.

It is not, gentlemen, due to any personal exertions or energy on my own part that we have arrived at this, the termination of another year of great prosperity and increased usefulness; but it is to those friends whose scientific efforts have been so well directed; and it is to you who have trodden so zealously the path of those few devoted men whom we may be proud to call our predecessors and the founders of this Institution. It is, I repeat, to your efforts that our increased prosperity must be attributed. It is a noble object that has invited us to these Halls of Science. We meet together to contemplate the teachings of God in Nature; and our mutual aim should be, and we hope has been, to decipher some new word in the pages of that great book, in order that we may the better learn the will and the workings of Him who ordereth all things well. We have sought to study the method of God's workings in nature; for in the vision of science there is nothing too minute for our notice, or unworthy of it. The means for the investigation of almost every branch of Natural Science are gradually extending; and the Montreal Natural History Society is not the least important of those institutions which are spreading over our country, and the world generally, scientific knowledge, for science is nothing more than knowledge reduced to order. But to say that science is worthy of your pursuit, is at best a waste of words. You know too well its importance; for by science we have converted the products of our forests and our fields into articles of commerce; we have by science abridged human labour to an immense extent; we have by science invented machines, some of immense power, all but surpassing human efforts at calculation, and others which almost rival the winds in swiftness, propelled on road-ways that have compassed our globe by their iron bands; and science, again, has nearly achieved a victory over the velocity of thought, light and sound, in the invention and application of our electrical telegraph.

Where shall I specially turn to contemplate the wondrous works of God, or to follow up the yearly march of science? Shall I dip with a Logan, a Dawson, a Hunt, and a Billings, beneath the

rocky covering of our globe, for a subject of discourse ? I dare
not. Their mantle would not fall with graceful folds upon one so
incompetent as myself. Our reports and journals bear ample
evidence of their united labours and individual researches. Or
shall I stroll through the deep forests or over the flowery sod,
where once trod the footsteps of a Holmes or a Barnston, one of
whom was removed from among us full of years and of honour;
while the other had scarcely entered upon the busy stage of science
ere he was called away? But why should I hesitate to find a
suitable theme in the vast domains of science? Why should I say
more? Ascend with me above the dust, ascend with me far above
those sure foundations that were laid in the ages of this our world,
far, far gone by; ascend with me above the clouds,—those cirrous
clouds, where the heavens are never obscured, where the atmosphere is pure and free from mist,—in the balmy but intensely cold
regions of space, where our earth, with its lofty mountains and
fertile valleys, with its noble mansions and its lovely cottages, is
only seen as a small planet; where our sun itself is dwindled
to a twinkling star; where the starry host is nearly lost from
vision,—merged, as it were, into a milky way;—and where the
great girdle of the heavens itself is but a faint nebulous mass.
Yet deep even into this immensity of space science has cast its
divining rod.

A Herschel discovered a world eighty times larger than our
own, which revolves in its circuit in a long period of time, corresponding to more than 80 of our years, ere its curved course is
run. Round this planet, thus removed some eighteen hundred
millions of miles, six moons revolving like our own accompany it
on its onward and extended course.

But from this distant world the shout of science was still
*Onward!* A Le Verrier and an Adams, with a colossal stride,
placed one foot, as it were, on our earth, and another on the surface of this distant globe, and pointed out the spot where Neptune
was to be found, a planet still further removed from us, and whose
period of revolution was more than double that of Uranus. But
even that planet appears near us when we measure the nearest
star that bedecks the vaulted canopy of heaven; for that is twenty
billions of miles distant from our sun.

If geology marks the progressive development of the rocks on
our globe, and counts its periods by millions of years, (for the rocks
are but incidents in the earth's history,) surely the astronomer

may well be lost in admiration by the contemplation of these
wondrous works that are manifest in

"the wide expanse,
Where stars, and suns, and systems shine."

The progress of astronomical science has shown us that our sun
can no longer be regarded as the centre of our solar system, but
that all the starry host is moving yearly in a grand procession
towards another, a far distant central sun, the great centre of our
universe; and we may well say, in the words of the poet,

"He sets the bright procession on its way,
And marshals on the order of the year."

Scarce a year has passed without adding to our list of the
Asteroids, until the number now reaches 85; while a very few
years ago it was but four. Are these asteroids the particles of a
larger planet? or are they new worlds opened up to human vision,
aided by science in the construction of the telescope? or have they
been for ever wanderers in the pathless regions of space? Here
science will one day, with a spectroscope, tell us if they are the
remains of a larger body. A short time will no doubt set this
question at rest, for if they are the particles of a larger planet,
which from any cause has burst asunder, the spectra will furnish
the same results for them all.

Modern investigations have shown that our sun possesses an
atmosphere, and that this atmosphere is disturbed by some action
that renders visible certain spots at different times, spots which
led Galileo to demonstrate the rotation of the sun upon its axis.
It is the opinion of modern observers that the photosphere, (our
sun's atmosphere) consists of solid or liquid bodies of a greater or
less magnitude, either slowly sinking, or suspended in equilibrio
in a gaseous medium; and that either the body of the sun itself
is older than the surrounding medium, or else that some chemical
or molecular changes have taken place where a spot is formed;
or that it is produced by matter coming from a colder region; or,
may be, by the solidification of its particles. But more recent
investigation would tend to show that the body of the sun itself is
hotter than the surrounding photosphere.

From the surface of the sun that imponderable fluid, light, is
diffused, shedding on this earth all the brilliancy of colour, and
tinting the landscape with an ever-varying degree of beauty.
What a glorious expanse of view, and what a vast field of know-

ledge has been revealed within even the few past months, bearing
on this subject of spectral analysis.

The immortal Newton, by means of the prism, resolved light
into its ultimate rays in the solar spectrum, a fitting rival to the
rainbow. Fraunhofer discovered that this spectrum was traversed by numerous dark lines or bands which gave no light or
colour, indicating that at the source from whence they emanated,
the rays of light were absorbed in their passage from the sun to
our earth, and probably some by the earth's atmosphere. More
probably some are absorbed in the atmosphere of the sun itself,
for the most recent investigations in this department of physical
research have shown that a glowing and gaseous atmosphere
surrounds the solid nucleus of the sun, which, possessing a still
higher temperature, approaching the intense heat of the brightest
whiteness.

The polarized rays of this light exhibit spectra still more beautiful and intense than the solar spectrum itself. Forms of the
most symmetrical order are constantly presented when a polarized
ray of light is passed through various substances; and these
appearances are constantly varied when we change, by means of
pressure, the molecular arrangement of these bodies.

And are we not, by the photographic art, able to preserve, in
unfading lines, the lineaments of those we love, of those that are
great, and wise, and good; as well as to transfer to paper, by this
process of *sun-painting*, those cherished spots on earth most dear
to us, every modulation of the landscape, the familiar dell, and the
rippling river by our homes of childhood?

But the progressive march of science has not stopped here.
The investigations by means of the spectral analysis have penetrated into those regions of space to which I have already alluded,
and the fixed stars have been the objects of intense interest. The
astronomers had well said that they were distant suns, like our
own, shining by their own light; and this opinion has been confirmed by the spectroscope. They are composed of the same matter
as our sun; and in the spectra of these stars, the dark lines are
wonderfully well brought out and defined.

Many of the stars of the first magnitude have been subjected to
direct experiment; and it has been shown that they possess in
their atmospheres many of our terrestrial elements. *Aldebaran*, a
star of the first magnitude, possesses sodium, magnesium, hydrogen, calcium, iron, bismult, tellurium, antimony and mercury,

besides others which give negative evidence only. *Alpha Orionis* has been carefully examined, and contains most of the above-named elements with the exception of hydrogen. The presence of hydrogen has been noticed in the sun, and in almost forty fixed stars, and is eminently characteristic, showing that its presence belongs to the atmospheres of the luminous bodies themselves, and not merely to our own atmosphere.

These investigations have confirmed and demonstrated, beyond the shadow of a doubt, that all the planets shine by light reflected from the sun, and that any variety differing from the solar spectrum may be attributed to the peculiar properties of the atmospheres that surround the planets themselves.

One of the most important and interesting deductions to be drawn from these researches, is in connection with the origin of the colour of the stars. That a difference of colour in the stars does exist, is too well known to require any comment : for " one star differeth from another star in glory." And it is now no longer a matter of conjecture that the brightest stars at least are, like our sun, giving energy and life to systems of worlds like our own, adapted for the abode of intelligent life. While yellow and red stars are the most frequent, in double stars the contrasted colours are green and blue. The source of the light of the stars must be a solid or liquid body in a state of incandescence, as only such bodies, when raised to a high temperature, give out a continuous spectrum. In the case of the fixed stars and the sun, this continuous spectrum becomes crossed by dark bands, which are produced by the absorbing power of the constituents, held in a vaporous form in the investing atmospheres. These atmospheres vary in chemical constitution, according to the elements composing the star ; and the dark lines are produced by the absorptive power of the vapours forming the stellar atmospheres. They correspond to the bright lines they would form in an incandescent state, and would be the strongest and most numerous in the more refrangible portions of the spectrum, consequently a star would have a red or orange tint should that part of the spectrum suffer least absorption : while, on the contrary, should the red and yellow portion have most lines, the blue and green rays would then predominate in the colour of the star.

In *Sirius*, the ' dog star,' which is of a brilliant white, there are no lines sufficiently intense, in any particular part of the spectrum, to interfere with our receiving the light in about the same proportion as to the quantity of the different coloured rays, to that which starts from the incandescent light-giving surface. Sodium, magnesium, hydrogen, and probably iron, have been found in this star ; and even a photograph on wet collodion has been obtained. In reference to double stars, observations on *Beta Cygni* and *Alpha Hercules* confirm these observations.

Various opinions have been ventured on the composition of the nebulæ. It has been affirmed that they are masses of minute stars, and only require higher optical powers to reduce them to distinct vision. The construction of Lord Rosse's telescope was looked forward to as tending to set the matter at rest ; but, instead of this, it seemed to involve the question in still greater difficulty. Its solution was not lost sight of during the past year, and the spectrum observation has been shown to have an important bearing on the nebular hypothesis of the cosmical origin of the universe. It shows that the elementary substances must have existed in different proportions at different points of the nebulous mass ; otherwise, by condensation, equal portions of the elements from the surrounding vapour would have been collected.

There is also an analogy to the manner in which the components of the earth's crust are distributed, for some of these elements are widely diffused through vegetable, animal, and mineral matter.

It has been further shown that it is only liquid and solid bodies that give out a continuous spectrum ; while gases alone, when rendered luminous by heat, give out light which, after dispersion by the prism, is found to consist of certain degrees of refrangibility only, and which appear as bright lines on a dark ground, contrary to the solar spectrum, which shows dark lines on a bright ground. This fact has shown that, in the nebulæ, large masses of gas exist, and they possess no resemblance whatever to stars or clusters of stars. The nebulæ, therefore, are not masses of stars removed to such a distance as to render them irresolvable, but consist, for the most part, of luminous gases.

This presents to us, at once, another instance of unity in nature, by recognizing each of the simple bodies held in suspension in the flame, whose rays are decomposed by the prism. The dispersion of the sun's rays by the prism forms the standard of observation ; any deviation will show either bright lines in the place of dark ones, or dark lines in the place of bright ones. Nickel, chromium, magnesium, iron, potassium, sodium, barium, copper, cobalt and zinc, are found always present in the sun's atmosphere in a state of vapour.

The possession of an atmosphere by the moon has been the subject of frequent investigation and conjecture; but, by the spectrum analysis, it is now rendered certain that the moon has no atmosphere, at least on that side presented to our view. This has been lately further confirmed by observing the different spectra shown by the occultation of a star by the moon at the moment of contact, by obtaining the two separate spectra at once in the field of view.

It may be thought that the few remarks on the branches of science to which I have more immediately alluded, do not fairly come within the scope of the Natural History Society. But as, in looking over the annual addresses for the past few years, I found no account of any of the progressive steps in the sciences generally, except in those of Geology and Botany, I deemed it not unworthy to allude to some of these more recent researches in other departments of physical science.

I ought not to close this short address without expressing my great regret that Montreal does not possess any adequate means, owing to the want of proper instruments, for prosecuting the science of Astronomy. A climate like that of Lower Canada, which furnishes, upon an average, 120 nights in a year suitable for celestial observation, offers a vast field for astronomical labours, and also for the investigations now being carried on in celestial chemistry, and the spectum analysis. Since our last annual meeting, many original papers on subjects more intimately connected with Natural History have been read before the Society, or printed in the *Canadian Naturalist*, the perusal of which will shew that many new and curious facts have been observed and recorded, bearing upon the geology, zoology and botany of British North America. These papers will furnish evidence that the members of this Society have not been idle during the past session, and that some of them have devoted a considerable time to the study of those objects which come more directly within its scope. Those who are more particularly engaged in the study of natural history in Canada, know further that investigations have been carried on during the past Summer, the results of which have not yet been recorded. Among the papers to which I may more particularly refer are: four on Geology and Palæontology, by Dawson, Billings, Packard and Whiteaves; four on Zoology, by

Stimpson, Parkes, Couper and Ritchie; two on Botany, one by Mr. Watt, and another from Dr. Gibb; and one on Geography, from Dr. Hunt.    I would refer to the pages of the *Canadian Naturalist* for more ample information on these points.

The pursuit of science, in its legitimate sense, is to endeavour to advance man's happiness, and to elevate and refine every human sentiment.  Associations of a like character to our own are intended to diffuse intelligence and the light of truth to man, to fit him for a higher state of existence.

The study of nature has formed the object of the most elevated and aspiring thoughts,—thoughts that have dwelt on the works and wonders of creation.  What is more beautiful or more elevating than those aspirations that direct us to contemplate the wisdom and goodness of God?  and what can be more pleasing than that kindred minds should associate in mutual harmony, and contribute each his small portion (though small) to the grand treasury of knowledge and of truth?  Nor is it possible to suppose that the onward progress of true science will ever operate to the disparagement of that devout homage we owe to Him in whose hands are held our daily wants and future destiny; but on the contrary science, if directed in the proper paths, will aid in fitting us, after a life devoted to its pleasures and its beauties, for the enjoyment of that intellectual intercourse which has ever been among the holiest and noblest aspirations of man.

I have not entered much (nor did I intend) into the business part of the Society's operations, properly so called, leaving it to your Council, Scientific Curator and Treasurer to present their reports, which, I have no doubt, will be very satisfactory.  But I must not forget to mention the eminent and efficient services of Mr. Whiteaves.  A look into our museum will, I am sure, convince any one of the amount of labour he has bestowed; and I feel sure that your Council will render also a good account of his recent visit to England.

For my own part, I am sorry to say that a lack of time has prevented me from filling the office of President so well as I could have wished.  In resigning the charge into your hands, I must be allowed to express a fervent wish that increasing prosperity may mark our way; and to say that we may congratulate ourselves on our increasing usefulness in spite of a Winter of more than ordinary excitement, owing to a most wicked and unheard-of threat of invasion of our country by strangers, many of our young men having taken up arms in defence of our homes. But I trust that now peace is again restored to us, and hope that war, with all its appalling features, may merge into the calmer pursuits of science; and that the Montreal Natural History Society may long continue to diffuse and spread knowledge; for

"There's beauty all around our paths, if but our watchful eyes
Can trace it 'midst familiar things, and through their lowly guise."

----

## ON THE VITAL STATISTICS OF MONTREAL.

By PHILIP P. CARPENTER, B.A., PH. D., Hon. Sec. of the Montreal Sanitary Association.

In the *Canadian Naturalist* for 1859, pp. 173-186, was published the first attempt to eliminate and explain the sanitary statistics of Canada. The facts and figures therein set forth were carefully scrutinized in this and other cities. As was to be expected, the conclusions arrived-at were frequently called in question; but the writer was charged with inaccuracies which belonged to the data, and not to the working-out of the materials. The figures were not set forth as accurate; but only as the *nearest approach to accuracy which was then attainable*.

The census of 1861 has now furnished elements for comparison with similar results in the previous decade; and the yearly tabulation of burials and baptisms in the city of Montreal and in the adjacent counties has added to the cumulative evidence of the peculiar unhealthiness of the city. It is proposed, in the present paper, to present the results of these two sources of information; and to compare them with a third source, viz. the weekly returns of interments at the city cemeteries, which were not accessible to the writer in 1859.

### A. CENSUS OF 1861.

It must be premised that the deaths are twice tabulated in the census returns, viz. under ages, and under diseases. On analyzing these in order to ascertain the proportions of deaths from zymotic diseases, of deaths under 5 years, and of deaths above 70 years, to the total deaths, it was found that in Quebec City, then the capital of Canada, there was no less a discrepancy than 296, in the total number of deaths recorded, between these two tabulations. Such a glaring inaccuracy in a work executed at considerable expense, and demanding the greatest care to make it of practical value, is not calculated to raise the character of the Canadian Executive; and throws considerable doubt on the value of the returns in general. Evidence is given in the 'Second Report of the Financial and Departmental Commission,' Feb. 1864, pp. 32 et seq., that "the irregularities in the returns themselves resulted from the ignorance of many of the enumerators as to the object of the different columns; and carelessness in leaving some of them blank, or filling them in a manner that was *manifestly absurd.* Where the addition of several columns should have agreed with the total given in some other column, it *often happened that irreconcilable differences occurred. . . .* Some mode of bringing these totals into harmony was necessary; and an arbitrary system of what I must call *cooking the figures* was resorted to for the purpose."

The returns for Montreal City are said to have been made with the greatest attainable accuracy; yet the deaths for the year are only stated as 2,038, while we know that 3,181 interments actually took place during the year at the two cemeteries, being a difference of 1,143, or *more than* 50 *per cent.* If it be supposed that this marvelous discrepancy arose from a different division of the year, the fact remains that the interments for 1860 were 3,171, and for 1862, 3,461; in neither case presenting a perceptibly lower rate.

If such be the manifest and gigantic untruth in the returns of the two largest cities of British America, it is hard to place any reliance on returns of places of less importance, least of all of country districts. Even if the figures had been accurately given, they would only have established facts for a single year, which might have exceptional: as it is, they must only be accepted for comparative, not for absolute results. Such as they are, they are presented in the following table, where the first two columns A and B give the actual population and mortality. Column C presents the average deaths among each thousand of the population. Column D shews the number of deaths, out of every hundred from all causes, which were due to xymotic diseases. When this proportion is permanently high, it is a sure sign of bad air outside or within the dwelling, or of polluted water: where it is exceptionally high (as, apparently, in Ottawa, Laval, Vaudreuil, Soulanges and Laprairie) it betokens an epidemic, which is probably due to cumulative corruptions: where it is remarkably low, it may be

taken as a very favourable sign of the sanitary conditions. Column
E gives the percentage of the total deaths which took place under
five years of age.   If accurate, unless there were some special
infantile epidemic, the high or low percentage in this column
ought to be a sure test of sanitary condition; but the high
rate in healthy Upper Canada, never falling below 35 p. c., and in
even the country districts of Lower Canada (with the exception of
Soulanges), needs some explanation not yet given.   Column F
gives the number, out of every hundred deaths, which were of
people above the allotted term of 70 years of age.   Contrary to
the previous columns, it ought to be highest in the most healthy
districts; but the numbers are so low that they could only be
trusted on an average of years, or for a large population.   Thus
the low rate for Three Rivers, and the very high rate for Soulanges
(*nearly five times that of Montreal*) are probably accidental.
Column G exhibits the proportion between the births and deaths
in the year; the figures representing the deaths in each district to
every hundred births.   If accurate, these ought to be lowest in the
most healthy districts, as we see in the case of Verchères which
presents only half the death-rate of Montreal.

The last column, II, representing the number of Catholics out
of every hundred in the population, has been added to test the
value of a suggestion made in certain quarters that the religious
customs of the French Canadians, who bring their infants to be
baptized in the church, even in the coldest weather, was a main
cause of the excessive infantile mortality of Montreal.   It will be
seen that the proportion of Catholics is *less in Montreal* than in
any other quoted district of Lower Canada, except Sherbrooke.

The returns may be regarded (subject to exceptions) as sufficiently correct to show the *comparative* mortalities of cities and
adjacent counties, and to compare these with the ratios worked-out
from the preceding census.   It is presumed that the causes of inaccuracy will affect the different returns in somewhat of the same
ratio.   They must also be taken (whether accurate or not) as our
only data for the actual population; and, by comparison with the
census of 1851, for the yearly average rate of increase.   There
was no temptation to "cook the figures" in this, the easiest part
of the work; least of all, to reduce the population below its actual
extent.

*In all the columns which include Quebec city, two sets of figures
are bracketed together* for the reason stated above.   Analogy proves

that the higher rate, assigning 1,111 deaths, is more likely to be correct.

### 1. *Sanitary Statistics of the Census of* 1861.

| | A | B | C | D | E | F | G | H |
|---|---|---|---|---|---|---|---|---|
| | Population. | Total Deaths. | Deaths per 1000 living. | Percentage of total deaths from Zymotic Diseases. | Percentage of total deaths under 5 years. | Percentage of total deaths above 70 years. | Number of Deaths to each 100 Births. | Percentage of Catholics. |
| ALL CANADA........ | 2,507,657 | { 23,088 / 23,384 | 9.2 / 9.3 | 22.3 / 21.9 | 49.7 / 48.9 | 7.0 / 6.8 | { 24 / 25 } | 48 |
| *Upper* Canada......... | 1,396,091 | 10.160 | 7.2 | 18.6 | 42.1 | 6.6 | 19 | 18 |
| Do. less 5 cities........... | 1,292,207 | 8,813 | 6.8 | 18.9 | 41.4 | 7.0 | 17 | 18 |
| Toronto................. | 44,821 | 727 | 16.2 | 14.2 | 48.7 | 3.5 | 45 | 27 |
| Hamilton................ | 19,096 | 217 | 11.3 | 18.9 | 49.9 | 3.7 | 31 | 25 |
| Ottawa.................. | 14.669 | 172 | 11.7 | 32.5 | 48.2 | 3.5 | 29 | 56 |
| Kingston ............... | 13,743 | 129 | 9.4 | 16.3 | 34.9 | 5.4 | 26 | 34 |
| London................. | 11,555 | 102 | 8.8 | 7.8 | 39.2 | 3.0 | 24 | 18 |
| *Lower* Canada......... | 1,111,566 | { 12,928 / 13,224 | 11.6 / 11.9 | 25. / 24.5 | 55.4 / 54.2 | 7.3 / 7.1 | { 31 / 32 } | 85 |
| Do. less 2 cities ......... | 970,134 | 10,075 | 10.3 | 25.1 | 53.3 | 8.2 | 29 | 86 |
| Do. less 4 cities ......... | 958,177 | 9,877 | | | | | | |
| Montreal ....... ..... | 90,323 | 2,038 | 22.5 | 23.5 | 66.0 | 3.4 | 55 | 73 |
| Quebec.............. | 51,109 | { 815 / 1,111 | 15.9 / 21.7 | 27.6 / 20.2 | 55.2 / 40.5 | 4.1 / 3.6 | { 46 / 63 } | 81 |
| Quebec *County*........ | 27,893 | 411 | 14.7 | 14.8 | 48.2 | 10.2 | 38 | 87 |
| Three Rivers ..... .... | 6,058 | 106 | 17.5 | 21.7 | 56.6 | 1.9 | 44 | 92 |
| Sherbrooke ............ | 5,899 | 92 | 15.6 | 27.1 | 42.1 | 6.5 | 39 | 46 |
| Hochelaga *County*..... | 16,474 | 226 | 13.7 | 23.0 | 74.3 | 2.6 | 40 | 88 |
| Jacques Cartier " ...... | 11,218 | 140 | 12.4 | 18.6 | 52.1 | 5.7 | 39 | 92 |
| Laval " ...... | 10,507 | 152 | 14.4 | 32.9 | 56.5 | 8.5 | 58 | 99 |
| Vaudreuil " ...... | 12,282 | 163 | 13.3 | 34.3 | 60.7 | 8.0 | 34 | 91 |
| Soulanges " ...... | 12,221 | 149 | 12.2 | 29.5 | 28.2 | 14.7 | 29 | 94 |
| Laprairie " ...... | 14,475 | 183 | 12.7 | 35.5 | 58.4 | 5.4 | 49 | 96 |
| Chambly " ...... | 13,132 | 121 | 9.9 | 17.3 | 43.0 | 6.6 | 28 | 96 |
| Verchères " ...... | 15,485 | 167 | 10.8 | 18.5 | 49.7 | 8.9 | 27 | 99 |
| Total of 8 Counties round Montreal..... | 105,794 | 1,301 | 12.3 | 26.2 | 52.8 | 7.5 | 36 | 94 |
| Montreal City.......... | 90,323 | 2,038 | 22.5 | 23.5 | 66.0 | 3.4 | 55 | 73 |
| Excess for Montreal..... | −15,471 | +737 | +10.2 | −2.7 | +13.2 | −4.1 | +19 | −21 |
| Total of 7 of the above Counties, leaving out Verchères.......... | 90,309 | 1,134 | 12.5 | 27.3 | 53.3 | 7.3 | 38 | 94 |
| Montreal City.......... | 90,323 | 2,038 | 22.5 | 23.5 | 66.0 | 3.4 | 55 | 73 |
| Excess for Montreal..... | +14 | +904 | +10.0 | −3.8 | +12.7 | −3.9 | +17 | −21 |
| Comparison of { London.............. | 11,555 | 102 | 8.8 | 7.8 | 39.2 | 3.0 | 24 | 18 |
| { Montreal.............. | 90,323 | 2,038 | 22.5 | 23.5 | 66.0 | 3.4 | 55 | 73 |
| Excess for Montreal..... | +78,768 | +1,936 | +13.7 | +15.7 | +26.8 | +.4 | +31 | +55 |

In the above schedule is first given the general average for the whole of Canada, from Gaspé to Essex, including the cities.

Next come the figures; 1. for the whole of Upper Canada; 2. for the same, excluding the five principal cities, but including all the others; and 3. for the five cities, in the order of their population. As compared with England, one cannot but be struck with the extremely low rate of mortality throughout. English insurance companies doing business in the province according to their home tables, may expect to gain considerably on life policies.

The third group presents the principal statistics for Lower Canada; first for the whole province; next for the same, leaving out the two unhealthy cities, Montreal and Quebec; next for the province, leaving out also Three Rivers and Sherbrooke; (these however, although as unhealthy as Toronto, do not affect the general average;) next for Montreal, and for Quebec with its double entry of "uncooked" figures; next for the county of Quebec, leaving out the city; and lastly for the two smaller towns, which, though healthy in comparison with their populous neighbours, are much more unhealthy than the larger cities of Upper Canada.

The next group contains the figures for the eight counties round Montreal, which were included in the registration district, and whose returns are preserved at the Protonotary's Office. Some of these display a high rate both of xymotic and of infantile mortality; yet when their total is added up, and the average taken and compared with that of the city repeated below, the *excess of deaths amounts to one citizen taken yearly out of every hundred,* who would have lived had he dwelt in the country, with the same climatal conditions, and a preponderating Catholic element.

The contrast is perhaps rendered more apparent by leaving out Verchères from the above total, and thus bringing the country population to an almost exact equality with that of the city. Although the abstraction of this healthy district somewhat raises the death-rate for the rural population; we find that in that year 904 *persons were killed by city life;* 12 per cent more of city than of rural deaths were of children under five years; less than half the number reached the age of 70; and there were 17 additional deaths to set against each hundred births. This was in spite of special epidemics which appear to have visited at least half of the rural districts, and which caused nearly 4 out of every hundred deaths more than in the city.

The last group of figures shews the contrast between Montreal, the most unhealthy, and London, the most healthy of Canadian cities, which presents a death-rate below that of the rural districts of Lower Canada. It appears that the *extra* mortality of Montreal amounts to 137 in every 10,000 persons; that for every 10 persons who die in London, 25 die in the older city; and that, out of every hundred deaths, more than 26 additional cases of children cut off under 5 years of age are found in Montreal.

The following is a comparison of the statistics of population and mortality between the census of 1851 and that of 1861. Some particulars from the report of the (English) Registrar General for 1857* are added.

## 2. *Comparative Sanitary Statistics of the Census of* 1851 *and of* 1861.

| | Population. | | Total Deaths. | | Deaths per 1000 living. | | Excess of Deaths in 1861 over rural districts of | |
|---|---|---|---|---|---|---|---|---|
| | | | | | | | Upper Canada | Lower Canada |
| | 1851. | 1861. | 1851. | 1861. | 1851 | 1861 | | |
| ALL CANADA | 1,842,265 | 2,507,657 | 19,449 | 23,384 | 10.5 | 9.3 | 6,269 | ... |
| *Upper* Canada.. | 952,004 | 1,396,091 | 7,775 | 10,160 | 8.2 | 7.2 | 558 | ... |
| Do. less 5 cities | 880,737 | 1,292,207 | 6,754 | 8,813 | 7.5 | 6.8 | ... | ... |
| Toronto........ | 30,775 | 44,821 | 474 | 727 | 15.4 | 16.2 | 411 | 263 |
| Hamilton....... | 14,112 | 19,096 | 172 | 217 | 12.2 | 11.3 | 86 | 19 |
| Ottawa ........ | 7,760 | 14,669 | 90 | 172 | 11.5 | 11.7 | 71 | 20 |
| Kingston ....... | 11,585 | 13,743 | 185 | 129 | 15.9 | 9.4 | 35 | .. |
| London........ | 7,035 | 11,555 | 100 | 102 | 14.2 | 8.8 | 23 | .. |
| *Lower* Canada.. | 890,261 | 1,111,566 | 11,674 | 13,224 | 13.1 | 11.9 | 5,668 | 1778 |
| Do. less 2 cities. | 790,494 | 970,134 | 8,632 | 10,075 | 10.9 | 10.3 | 3,395 | .... |
| Montreal....... | 57,715 | 90,323 | 1,978 | 2,038 | 34.4 | 22.5 | 1,417 | 1101 |
| Quebec...... .. | 42,052 | 51,109 | 1,064 | 1,111 | 25.3 | 21.7 | 761 | 582 |

| | |
|---|---|
| All England................................................ | 22.0 |
| London.................................................... | 25.0 |
| Eastbourne, Sussex........................................ | 15.0 |
| Liverpool................................................. | 36.0 |
| Average Deaths in all England from xymotic diseases, out of every hundred deaths........................................ | 22.0 |
| Do. under five years...................................... | 39.1 |

If these returns could be relied upon, they would present an extremely flattering picture of Canada in general, and even of the cities in particular, as compared with the rural districts and cities of England, and as compared with its own condition ten years previously; Toronto and Ottawa being the only cities in which

---

* This is the latest return accessible at the free library in the Mechanics' Institution. It represents an average of many years. Not a single district in England is found to have a mortality less than 15 per 1000, or more than 36.

the mortality has increased. But as we know that the deaths for Montreal are glaringly understated, we are obliged to doubt the accuracy of the returns in other districts also. As the registers of interments at cemeteries and churchyards must be always accessible to the enumerators, it is hoped that the authorities will take the necessary steps to insure accuracy at the next decennial census.

The following table has been calculated in order to estimate the proportion borne between the interments at different ages, and the number living at the same age. The "total deaths" are probably much below the real numbers, but the *ratio between the ages* may be sufficiently near the truth.

3. *Population and Deaths in Montreal at different ages : from the Census of* 1861.

| 1861. | Number living. | Total Deaths. | Deaths per 1,000 living at the same age. | Quebec. Do. | Lower Canada, less 4 cities. |
|---|---|---|---|---|---|
| Under 1 year...................... | 3,700 | 1,006 | 271.3 | 161.9 | 82.6 |
| From 1 to 2 years................. | 3,183 | 179 | 56.2 | 48.8 | 43.8 |
| "  2 to 3  " ................ | 2,883 | 70 | 24.3 | 33.2 | 16.0 |
| "  3 to 4  " ................ | 2,821 | 46 | 16.3 | 17.9 | 10.2 |
| "  4 to 5  " ................ | 2,609 | 44 | 16.5 | 11.6 | 7.2 |
| "  0 to 5  " ................ | 15,196 | 1,345 | 88.5 | 58.4 | 32.3 |
| "  5 to 10  " ................ | 10,363 | 86 | 8.3 | | |
| "  10 to 15  " ................ | 9,200 | 37 | 4.0 | | |
| "  15 to 20  " ................ | 10,890 | 55 | 5.5 | | |
| "  0 to 10  " ................ | 25,559 | 1,431 | 55.9 | (It was not judged | |
| "  10 to 20  " ................ | 20,030 | 92 | 4.5 | necessary to | |
| "  20 to 30  " ................ | 18,174 | 119 | 6.5 | complete the | |
| "  30 to 40  " ................ | 11,044 | 89 | 8.6 | table for adult | |
| "  40 to 50  " ................ | 7,248 | 50 | 6.9 | deaths in Que- | |
| "  50 to 60  " ................ | 4,476 | 72 | 16.0 | bec and the | |
| "  60 to 70  " ................ | 2,460 | 56 | 22.8 | rural districts.) | |
| Above 70 and unknown............. | 1,272 | 129 | 101.4 | | |
| All Ages........................... | 90,323 | 2,038 | 22.5 | | |

It appears, therefore, that for every hundred children who die under one year in Montreal, only *sixty* die in Quebec, and *thirty* in the country districts. For every hundred who die under five years in Montreal, *sixty* die in Quebec, and only *thirty-six* in the country districts.

### B. PROTONOTARY'S RETURNS.

It appears, by the rate of increase ascertained from the census of 1861, that the population of Montreal City must have been greater than that assumed in the table printed in the *Canadian Naturalist*, 1859, p. 176, so far as the later years are concerned. Subtracting that rate, viz., 3,260 annually, to find the population before 1861, and adding it for the subsequent years,* we are able to present a table approximately correct, as follows :

### 4. *Montreal City: Returns of Baptisms and Funeral Services.*

| Year. | Supposed Population | Births. | Deaths. | Excess of Births over Deaths. | Deaths per 1.000 living. | Deaths per 100 Births. |
|---|---|---|---|---|---|---|
| 1859................. | 83,803 | 4,238 | 2,581 | +1,657 | 30.8 | 60 |
| 1860................. | 87,063 | 4,438 | 3,016 | +1,422 | 34.7 | 68 |
| 1861................. | 90,323 | 4,579 | 3,005 | +1,574 | 33.2 | 65 |
| 1862 ................ | 93,583 | 4,811 | 3,222 | +1,589 | 34.4 | 67 |
| 1863................. | 96,843 | 5,388 | 3,510 | +1,878 | 36.2 | 65 |
| 1864................. | 100,103 | 4,024 | 4,306 | — 282 | 43.0 | 107 |
| 1865................. | 103,363 | 4,339 | 3,732 | + 607 | 36.1 | 86 |
| Average of 7 years........... | 93,583 | 4,545 | 3,390 | +1,155 | 36.2 | 74 |
| Average of 6 years(—1864)... | 92,496 | 4,632 | 3,177 | +1,455 | 34.3 | 68 |

The returns from which this table is constructed were the most accurate known at the time the former article was written. They are now known to be consideraby below the truth. They only profess to register religious services at birth and death; so that many children are born, and some corpses perhaps interred, without the names appearing in the clerical registers. The returns are not always sent in with becoming punctuality; and none are yet accessible for the year 1866. Their chief use is in furnishing data for the comparison of births and deaths; and of the city with the country districts. These last consisted, from 1859–1861, of the following counties, viz.: Hochelaga, Jacques Cartier, Laval, Vaudreuil, Soulanges, Laprairie, Chambly and Verchères. In 1862 Vaudreuil, and in 1863 Soulanges, were removed to another registration district; but their averages have been added in, to make the returns for the different years correspond. The population in 1861 is taken from the census; a comparison of this with the census of 1861 gives 3817 as the average yearly rate of increase. It is probable that these country returns are more accurate than those of the city; the population being less affected by immigration; and the proportion who are careless as to religious observances being much smaller. It will be specially noticed that there is no remarkable fluctuation in births in 1863–4, nor extra mortality in 1864.

---

\* This simple mode is not exact, being less than the real rate. But as the recorded deaths are also below the real numbers, the lower totals of population make the averages more near the truth.

## 5. *Eight Adjacent Counties : Returns of Baptisms and Funeral Services.*

| Year. | Supposed Population | Births. | Deaths. | Excess of Births over Deaths. | Deaths per 1,000 living. | Deaths per 100 Births. |
|---|---|---|---|---|---|---|
| 1859........................ | 98,160 | 4,087 | 1,881 | +2,206 | 19.1 | 46 |
| 1860........................ | 101,977 | 4,013 | 1,787 | +2,226 | 17.5 | 44 |
| 1861........................ | 105,794 | 3,935 | 1,799 | +2,136 | 17.0 | 45 |
| 1862........................ | 109,611 | 3,882 | 2,020 | +1,862 | 18.4 | 52 |
| 1863........................ | 113,428 | 3,895 | 1,823 | +2,072 | 16.0 | 47 |
| 1864........................ | 117,245 | 3,712 | 2,019 | +1,693 | 17.1 | 54 |
| 1865........................ | 121,062 | 3,943 | 2,045 | +1,898 | 16.9 | 52 |
| Average of 7 years.......... | 109,611 | 3,923 | 1,911 | +2,012 | 17.4 | 48 |
| Do.  Montreal............... | 93,583 | 4,545 | 3,390 | +1,155 | 36.2 | 74 |
| Balance for the city, + and— | —16,028 | + 622 | +1,479 | — 857 | +18.8 | + 26 |

It appears, therefore, that although the average population of Montreal is more than *sixteen thousand less* than that of the eight counties, (making a difference greater than the whole population of Verchères,) it furnishes yearly 1479 *more deaths*, being at the rate of 188 additional yearly deaths among each myriad of the living population, which is *more than double the country rate of dying.*

It is found to be a standard fact in sanitary statistics, that, by a compensating power in nature, extra deaths are accompanied by extra births, so that if a city has above the normal number of *births* in proportion to the population, it will be found to have also an abnormal number of *deaths*. We find therefore that, for the *smaller population* of Montreal, there is yet a yearly *excess of 622 births*; yet in spite of this, there is a yearly loss to the city, on comparing the balance of births and deaths with that of the country, amounting to 857 souls, or 26 extra deaths out of every hundred births. Such is the contrast presented, not by a single year, as in the census returns, but by the average of seven years, between the city and the country, both having the same climatal conditions, and the balance of comforts and the means of living being decidedly in favour of Montreal.

### C. INTERMENTS AT THE CEMETERIES.

We have been obliged to express doubts as to the accuracy of the previous returns. Those of the census, even if correct, apply to one year only. Those of the clergy apply only to religious services; and among them may be some which are not accurately registered. But of the graves dug, and the coffins actually interred, there can be no mistake. That the name, age, and other circumstances attending the death of a citizen should be actually entered in the register, without that person actually having died, cannot be believed. Citizens may have died, and been interred elsewhere; they may have been interred at the cemeteries, and by bare possibility an entry not have been made; the returns may not therefore be complete, but they cannot be gainsaid so far as they go. That such and such numbers of persons were interred at Cote des Neiges and at Mount Royal Cemeteries on such and such dates, is recorded in black and white, and forms a record of human life prematurely cut off, truly fearful to contemplate.

It is no doubt true that several interments are made of country residents: but the suburban districts are not populous enough materially to affect the averages; and the number of countrymen buried from them is probably balanced by citizens who die or are interred elsewhere. The census returns of population may indeed be incorrect; and therefore the assumed yearly increase, and the actual rate of mortality per thousand. But there are three classes of facts which are not affected by these chances of error, and which are of the highest importance; viz.: 1. the comparative mortality from one year to another; 2. the comparative mortality at different seasons of the year; and 3. the comparative mortality of children and adults.

In accordance with a Municipal Bye-Law, weekly returns are tabulated, at the office of the City Clerk, of all interments in the burial grounds of the City of Montreal. They are compiled from sheets sent from the "Catholic Cemetery;" and from the "Protestant Vaults or Burial-ground." The latter is said to include all interments made elsewhere than in the Cote des Neiges Cemetery. These sheets are ruled to contain the

| No. | Name. | Date of Decease. | | | MALES. | | |
|-----|-------|------------------|---|---|--------|---|---|
| | | | | | Children. | Married Men. | Widowers. |

| | | | FEMALES. | | | |
|---|---|---|----------|---|---|---|
| Bachelors. | | | Children. | Married Women. | Widows. | Unmarried Women. |

| AGE. | | | PLACE OF RESIDENCE. | | | |
|------|---|---|---------------------|---|---|---|
| Years. | Months. | Days. | Street. | Ward. | | Country. | Disease. |

The last two columns, in the Catholic sheet examined as a specimen, and even the previous ones of place of residence, are imperfectly filled up. With more care in the registration, and with accurate tabulation extending over a series of years, these sheets might afford materials for *fixing the special localities of extra mortality*, which might produce most important results. Many of the streets being extremely long, and containing houses, even in the same ward, differing very greatly in sanitary condition, the *number* of the house ought *in every case* to be recorded. As in England, no interment ought to be allowed, without the production of a duly authorized medical certificate, assigning both the *proximate* and the *remote* cause of death, *both of which* should be recorded.

The only items tabulated in the City Clerk's register are the *numbers* in the columns for males and females, and the *totals for each week*. There are two columns for disease, simply divided between 'epidemic' and 'others;' but the epidemic of cholera, which caused this return to be instituted, (on July 16, 1854,) having terminated in November, no returns have been entered under the disease columns since that date. The columns for 'children' include all deaths under twelve years of age.

The returns for 1854 are of course incomplete. There is an entry of 274 deaths from cholera, from June 28 to July 11; and of the total deaths registered from cholera being 1067, principally in July. The greatest mortality was in the week ending July 23rd, viz.: 281; the least, Nov. 25, viz.: 33. The totals are as follows:

### 6. *Partial Returns of Deaths in Montreal, for the Cholera year*, 1854.

| 1854. | Children. | Adults. | Total. | Weekly Average. |
|---|---|---|---|---|
| July,  3 weeks.............. | 414 | 396 | 810 | 270.0 |
| Aug., 4    "   .............. | 262 | 278 | 540 | 135.0 |
| Sept., 5    "   .............. | 211 | 93 | 304 | 60.8 |
| Oct., 4    "   .............. | 103 | 60 | 163 | 40.7 |
| Nov., 4    "   .............. | 99 | 74 | 173 | 43.2 |
| Total...................... | 1,089 | 901 | 1,990 | 99.5 |

The cemetery tables enable us to present the complete returns for twelve years, from Jan. 1, 1855, to Dec. 31, 1866, inclusive; and to divide them between 'children' and adults.

The population for each year has been calculated, as exactly as possible, not by adding and subtracting a fixed quantity, as in tables 4 and 5, but according to the *average rate of increase*, which is found to be very nearly 4·7 per cent.; (that of all England being somewhat under 2 p. c.) Of course a considerable part of this large increase is due to immigration, and is a fluctuating element. This was probably greatest during the American war, and least in 1866, when the nominally high wages in the United States tempted many to emigrate. Due allowance is made for the excess of deaths over births in 1854 and 1864.

The following table presents the total population; the total deaths; the deaths of all above 12 years of age, called *adults;* and of those under 12, classed as children. Corresponding columns exhibit the proportion of each entry of death to 1000 living persons *of all ages.* A separate column exhibits the proportion between every 100 deaths of *persons of all ages above* 12, and the corresponding deaths in the same year below 12. *In every year except* 1866, *the latter are more than double.*—In order to render more conspicuous the high death-rate of the city, a tenth column shews the average group of individuals among whom a *single death* occurs, viz.: among every 30 in the healthier years, every 28 in the balance of years, every 22 in 1864, and every 17 in the cholera year. The eleventh column shews the actual number of deaths which occurred in the city above the rural average; that is, of lives which might have been saved, had the people been scattered over the neighbouring counties. The last column presents the same *excess* of city death, as compared with each 1000 living.

It will be observed that although so large a proportion of the moribund population were killed off in the cholera year, the succeeding year, 1855, was still unhealthy. From 1856-1859, the mortality, though frightfully great, was below the average. The six years from 1860-1865 march on with steady course, presenting a death-rate only equalled, in the worst English cities, during periods of special pestilence. In 1866, there is a marvellous and sudden rebound to the death-rate of the least unhealthy year, 1858. During 1864, there was a terribly fatal epidemic of scarlatina, its virulence being no doubt caused by the accumulations of xymotic poison, which then attained their maximum. These fluctuations are brought out most strongly in the column for children's deaths: they are much slower in affecting adults. With them the rise does not begin till 1863; it is even somewhat lower in 1864; and there is no change for the better in 1866.

## 7. Rate of Mortality for the City of Montreal, 1855–1866 : from the Cemetery Returns.*

| Year. | Population. | Deaths of Adults. | Or per 1000 of all ages living. | Deaths of Children. | Or per 1000 of all ages living. | Deaths of Children to every 100 deaths of adults. | Total Deaths. | Or per 1000 all ages living. | Or one Death in every | Yearly excess of deaths of all ages as compared with adjacent counties. | Or per 1000 living. |
|---|---|---|---|---|---|---|---|---|---|---|---|
| 1855...... | 68,347 | 712 | 10.4 | 1,704 | 24.6 | 236 | 2,416 | 35.2 | 28 living. | 1,250 | 18.3 |
| 1856...... | 71,581 | 743 | 11.1 | 1,617 | 22.6 | 204 | 2,360 | 32.9 | 30 " | 1,270 | 17.9 |
| 1857...... | 74,951 | 796 | 10.6 | 1,694 | 22.6 | 213 | 2,490 | 33.3 | 30 " | 1,442 | 18.3 |
| 1858...... | 78,483 | 771 | 9.8 | 1,739 | 22.1 | 226 | 2,510 | 32.0 | 31 " | 1,334 | 17.0 |
| 1859...... | 82,180 | 847 | 10.3 | 1,919 | 23.2 | 225 | 2,766 | 33.7 | 29 " | 1,197 | 14.6 |
| 1860...... | 86,040 | 922 | 10.7 | 2,249 | 26.1 | 244 | 3,171 | 36.8 | 27 " | 1,659 | 19.3 |
| 1861...... | 90,323 | 945 | 10.4 | 2,236 | 24.7 | 237 | 3,181 | 35.2 | 28 " | 1,643 | 18.2 |
| 1862...... | 94,568 | 996 | 10.6 | 2,465 | 26.0 | 245 | 3,461 | 36.6 | 27 " | 1,811 | 19.2 |
| 1863...... | 99,011 | 1,071 | 11.8 | 2,535 | 25.6 | 217 | 3,606 | 36.4 | 27 " | 2,019 | 20.4 |
| 1864...... | 103,664 | 1,165 | 11.2 | 3,536 | 34.1 | 305 | 4,701 | 45.3 | 22 " | 2,921 | 28.2 |
| 1865...... | 106,375 | 1,171 | 11.0 | 2,854 | 26.8 | 243 | 4,025 | 37.8 | 26 " | 2,216 | 20.9 |
| 1866...... | 111,374 | 1,226 | 11.0 | 2,384 | 21.4 | 194 | 3,610 | 32.2 | 10 " | 1,669 | 15.0 |
| Total...... | | 11,365 | | 26,932 | | | 38,297 | | | 20,440 | |
| Average of 12 years.. | 89,860 | 947 out of 61,224 adults living. | 10.5 or 15.4 per 1000 adults living, or 1 in every 65 adults, yearly. | 2,224 (out of 29,099 or living.) | 25.0 or 77.2 per 1000 children, or 1 in every 13 children every year. | 238 | 3,191 | 35.5 | 28 " | 1,696 | 18.9 |
| Estimated population of the cholera year 1854. Do of 1867.. | 66,097  116,608 | | | | | | 3,878 | 58.7 | 17 " | 2,624 | 39.7 |

* The table may be thus read :—" In 1864 the population of the city being 103,664, 4701 corpses were interred ; that is, 453 out of every myriad living (98 above the average), or one out of every 22 persons in the city. If the same people had lived in the country round, 2,921 need not have died, or 282 lives would have been saved in every myriad. But of the corpses interred, only 1165 were above twelve years old : that is, 112 out of every myriad living, which is only seven above the average. All the other corpses were of *children under twelve*, viz., 3,536, or 341 out of every myriad people in the city, which is 91 above the average. In that year, *to every hundred who died from twelve years old upwards, there were no less than 305 infants and young children under twelve*." In the same way, all the tables may be read without decimals, by quoting 1,000 for 100, and 10,000 for 1,000 : or, for common purposes, the 1,000 may be retained, and the decimal figure omitted.

We are now in a position to judge of the statistics recorded under sections A & B. The following table exhibits these in comparison with the totals from the cemeteries. It appears that during the eleven years *no fewer than 2,134 deaths have escaped registration by the clergy;* being never less than 76 in a year; on the average 194; and, in the deadly year, actually 395. The average equals 6 per cent of the total deaths; or 22 unrecorded deaths to every 10,000 living.

In the case of the census returns, the deficiency is still more startling; *no fewer than 36 per cent of the total deaths* having escaped recording.

## 8. *Comparison of 3 returns of Deaths in Montreal, 1855-1865.*

| Year. | Cemetery Returns. | Clergy Returns | Not entered in Clergy returns. | Census Return | Not entered in Census Return |
|-------|-------------------|----------------|-------------------------------|---------------|------------------------------|
| 1855 | 2,416 | 2,231 | 185 | | |
| 1856 | 2,360 | 2,284 | 76 | | |
| 1857 | 2,490 | 2,367 | 123 | | |
| 1858 | 2,510 | 2,299 | 211 | | |
| 1859 | 2,766 | 2,581 | 185 | | |
| 1860 | 3,171 | 3,016 | 155 | | |
| 1861 | 3,181 | 3,005 | 176 | 2,038 | 1,143 |
| 1862 | 3,461 | 3,222 | 239 | | |
| 1863 | 3,606 | 3,510 | 96 | | |
| 1864 | 4,701 | 4,306 | 395 | | |
| 1865 | 4,025 | 3,732 | 293 | | |
| Total.... | 34,687 | 32,553 | 2,134 | | |

MORTALITY OF 1861.—Cemetery....................35.2 per 1000 living.
Protonotary.................33.2 "
Census........................22.5 "
Not registered by the clergy... 2.0 "
Not recorded in census.........12.7 "

These facts are surely sufficient to convince the most sceptical of the importance of a compulsory civil registration of births and deaths. In addition to the usual details, it is very necessary to provide that no death be registered without the production of a medical certificate, declaring the remote as well as the proximate cause of death. There should be heavy penalties for any interment without previous registration.

The next step in our analysis leads to very important results: it is, to distribute the total deaths for each year under the *months* in which they occur. This is done in table 9 for all ages; in table 10, for children under 12; and in table 11, for children above 12 and adults. The numbers which include five weeks instead of four are distinguished by large-faced figures. The totals for each year are added at the bottom; *for the same month in the twelve years*, in the last column.

## 9. Total Deaths in Montreal, of all ages, for each month from January, 1855, to December, 1866.

| Year. | 1855. | 1856. | 1857. | 1858. | 1859. | 1860. | 1861. | 1862. | 1863. | 1864. | 1865. | 1866. | Total of each month, for 12 years. |
|---|---|---|---|---|---|---|---|---|---|---|---|---|---|
| January .. | 138 | 135 | 217 | 194 | 206 | 209 | 215 | 262 | 287 | 411 | 291 | 227 | 2,792 |
| February. | 203 | 150 | 184 | 188 | 188 | 204 | 204 | 215 | 186 | 308 | 275 | 234 | 2,539 |
| March ... | 200 | 227 | 196 | 148 | 182 | 203 | 233 | 277 | 218 | 360 | 259 | 207 | 2,920 |
| April..... | 235 | 176 | 193 | 187 | 243 | 223 | 182 | 199 | 231 | 521 | 385 | 293 | 3,068 |
| May..... | 187 | 215 | 250 | 215 | 177 | 239 | 212 | 336 | 294 | 366 | 303 | 258 | 3,051 |
| June.... | 252 | 183 | 191 | 180 | 238 | 399 | 272 | 339 | 304 | 413 | 326 | 284 | 3,381 |
| July...... | 245 | 240 | 256 | 340 | 451 | 383 | 407 | 457 | 462 | 637 | 556 | 415 | 4,858 |
| August ... | 271 | 348 | 312 | 217 | 286 | 309 | 378 | 440 | 504 | 449 | 420 | 387 | 4,321 |
| September | 218 | 194 | 189 | 243 | 188 | 300 | 270 | 243 | 280 | 300 | 426 | 394 | 3,245 |
| October .. | 135 | 155 | 202 | 224 | 217 | 210 | 256 | 216 | 273 | 317 | 266 | 265 | 2,741 |
| November | 91 | 172 | 156 | 167 | 168 | 181 | 317 | 271 | 283 | 252 | 249 | 260 | 2,567 |
| December. | 182 | 165 | 144 | 193 | 222 | 251 | 235 | 206 | 284 | 367 | 269 | 296 | 2,814 |
| Total of each year. | 2,416 | 2,360 | 2,490 | 2,510 | 2,766 | 3,171 | 3,181 | 3,461 | 3,606 | 4,701 | 4,025 | 3,610 | 38,297 |

## 10. Deaths of Children under 12 in Montreal, for each month, from 1855–1866.

| Year. | 1855. | 1856. | 1857. | 1858. | 1859. | 1860. | 1861. | 1862. | 1863. | 1864. | 1865. | 1866. | Total of each month, for 12 years. |
|---|---|---|---|---|---|---|---|---|---|---|---|---|---|
| January... | 119 | 85 | 159 | 134 | 133 | 125 | 129 | 174 | 176 | 312 | 201 | 150 | 1,897 |
| February . | 117 | 87 | 133 | 127 | 126 | 135 | 124 | 154 | 108 | 235 | 187 | 146 | 1,679 |
| March ... | 180 | 151 | 127 | 97 | 113 | 185 | 163 | 186 | 138 | 273 | 184 | 183 | 1,980 |
| April..... | 164 | 116 | 123 | 119 | 165 | 144 | 114 | 131 | 140 | 387 | 254 | 183 | 2,040 |
| May...... | 121 | 140 | 176 | 140 | 133 | 171 | 144 | 228 | 197 | 266 | 206 | 152 | 2,074 |
| June...... | 180 | 124 | 128 | 114 | 176 | 325 | 180 | 263 | 218 | 317 | 234 | 181 | 2,440 |
| July...... | 203 | 179 | 188 | 257 | 351 | 302 | 337 | 391 | 376 | 519 | 453 | 341 | 3,927 |
| August... | 213 | 281 | 237 | 158 | 224 | 242 | 294 | 338 | 397 | 365 | 320 | 289 | 3,358 |
| September | 161 | 134 | 125 | 166 | 94 | 216 | 197 | 176 | 195 | 241 | 335 | 280 | 2,320 |
| October .. | 79 | 105 | 115 | 148 | 143 | 130 | 180 | 139 | 197 | 210 | 174 | 157 | 1,777 |
| November | 48 | 106 | 89 | 114 | 108 | 117 | 216 | 156 | 193 | 173 | 154 | 156 | 1,630 |
| December. | 119 | 109 | 94 | 135 | 153 | 157 | 158 | 129 | 200 | 238 | 152 | 166 | 1,810 |
| Total of each year. | 1,704 | 1,617 | 1,694 | 1,739 | 1,919 | 2,249 | 2,236 | 2,465 | 2,535 | 3,536 | 2,854 | 2,384 | 26,932 |

## 11. Deaths of Adults and Children above 12 in Montreal, for each month, from 1855–1866.

| Year. | 1855. | 1856. | 1857. | 1858. | 1859. | 1860. | 1861. | 1862. | 1863. | 1864. | 1865. | 1866. | Total of each month, for 12 years. |
|---|---|---|---|---|---|---|---|---|---|---|---|---|---|
| January... | 19 | 50 | 58 | 60 | 73 | 84 | 86 | 88 | 111 | 99 | 90 | 77 | 895 |
| February . | 86 | 63 | 51 | 61 | 62 | 69 | 80 | 61 | 78 | 73 | 88 | 88 | 860 |
| March. ... | 80 | 79 | 69 | 51 | 69 | 78 | 70 | 91 | 80 | 87 | 75 | 114 | 940 |
| April..... | 71 | 60 | 70 | 68 | 78 | 79 | 68 | 68 | 91 | 134 | 131 | 110 | 1,028 |
| May...... | 65 | 75 | 74 | 75 | 44 | 68 | 68 | 108 | 97 | 100 | 97 | 106 | 977 |
| June...... | 72 | 59 | 63 | 66 | 62 | 74 | 92 | 76 | 86 | 96 | 92 | 103 | 941 |
| July...... | 42 | 61 | 68 | 62 | 100 | 81 | 70 | 66 | 86 | 118 | 103 | 74 | 931 |
| August . . | 58 | 67 | 75 | 59 | 62 | 69 | 84 | 102 | 107 | 84 | 100 | 98 | 963 |
| September | 57 | 60 | 64 | 77 | 94 | 84 | 73 | 67 | 85 | 59 | 91 | 114 | 925 |
| October... | 56 | 50 | 87 | 81 | 74 | 80 | 76 | 77 | 76 | 107 | 92 | 108 | 964 |
| November | 43 | 66 | 67 | 53 | 60 | 64 | 101 | 115 | 90 | 79 | 95 | 104 | 937 |
| December. | 63 | 56 | 50 | 58 | 69 | 94 | 77 | 77 | 84 | 129 | 117 | 130 | 1,004 |
| Total of each year. | 712 | 743 | 796 | 771 | 847 | 922 | 945 | 996 | 1,071 | 1,165 | 1,171 | 1,226 | 11,365 |

In order to bring out more vividly the startling differences exhibited by the foregoing tables, not in one year only, nor in many, but in *each one of a long series*, a fresh series of tables has been constructed, nos. 12-14, exhibiting the average *weekly* mortality of each class during each month. This is done by dividing the previous items by 4 or by 5; fractions below one-tenth being omitted. The averages for each year, and for the sum of years, are in each case *constructed from the totals*, and not by the mere addition of the previous items, which would involve error from the disregarded hundredths.

## 12. *Average Weekly Mortality, of all ages, for each month from January, 1855, to December, 1866.*

| Year. | 1855 | 1856 | 1857 | 1858 | 1859 | 1860 | 1861 | 1862 | 1863 | 1864 | 1865 | 1866 | Average per week in each month, for 12 years. |
|---|---|---|---|---|---|---|---|---|---|---|---|---|---|
| January.... | 34.5 | 33.7 | 43.4 | 38.8 | 41.2 | 52.2 | 53.7 | 65.5 | 57.4 | 82.2 | 72.7 | 56.7 | 52.7 |
| February... | 50.7 | 37.5 | 46.0 | 47.0 | 47.0 | 51.0 | 51.0 | 53.7 | 46.5 | 77.0 | 68.7 | 58.5 | 52.9 |
| March...... | 52.0 | 45.4 | 49.0 | 37.0 | 45.5 | 52.6 | 46.6 | 55.4 | 54.5 | 90.0 | 64.7 | 59.4 | 54.1 |
| April....... | 58.7 | 44.0 | 48.2 | 46.7 | 48.6 | 55.7 | 45.5 | 49.7 | 57.7 | 104.2 | 76.6 | 73.2 | 60.1 |
| May........ | 46.5 | 43.0 | 50.0 | 43.0 | 44.2 | 59.7 | 53.0 | 67.2 | 58.8 | 91.5 | 75.7 | 64.5 | 57.6 |
| June........ | 50.4 | 45.7 | 47.7 | 45.0 | 59.5 | 79.8 | 54.4 | 84.7 | 76.0 | 103.2 | 81.5 | 56.8 | 65.0 |
| July........ | 61.2 | 60.0 | 64.0 | 69.8 | 90.2 | 95.7 | 101.7 | 114.2 | 115.5 | 127.4 | 111.2 | 103.7 | 93.4 |
| August...... | 67.7 | 69.6 | 62.4 | 54.2 | 71.5 | 77.2 | 75.6 | 88.0 | 101.0 | 112.2 | 105.0 | 96.7 | 82.1 |
| September.. | 43.6 | 48.5 | 47.2 | 60.7 | 47.0 | 60.0 | 67.5 | 60.7 | 70.0 | 75.0 | 85.2 | 78.8 | 62.4 |
| October.... | 33.7 | 38.7 | 40.4 | 45.8 | 43.4 | 52.5 | 64.0 | 54.0 | 54.6 | 63.4 | 66.5 | 66.2 | 51.7 |
| November.. | 22.7 | 34.4 | 39.0 | 41.7 | 42.0 | 45.2 | 63.4 | 54.2 | 70.7 | 63.0 | 62.2 | 65.0 | 50.3 |
| December.. | 36.4 | 41.2 | 36.0 | 48.2 | 44.4 | 50.2 | 58.7 | 51.5 | 71.0 | 73.4 | 53.8 | 59.2 | 52.1 |
| Average week for 12 months. | 46.4 | 45.4 | 47.9 | 48.2 | 52.2 | 60.9 | 61.1 | 66.5 | 69.3 | 88.7 | 77.4 | 69.4 | 61.2 |

## 13. *Average Weekly Mortality of Children under 12, for each month, from Jan. 1855, to Dec. 1866.*

| Year. | 1855 | 1856 | 1857 | 1858 | 1859 | 1860 | 1861 | 1862 | 1863 | 1864 | 1865 | 1866 | Average per week in each month, for 12 years. |
|---|---|---|---|---|---|---|---|---|---|---|---|---|---|
| January........ | 29.7 | 21.2 | 31.8 | 26.8 | 26.6 | 31.2 | 32.2 | 43.5 | 35.2 | 62.4 | 50.2 | 37.5 | 35.8 |
| February....... | 29.2 | 21.7 | 33.2 | 31.7 | 31.5 | 33.7 | 31.0 | 38.5 | 27.0 | 58.7 | 46.7 | 36.5 | 34.9 |
| March.......... | 36.0 | 30.2 | 31.7 | 24.2 | 28.2 | 37.0 | 32.6 | 37.2 | 34.5 | 68.2 | 46.0 | 36.6 | 36.6 |
| April.......... | 41.0 | 29.0 | 30.7 | 29.7 | 33.0 | 36.0 | 28.5 | 32.7 | 35.0 | 77.4 | 50.8 | 45.7 | 40.0 |
| May........... | 30.2 | 28.0 | 35.2 | 28.0 | 33.2 | 42.7 | 36.0 | 45.6 | 39.4 | 66.5 | 51.5 | 38.0 | 38.9 |
| June........... | 36.0 | 31.0 | 32.0 | 28.5 | 44.0 | 65.0 | 36.0 | 65.7 | 54.5 | 79.2 | 58.5 | 36.2 | 46.9 |
| July........... | 50.7 | 44.7 | 47.0 | 57.4 | 70.2 | 75.5 | 84.2 | 97.7 | 94.0 | 103.8 | 90.6 | 85.2 | 75.5 |
| August........ | 53.2 | 46.2 | 47.4 | 39.5 | 56.0 | 60.5 | 58.9 | 67.6 | 79.4 | 91.2 | 80.0 | 72.2 | 63.3 |
| September..... | 32.2 | 33.5 | 31.2 | 41.5 | 23.5 | 43.2 | 49.2 | 44.0 | 48.7 | 60.2 | 67.0 | 56.0 | 44.6 |
| October....... | 19.7 | 26.2 | 23.0 | 29.6 | 28.6 | 32.5 | 45.0 | 34.7 | 39.4 | 42.0 | 43.5 | 39.2 | 33.5 |
| November..... | 12.0 | 21.2 | 22.2 | 28.5 | 27.0 | 29.2 | 43.2 | 31.4 | 48.2 | 43.2 | 38.5 | 39.0 | 31.9 |
| December..... | 23.8 | 27.2 | 23.5 | 33.7 | 30.6 | 31.4 | 39.5 | 32.2 | 50.0 | 47.6 | 30.4 | 33.2 | 33.5 |
| Average week for 12 months... | 32.8 | 31.1 | 32.6 | 33.4 | 36.2 | 43.2 | 43.0 | 47.4 | 48.7 | 66.6 | 54.9 | 45.9 | 43.0 |

## 14. Average Weekly Mortality of Adults and Children above 12, for each month from Jan., 1855, to Dec., 1866

| Year. | 1855 | 1856 | 1857 | 1858 | 1859 | 1860 | 1861 | 1862 | 1863 | 1864 | 1865 | 1866 | Average per week, in each month, for 12 years. |
|---|---|---|---|---|---|---|---|---|---|---|---|---|---|
| January........ | 4.7 | 12.5 | 11.6 | 12.0 | 14.6 | 21.0 | 21.5 | 22.0 | 22.2 | 19.8 | 22.5 | 19.2 | 16.9 |
| February........ | 21.5 | 15.7 | 12.7 | 15.2 | 15.5 | 17.2 | 20.0 | 15.2 | 19.5 | 18.2 | 22.0 | 22.0 | 17.9 |
| March.......... | 16.0 | 15.2 | 17.2 | 12.7 | 17.2 | 15.6 | 14.0 | 18.2 | 20.0 | 21.7 | 18.7 | 22.8 | 17.4 |
| April.......... | 17.7 | 15.0 | 17.5 | 17.0 | 15.6 | 19.7 | 17.0 | 17.0 | 22.7 | 26.8 | 26.2 | 27.5 | 20.1 |
| May........... | 16.2 | 15.0 | 14.8 | 15.0 | 11.0 | 17.0 | 17.0 | 21.6 | 19.4 | 25.0 | 24.2 | 26.5 | 18.5 |
| June.......... | 14.4 | 14.7 | 15.7 | 16.5 | 15.5 | 14.8 | 18.4 | 19.0 | 21.5 | 24.0 | 23.0 | 20.6 | 18.1 |
| July.......... | 10.5 | 15.2 | 17.0 | 12.4 | 20.0 | 20.2 | 17.5 | 16.5 | 21.5 | 23.6 | 20.6 | 18.5 | 17.9 |
| August........ | 14.5 | 13.4 | 15.0 | 14.7 | 15.5 | 16.7 | 16.8 | 20.4 | 21.4 | 21.0 | 25.0 | 24.5 | 18.1 |
| September...... | 11.4 | 15.0 | 16.0 | 19.2 | 23.5 | 16.8 | 18.2 | 16.7 | 21.2 | 14.7 | 18.2 | 22.8 | 17.8 |
| October........ | 14.0 | 12.5 | 17.4 | 16.2 | 14.8 | 20.0 | 19.0 | 19.2 | 15.2 | 21.4 | 23.0 | 27.0 | 18.2 |
| November...... | 10.7 | 13.2 | 16.7 | 13.2 | 15.0 | 16.0 | 20.2 | 23.0 | 22.5 | 19.7 | 23.7 | 26.0 | 18.3 |
| December...... | 12.6 | 14.0 | 12.5 | 14.5 | 13.8 | 18.8 | 19.2 | 19.3 | 21.0 | 25.8 | 23.4 | 26.0 | 18.6 |
| Average week for 12 months... | 13.7 | 14.3 | 15.3 | 14.8 | 15.9 | 17.7 | 18.1 | 19.1 | 20.6 | 21.9 | 22.5 | 23.6 | 18.1 |

It was natural to expect that there should be some difference between the mortality at different seasons of the year. It is found in England, on the average of 10 years, that this difference does not affect in the same degree the town and the country population.

## 15. English Seasonal Variations between Town and Country Mortality.

| | Large Towns. | Country. | Town Excess. |
|---|---|---|---|
| Deaths in an average quarter, for every 1000 living...................... | 25.9 | 20.0 | 5.9 |
| Do Winter quarter................ | 27.5 | 22.8 | 4.7 |
| Do Spring " ............... | 24.6 | 20.8 | 3.8 |
| Do Summer " ............... | 26.2 | 17.8 | 8.4 |
| Do Autumn " ............... | 25.4 | 18.7 | 6.7 |

The town excess is thus shown to be intensified most in summer, and next in autumn; no doubt because the zymotic poisons are rendered most active in the hottest weather, and their influence continues till the frosts of winter. The effect of the heat in the five *plague years* of London which have been recorded in history is very noteworthy. The bills of mortality show the following average for every 1000 persons living.

## 16. Plague Years in London.

| | | |
|---|---|---|
| Winter Quarter : January, February, March.................... | 17 | per 1000 living. |
| Spring " April, May, June............................ | 20 | " |
| Summer " July, August, September...................... | 163 | " |
| Autumn " October, November, December.............. ..... | 50 | " |
| Total..................................................... | 250 | " or 1 in 4 |

But if there are no special stenches to be drawn-out into virulence by the summer sun, the cold of winter renders it the most
unhealthy of the seasons; as shown by the following table for a
year in which the minimum temperature was 11°.

### 17. *Mortality of London Seasons in* 1830.

| Winter Quarter ........ | Average Temperature 36° | Total Deaths 8.5 per 1000 living. |
|---|---|---|
| Spring   "   ........ | "   "   53° | "   7.0   " |
| Summer   "   ........ | "   "   61° | "   6.0   " |
| Autumn   "   ........ | "   "   45° | "   6.6   " |
| Total of the year........ | Mean   "   48.9° | "   28.1   " |

The same is shown in the average of all England for 1857;
when, the average quarter being assumed as 1000 deaths, winter
furnished 1050, autumn 1045, spring 955 and summer 950. A
long series of observations has led to such uniform results in England
that the Registrar General is able to predict a definite excess of
mortality for every considerable fall in the thermometer. The
severe frost of Jan. 1867, caused an excess of 732 deaths in a
fortnight in London alone; of which only 50 were of young persons under 20, and 411 were of old people about 60. The same
frost raised the death-rate in the 13 large towns to 31 per 100.

It would therefore be naturally expected that in the extreme
cold of a Lower Canadian winter, the death-rate would rise proportionally. But it is not so. For adults there is a marvelous uniformity between the different months of the year. Old people, and
indeed all above 12, do not appear to be rendered moribund either
by the intense frosts of winter or the unhealthy heats of summer.
On the average of 12 years, it does not appear that their mortality
varies more than 9 out of every 10,000 living at all ages; or as 10
to 12 between January, the most healthy, and April, the least
healthy of the months. The *lowest* recorded mortality was in
January, 1855, (many of the moribund adults having been cut off
by cholera in the previous summer); and the contrast of the year
is consequently the greatest, being 16.8 between that month and
February. The *highest* recorded mortality of adults was in
April, 1866, when the thawed stenches of an unusually severe
winter were precipitated on the putrifying corruptions of previous
years; the contrast of the year between April and July being 9.0.
The year of death, 1864, affords a somewhat greater contrast, viz.,
12.1 between April and September; but those above twelve years
old do not appear to have been more unhealthy than usual.

If winter cold does not specially kill the aged, we are not surprised to find that it appears by no means unhealthy to children.

The five coldest months are uniformly the most healthy; the two hottest, not only uniformly unhealthy, but so frightfully destructive that *July kills off* 247 *children out of every* 10,000 *of all ages living, in addition to the* 184 *who die in November;* which is as 23 to 10, or *more than double.* This is *nearly double* the excess of the terrible year of death 1864 over the most healthy of the years 1858. These facts are brought out in fearful contrast in the following table.

17. *Comparative Weekly Mortality of each Month, on the average of* 12 *years,* 1855–1866.

| Deaths of Children. Yearly average to 1000 of all ages living | Deaths of Adults. Yearly average to 1000 of all ages living | Deaths of all ages. Yearly average to 1000 of all ages living | Total yearly mortality to 1000 of all ages living. |
|---|---|---|---|
| November......18.4 | January........ 9.7 | November......29.0 | 1858............32.0 |
| October.........19.3 | March...........10.0 | October.........29.8 | 1866............32.4 |
| December......19.3 | September......10.3 | December......30.1 | 1856............32.9 |
| February.......20.2 | February.......10.4 | January......,...30.5 | 1857............33.3 |
| January........20.6 | July ...........10.4 | February.......30.6 | 1859............33.7 |
| March..........21.2 | June ...  ......10.5 | March..........31.3 | 1861............35.2 |
| May............22.5 | August.........10.5 | May ...........33.3 | 1855............35.3 |
| April... ......23.2 | October.........10.5 | April...........34.8 | 1863............36.4 |
| September......25.3 | November ......10.6 | September......36.1 | 1862............36.6 |
| June ...........27.1 | May ............10.7 | June............37.6 | 1860............36.8 |
| August.........30.8 | December......10.8 | August.........47.5 | 1865............37.8 |
| July..............43.1 | April............11.6 | July.............54.0 | 1864............45.3 |
| Average ........24.8 | Average........10.5 | Average.........35.5 | Average .........35.6 |
| Excess of July } over Nov .. } 24.7 | Excess of April } over Jan.... } .9 | Excess of July } over Nov... } 25.0 | Excess of 1864 } over 1858..... } 13.3 |
| Or as one to.... 2.3 | Or as one to.... 1.2 | Or as one to.... 1.9 | Or as one to...... 1.4 |

But this is not all the contrast. It is rendered even more marked by comparing not the months but the *weeks* of greatest and least mortality. This is done for each year in table 18. It will be noticed that the maximum is UNIFORMLY in *July or the first week in August.* The minimum is always in one of the cold months; or at least, as shown in the notes, a cold week appears with nearly as low a rate. There is one distinct exception for the minimum of 1866, which appears in June: for this there is a clear reason, which will presently be shown to add a striking confirmation to the general rule. In the year of mother's woe, 1864, there is an excess in July of 101 deaths over the 44 of October; which is the same as adding 51 per 1000 to the death rate of the city. In the cholera year, the deaths rose from 33 to 281; which last, if continued, would have added 195 per 1000 to the death rate of the city.—a mortality which only admits of parallel with the plague years of London before the fire. In this table, the extremes are of total mortality; as we have seen but little change in that of adults, there is no doubt that if the maxima and minima of children's deaths had been eliminated, the result would have appeared even more appalling.

18. *Weeks of Maximum and Minimum Mortality in Montreal, 1855–1866.*

| Year. | Highest Mortality in week ending | Lowest Mortality in week ending | Which is at the yearly rate, per 1,000 of the living inhabitants, of Maximum Minimum | | Range of variation at yearly rate per 1,000 living. | Actual Range of variation between max. and min. weeks. | General Average of year per 1,000 living. |
|---|---|---|---|---|---|---|---|
| 1854 | July 23..281 | Nov. 25..33 | 221 | 26 | 195 | 248 | 61·4 |
| 1855 | Aug. 4.. 78 | " 17..18 | 59 | 13 | 46 | 60 | 35·3 |
| 1856 | " 2.. 93 | " 22..19 | 67 | 14 | 53 | 74 | 32·9 |
| 1857 | July 18.. 79 | Dec. 19..25 | 55 | 17 | 38 | 54 | 33·3 |
| 1858 | " 17.. 81 | Nov. 13..29 | 53 | 19 | 34 | 52 | 32·0 |
| 1859 | " 9.. 97 | *May 7..30 | 61 | 19 | 42 | 67 | 33·7 |
| 1860 | " 7..106 | Nov. 17..36 | 64 | 27 | 37 | 70 | 36·8 |
| 1861 | " 20..118 | †Mar. 9..31 | 67 | 18 | 49 | 87 | 35·2 |
| 1862 | " 19..123 | Dec. 6..43 | 68 | 23 | 45 | 80 | 36·6 |
| 1863 | " 25..124 | ‡Feb. 7..44 | 65 | 23 | 42 | 80 | 36·4 |
| 1864 | " 2..145 | Oct. 22..44 | 73 | 22 | 51 | 101 | 45·3 |
| 1865 | " 1..127 | " 28..45 | 59 | 21 | 38 | 82 | 37·8 |
| 1866 | " 21..121 | §June 9..44 | 54 | 19 | 35‖ | 77¶ | 32·4 |

\* Nov. 5 and 19 are each quoted at 33 ; Oct. 8 at 32 ; and Jan. 8 at 33. All other weeks in the year are 40 or above.
† December 21 is quoted at 55.
‡ October 17 is quoted at 45.
§ Jan. 20 and Dec. 1 are each quoted at 45.
‖ Average range per 1,000, without cholera year, 42.
¶ Actual range of variation, on the average of 12 years, (leaving out 1854,) 72.

The number of living children in Montreal under 12 is to the total population as 29,249 is to 90,323: those of children under 5 years to the total children as 15,196 is to 29,249; those under 1 year to those under 5 as 3,700 to 15,196. From these elements, furnished by the census of 1861, and from the corresponding totals of deaths, the deaths of Montreal children under 12 years may be calculated *in proportion to those living, of the same ages.*

19. *Death-rate of Montreal Children under* 12, *as compared with* 1000 *children living at the same age.*

| Average of years, months, and ages. | Estimated number of living children. | Total Deaths of Children. | Deaths per 1000 living children. | Or, one death out of every |
|---|---|---|---|---|
| Average of 12 years, 1855-1866, for all children under 12..... | 29,099 | 2,244 | 77·2 | 13 children living. |
| The child-killing year, 1864, for all children under 12..... | 33,591 | 3,536 | 105.2 | 9½ " " |
| The least unhealthy year, 1866, for all children under 12..... | 36,066 | 2,384 | 66.1 | 15 " " |
| The most unhealthy month, July, 1855 to 1866, for all children under 12......... } | 29,099 | 3,927 | 134·9 | 7½ " " |
| The least unhealthy month, November,1855-1866, for all children under 12......... } | 29,099 | 1,630 | 56.3 | 18 " " |
| Lower Canada, less 4 cities, 1861, for all children under 12 | 293,579 | 10,796 | 36.8 | 27 " " |
| Average of Montreal children under 5 years, 1855-1866...... | 15,119 | 2,139 | 141.5 | 7 " " |
| Average of Montreal children under 1 year, 1855-1866....... | 3,681 | 1,599 | 434·1 | 2¼ " " |

That is, three out of every seven children born in Montreal, die before they are one year old ! !  Or, out of every 7 children under five years of age, living at the beginning of the year, one (on the average) will die before its close.  Or, out of every 13 children, of all ages under 12, living in the city, on the average one will die during the year.  It appears from the census returns, that even of the children living on the Island outside the city limits, or in any country district from Soulanges to Gaspé, out of every group of 27 one must expect to lose his life within the year; but if those children had been taken to live in Montreal in 1864, *two* out of 19 would have been seized by the destroyer; even if they had lived amongst us last year, when children had a better chance of life than ever before, death would have seized one in every fifteen. Should these children spend July with their friends in the city, · for twelve consecutive years, they must expect to follow to the cemetery twice that number of their companions.

Lastly let us compare the slaughter of the innocents in Montreal with their condition in different parts of England.  Table 20 compares the deaths of children of different ages with the *total deaths at all ages* during the same year.

20. *Death-rate of Children living in Montreal and in England, compared with every* 1000 *deaths at all ages.*

|  | Deaths under 1 year. | Deaths under 5 years. | Deaths under 12 years. |
|---|---|---|---|
| North Lancashire........................................ | 174.3 | 318.7 | 377.3 |
| All England............................................ | 214.5 | 391.0 | 447.4 |
| London................................................ | 190.3 | 404.2 | 453.4 |
| Liverpool.............................................. | 256.9 | 482.6 | 528.6 |
| Montreal............... ............................... | 501.1 | 670.3 | 703.2 |
| | | | |
| Excess of Montreal over Liverpool................... | 244.2 | 187.7 | 174.6 |
|    Do      do    North Lancashire............ | 326.8 | 351.6 | 325.9 |

The London death-rate of children is below the average, because of the large immigration of adults.  There is perhaps a proportionate immigration into Montreal, for similar reasons. Liverpool is a commercial city, like our own with great natural advantages, but cursed with a neglect of the sanitary laws.  It is cursed also by drink and by debauchery, to a greater extent than any other town in England.  Being the most criminal as well as the most unhealthy city in the island, it is called the *Plague-spot on the Mersey.*  Yet the plague-spot on the St. Lawrence is *nearly twice as fatal,* in the first year of being, as the polluted queen of the Mersey, with its cul-de-sac courts and tide-backed sewers; while round the sands of Morecambe Bay (within a fraction) only *one* of the coffins contains an infant of days to *three* which are laid within the bosom of our mountain forests, because the city rulers, and the owners and occupiers of their dwellings, denied them the right to breathe, even for one short year, the pure air that nature is for ever wafting to our otherwise favoured city.

It was well said, in the Sanitary Report presented to the imperial parliament in 1858, pp. xxvii. that " 1. The lives of young children, as compared with the more hardened and acclimatized lives of the adult population, furnish a *very sensitive test of sanitary circumstances*, so that differences in the infantine death-rates, are, under certain qualifications, the *best proof of differences of household condition* in any number of compared districts. 2. Those places where infants are most apt to die, are necessarily the places where survivors are most apt to be sickly ; and where, if they struggle through a scrofulous childhood to realize an abortive puberty, *they beget a sicklier brood than themselves*. A high local mortality of children must almost necessarily denote a *high local prevalence of those causes which determine a degeneration of race*." These words are prompted by long experience, built on facts which cannot be gainsaid. If they are true of all high rates of infantile mortality, how awful must be their truth in this city where the rate is *the highest yet presented!* And if the number of graves in our cemeteries prove these things to be true on the average of the whole city, what must be the harvest of death if we subtract the population living on the healthy mountain-side, and mark the coffins from the houses in Griffintown ! Surely a fearful responsibility rests on the members of the City Council, and especially on the members of the Health and Road Committees, as well as on all owners of property and householders in the city. Has any man a right to draw money from the rents of houses, by living in which children cannot but be killed ? Has the Council a right to compel owners and tenants to cleanse their premises, while it leaves the streets, over which it assumes the entire control, unsewered and even *reeking with the surface filth of years ? *

---

* Instances were recorded by the Sanitary Association, of women who were compelled last summer to open their windows over the reeking fumes of the back courts, *because they could not bear the still greater stenches of the street.*

During the year 1864, without any known special predisposing cause, but apparently through the cumulative virulence of the deadly agencies *always at work*, the fearful scourge of mortal disease carried off 3,516 of our children, or 341 out of every myriad of our population, *which exceeded even the abnormal number of our births by* 280. It does not appear that the legal guardians of the public health took any steps to mitigate this frightful calamity; and again in 1865, the mortality of children (as well as of adults) was above even the high average of twelve years.

But in the spring of 1866, owing to a wholesome dread of cholera, a strong public opinion, an Order in Council, and the labours of the Sanitary Association (then first formed), the Corporation appointed two Health Officers *for three months*, and detailed police to act as inspectors. *Only a very partial surface cleansing of the yards was the result;* the streets remaining as before, the subsoil retaining all its pollutions, and the production of fresh poisons unchecked; and yet what was the result of this, aided probably by the unusually cold, wet, and windy season? *Four hundred and seventy lives of children were saved* as compared with the previous year; and June, which on the average is the most unhealthy month except July or August, *actually furnished the week of lowest deaths.* Yet, no sooner was the cleansing finished, and the July sun drew forth to the surface the substratum of xymotic poison, than the death-rate of the children rose at once from 362 per myriad to 852; and the *deaths of adults in the whole year exceeded those of* 1865 *by fifty-five.*

But if this minute instalment of what ought to be done, produced at once such a marvellous benefit as the saving of 470 children's lives, what might not be expected, were councillors, owners of property and householders to perform their manifest duties? And if they are not willing, for the love of God and the good of their brethren, to obey the plain laws of health and remove the causes of disease and death, ought not the power of the law to protect the helpless, and prevent the selfish from robbing their neighbours of their happiness, and the very lives of themselves and their children?

---

EDITOR'S NOTE.—The present number of this journal is published April 26, 1867.

## REVIEW.

FERNS : British and Foreign ; By John Smith, A. L. S.

The well-known ex-curator of the Royal gardens, Kew, has lately published this most useful fine manual; intended primarily to assist fern cultivators, it is nevertheless valuable also to botanists. He gives a very interesting history of the introduction of exotic ferns into European gardens; an essay on the genera of ferns and their classification; an enumeration of the ferns at present cultivated, and very full instructions on their cultivation. Mr. Smith's mode of classification aims to be natural and his tendency is to multiply genera unduly. His enumeration extends to 1084 species (nearly half of those known to science) ; he gives many synonyms, a reference to the best descriptions and engravings in standard works, and wood-cut illustrations of the genera. Sir William Hooker recently said of our author :—

" The formation of this fine collection [of cultivated ferns in " Kew gardens] is mainly due to the exertions and ability of Mr. " John Smith. His knowledge of ferns and his writings upon " them, justly entitle him to rank among the most distinguished " Pteridologists of the present day."

Mr. Smith gives us, northern North Americans, no credit for having the following ferns in our native flora :

Phegopteris rhætica (the Polypodium alpestre of British botanists), which is found on the eastern side of the Rocky Mountains ; Dryopteris Thelypteris, one of our commonest ferns ; Polystichum Lonchitis, which has a wide range and is locally plentiful; Scolopendrium vulgare, which is local but also abundant ; Asplenium Ruta-muraria, which is found in all the neighboring States, as far west as Michigan and further south than Virginia; A. viride, which ranges from Newfoundland to the Rocky Mountains, and perhaps thence to the Pacific Ocean; and A. septentrionale, not uncommon on the Rocky Mountains. We learn nothing of our author's views on Woodsia; he gives only two species, Ilvensis and hyperborea, and gives North America credit for neither of them; moreover his wood-cut, which is said to be a frond of Ilvensis, is unmistakeably hyperborea, as we understand that species.

We believe the following to be bad species : — Asplenium Michauxii is A. Filix-fœmina, one of the most variable of ferns ; Cystea tenuis is merely a form of the protean C. fragilis ; Aspidium atomarium should have been referred to C. bulbifera ; Osmunda spectabilis is not separable from O. regalis, nor does our Onoclea Struthiopteris differ from the European form. Onoclea gracilis, and Ophioglossum pedunculosum are unknown to us. Mr. Smith's arrangement of the following species of the genus Dryopteris (or Lastrea) is not understood by us. He places American plants thus : Filix-mas, remota, rigida, marginalis, Goldiana, dilatata, cristata, intermedia, spinulosa. We look on their affinities in a different light, and would arrange them as shewn below. Four of these forms we consider to be unquestionably one species ; dilatata is our more common form northward, and is well-marked as a variety, intermedia is identical with spinulosa and remota (as we understand it) hardly separable from it, while cristata is more closely allied to Goldiana than to any of the forms of spinulosa.

The publisher has done his part well, the book is neatly got up, well printed and remarkably cheap.

The question,—under what circumstances is the author or emendator of a genus justified in writing his own name after such old species as he chosen to place in it ? has lately been discussed; we incline to answer, " under no circumstances," being of opinion that a specific name should never be changed, and that the original author's name should always be affixed to it. We append a catalogue of northern North American ferns, giving our views of the nomenclature and classification of this order; it includes all the species mentioned by Michaux and by Dr. Gray, and most of those mentioned by Pursh and by Hooker. The classification is based principally on that of Dr. Mettenius. A few species known to us only by name are omitted. W.

## Suborder POLYPODINEÆ.

### Tribe ACROSTICHEÆ.

*Chrysodium*, Fée.

1. C. aureum (Linn. 1525).

Metten. Fil. Lips. 21; Acrostichum a. Linn. Sp. Pl.; Michx. Fl. Bor.-Am. ii. (1820) 272.

### Tribe POLYPODIEÆ.

*Vittaria*, Smith.

1. V. lineata (Linn. 1530).

Swartz, Syn. Fil. 109; V. angustifrons, Michx. 261.

*Polypodium*, Linn. in part.

1. P. vulgare, Linn. 1544.

Willd. Sp. Pl. v., 172.

2. P. polypodioides (Linn. 1525).

P. ceteraccinum, Michx. 271 ; P. incanum, Swartz 35, Pursh 659, Gray's Manual, ed. 2nd, 590.

*Gymnogramme*, Desvaux.

1. G. triangularis, Kaulfuss,

Enum. Fil. 75. Found on Vancouver Island by Mrs. Miles.

*Cheilanthes*, Swartz.

There are three well-defined species of this genus within Gray's limits; but as they have been sadly confused by some authorities, I am unable to give synonyms, nor do I know to which of the three Michaux's *Nephrodium lanosum* should be referred.

1. C. vestita, Swartz 128.

Willd. 458; Gray's Manual, 592.

2. C. tomentosa, Gray's Man.

Link, Fil. Hort. Berol, ii., 42? Hook. Sp. Fil. 65 ?

3. C. lanuginosa, Nuttall.

C. gracilis, Metten. Cheil. 36.

*Cryptogramme*, R. Brown.

1. C. crispa (Linn. 1522).

R. Brown, App. Frank. Journ. 754. Osmunda, Linn. Allosorus, Bernhardi. "Isle Royal in Lake Superior;"—Moore: probably the following.

2. C. acrostichoides R. Br. 767.

Hooker considers these two plants to be specifically identical, which is probably correct. Mr. Moore considers them generically distinct.

*Pellæa*, Link.

1. P. gracilis (Michx. 262 .

H. ok. Sp. Fil. ii., 138. Pteris g. Michx. 262, Pursh 668. Ledebour and Moore refer *Pteris Stelleri* (Gmelin) here, while Swartz and Hooker refer it to *C. crispus*; should the former prove to be correct, this plant must be named *Pellæa Stelleri*.

2 P. atropurpurea (Linn. 1534).

Link, Fil. Hort. Berol, 59. Pteris a. Linn. Michx. 261, Pursh 668.

*Pteris*, Linn. in part.

1. P. aquilina, Linn. 1533.

P. caudata, Linn. 1533, Pursh 668 is a variety found in the Southern U. S. and elsewhere.

*Adiantum*, Linn.

1. A. pedatum, Linn. 1557.

### Tribe ASPLENIEÆ.

*Blechnum*, Linn., Presl.

1. B. Spicant (Linn. 1522).

Smith, Turin Trans. v. 411 Osmunda, Linn.; Lomaria, Desv.; B. boreale, Swartz 115, Pursh 669.

2. B. serrulatum, Rich.

Michx. 264; Pursh 669.

*Woodwardia*, Smith.

1. W. areolata (Linn. 1526).

Lowe's Ferns, iv. t. 46. W. angustifolia, Smith ; Onoclea nodulosa, Michx. 272; W. onocleoides, Willd.; Pursh 669.

## 2. W. Virginica (Linn. Mant. 307).

Smith, 1. c. 412; W. Banisteriana, Michx. 263.

### Scolopendrium (Smith) Hook.

§ *vera.*

### 1. S. vulgare, Smith 421.

Asplenium Scolopendrium, Linn. 1537; S. officinarum, Swartz 89.

§ *Camptosorus*, Link.

### 1. C. rhizophyllus (Linn. 1536).

Link, Fil. Hort. Berol, ii. 69.

### Asplenium, Linn.

### 1. A. pinnatifidum, Nuttall,

Gen. N. A. Plants, ii. 251.

### 2. A. montanum, Willd. 342.

A. Adiantum-nigrum, Michx. 265.

### 3. A. Ruta-muraria, Linn. 1541.

### 4. A. septentrionale (Linn. 1524).

Hoffman, Deuts. Fl. ii. 12.

### 5. A. viride, Hudson,

Fl. Ang. 385; A. Tri.-ramosm Linn. 1511.

### 6. A. Trichomanes, Linn. 1540.

A. melanocaulon, Willd.; Pursh 666.

### 7. A. ebeneum, Aiton,

Hort. Kew. iii. 462; A. trichomanoides, Michx. 265.

### 8. A. marinum, Linn. 1540.

Attributed to the Lower Provinces by Sir Wm. Hooker,—probably in error.

### 9. A. angustifolium, Michx. 265.

### 10. A. thelypteroides, Michx. 265.

### Athyrium, Roth.

### 1. A. Filix-fœmina (Linn. 1551).

Roth, Fl. Germ. iii. 65 ; N. filix-f. and N. asplenioides, Michx. 268 ; also Aspd. angustum, Willd., Pursh 664. Perhaps an Asplenium.

## Tribe ASPIDIEÆ.

### Phegopteris, Fée.

### 1. P. Dryopteris (Linn. 1555).

Fée, Gen. Fil. 243. Nephrodium D., Michx. 270. (Mr. Moore refers Michaux's plant to the next species.)

### 2. P. Robertiana (Hoffm.).

P. calcarea, Fée, 1. c. 243; Polypodium calcareum, Smith ; doubtfully distinct from *P. Dryopteris.* Universally but erroneously attributed to North America.

### 3. P. connectile (Michx. 271).

Polypodium Phegopteris, Linn. 1550; P. connectile, Willd. 200, Pursh 659. Michaux's name ought to be restored to this plant; it has priority over those of Fée or Mettenius.

### 4. P. hexagonoptera (Michx. 271).

Fée, Genera Filicum, 243.

### 5. P. rhætica (Linn. 1552).

P. alpestris, Mettenius ; Polypodium alpestre, Hoppe ; Aspidium rhæticum, Swartz 59. Cascades ; Rocky Mts. 49° N. Lat., Dr. Lyall.

[P. montana (Volger).

More properly *Aspidium montanum;* though it has been placed here by Fée.]

### Aspidium, Swartz.

Polystichum, Roth. ; Dryopteris, Adanson.

§ *Dryopteris* (Schott) A. Gray.

Lastrea, Presl ; Nephrodium, Richards, R. Brown, Hooker; Polystichum, D.C., Koch, Ledebour.

### 1. D. Thelypteris (Linn. 1528).

Gray's Manual, Ed. 1st. 630.

### 2. D. Nov-Eboracensis (Linn. 1552).

Gray, 1. c. 630, N. thelypteroides, Michx. 267.

### 3. D. montana (Volger).

Aspd. Oreopteris (Ehrhart) Swartz 50. Mr. Moore says that *Aspidium montanum* has been found in Vermont—certainly an error.

### 4. D. spinulosa-dilatata.

Polypodium dilatatum, Hoffman ; Aspd. dilatatum, Swartz 420 ; and A. dumetorum, Willd. 263. D. dilatata, Gray, 1. c. 631. Dr. Gray justly considers this fern (which is common in eastern C. E.) to be merely a variety of *Aspidium spinulosum* Swartz.

### 5. D. spinulosa-vera.

Polypodium spinulosum Retzius; Aspd. s. Swartz 54, 520 ; A. intermedium, Willd. 262. Common west of Quebec.

### 6. D. spinulosa-remota.

Aspd. remotum, A. Br.; Nephrodium r. Hook. Br. Ferns, t. 22; Aspd. Boottii, Tuckerman. Dr. Gray refers *Dryopteris remota* here (as *A. spinulosum* var. *Boottii*)—it may prove to be a distinct species; it is not well known to me.

### 7. D. cristata (Linn. 1551).

Gray, 1. c. 631; A. Lancastriense, Sprengel, Swartz 52.

### 8.         var. majus (Eaton).

A. filix-mas, Pursh 667 ?

### 9. D. Goldiana, Hook.

Gray, 1. c. 631; A. filix-mas, Pursh 662?

### 10. D. Filix-mas (Linn. 1551).

Schott, Gen. Fil. t. 9. Rocky Mts.

### 11. D. marginale (Linn. 1552).

Gray, 1. c. 632.

### 12. D. arguta (Kaulf. 242).

N. rigidum var. Americanum, Hook. Sp. Fil. 60.

### 13. D. rigida (Hoffm.).

Not of Gray, 1. c. 631. A. rigidum, Swartz 53. Attributed to North America by Mr. Bentham—doubtless in error.

§ *Polystichum*, Schott.

Presl, A. Gray ; Aspidium, Richards, R. Brown, Ledebour.

### 1. P. fragrans (Linn. 1550).

A. fragrans, Swartz, 51. In technical characters this plant is more properly *Dryopteris fragrans*, and is so considered by Hooker, Ledebour, etc. I agree with Dr. Gray in considering that its natural affinity places it here.

### P. aculeatum (Linn. 1552).

A. aculeatum and A. lobatum (Alton) Swartz 53, and A. angulare, Willd. 257. The typical form (A. aculeatum, Willd. etc.) has not been found in North America. Mr. Moore's remark—"extends "from the eastern U. S. to Columbia on the north-"west coast"—is certainly an error. We have, however, two well-marked and constant varieties.

### 2.        var. Braunii (Koch).

A. Braunii, Spenner; P. Braunii, Fée; which is allied to the European *Aspidium aculeatum* var. *angulare.*

3.        var. lobatum, Deakin.

A. lobatum (Aiton) Swartz, Willd. 260. *Aspidium aculeatum* var. *lobatum* was found by Mrs. Girdwood during the past summer on Ile Perrot, near Ste. Anne.

4.   P. Lonchitis (Linn. 1548).

Schott, Gen. Fil. t. 9.

5.   P. acrostichoides (Michx. 267).

Schott, Gen. Fil. t. 9.

6.   P. munitum (Kaulf. 230).

Referred by Mr. Moore to *A. falcinellum*, Swartz 46. Vancouver Island, and 49° N. Lat., Dr. Lyall.

### Cystea, Smith.

I adopt Sir J. E. Smith's characteristic name for this genus, as I do not consider Bernhardi's genera to be of much value.—Eng. Fl. iv. 260, 264.

1.   C. bulbifera (Linn. 1553).

Aspidium b., Swartz 59. "A. atomarium Muhl.", Gray!

2.   C. fragilis (Linn. 1553).

Smith, l. c. 285 ; N. tenue, Michx. 269 ; A. atomarium and A. tenue, Pursh 665.

3.   C. montana (Lamarck).

Aspidium, Swartz 61. Said to be found in north-western America.

### Woodsia, R. Br.

1.   W. Ilvensis (Linn. 1528).

R. Br. Linn. Trans. xi. 173; Neph. rufidulum, Michx. 269; W. Ilvensis and W. hyperborea, Pursh 660.

2.   W. alpina (Bolton).

W. hyperborea, R. Br. l. c. t. 11; Hook. Br. Ferns, t. 9; W. alpina, Moore, Nat. pr. Br. Ferns, t. 106. More properly *W. Ilvensis* var. *alpina*. Scarcely distinct from No. 1—from which, however, it may usually be distinguished by its smoothness, shorter pinnas, more rounded lobes, and darker (often almost ebeneous) sti, es which have fewer scales.

3.   W. hyperborea (Liljeb.)

Newfoundland, per Geological Survey. I regard the *Acrostichum hyperboreum* of Liljeblad as quite distinct from the *A. alpinum* of Bolton, (Fil. Brit. t. 42), and as very closely allied to No. 4.

4.   W. glabella, R. Brown.

Rich. App. 39; Hook. Fl. Bor.-Am. t. 237. Probably identical with No. 3 and thus *W. hyperborea* var. *glabella*, but very distinct from Nos. 1 and 2.

5.   W Oregona, Eaton.

In Can. Nat. (1865) 90.

6   W. scopulina, Eaton.

l. c. 91.

7.   W. obtusa (Sprengel).

Torrey, Cat. Pl. 1840; Aspidium, Swartz 420, Pursh 662.

### Onoclea, Linn.

1.   O. sensibilis, Linn. 1517.

O. *obtusilobata* is merely an abnormal form having semi-fertile fronds.

2.   O. Struthiopteris (Linn. 1522).

Swartz 111. ; Struthiopteris Pennsylvanica, Willd. 289; Pursh 666. Hardly generically distinct from Onoclea.

### Tribe DAVALLIEÆ.

*Dicksonia,* L'Heritier.

1.   D ? punctilobula (Michx. 268).

Kunze in Silliman's Journal, Nov. (1848) 88. D. pilosiuscula (Muhl.) Willd. 484; Pursh 671.

### Sub. HYMENOPHYLLEÆ.

*Hymenophyllum,* Smith.

1.   H. ciliatum, Swartz 147.

Pursh 671. Doubtless an error of Pursh; he may have collected *Trichomanes radicans*, which is found in the Southern States.

### Suborder SCHIZÆINEÆ.

*Schizæa,* Smith.

1.   S. pusilla, Pursh 657.

*Lygodium,* Swartz.

1.   L. palmatum (Linn. 1518).

Swartz 154 ; Cteisium paniculatum, Michx. 275. Hydroglossum, Willd. 84, Pursh 656.

### Suborder OSMUNDINEÆ.

*Osmunda,* Linn.

1.   O. regalis, β. Linn. 1521.

O. spectablis, Willd. 98, Pursh 658.

2.   O. Claytoniana, Linn. 1521.

Pursh 657; O. interrupta, Michx. 273, Pursh 657.

3.   O. cinnamomea, Linn. 1522.

### Suborder OPHIOGLOSSEÆ.

*Botrychium,* Swartz.

1.   B. Lunaria (Linn. 1519).

Swartz 171. Osmunda, Linn.

2.        var. simplex.

B. simplex Hitchcock.

3.   B. matricariæfolium, A. Braun.

Osmunda matricaria, Breyn. B. rutaceum, Swartz 171. Possibly identical with No. 1. Doubtfully North American.

4.        var. lanceolatum.

Osmunda lanceolata Gmel. B. lanceolatum, Angstrom. Possibly a distinct species.

5.   B. virginianum (Linn. 1519).

Swartz 171; Pursh 656; B. gracile Pursh 656; Botrypus, Michx. 274.

6.   B. lunaroides (Michx. 274).

Swartz 172; B. fumarioides, Pursh 655.

7.        var. obliquum, Gray.

B. obliquum (Poir.) Muhl., Pursh 655.

8.        var. dissectum, Gray.

B. dissectum (Poir.) Muhl., Pursh 656.

*Ophioglossum,* Linn.

1.   O. vulgatum, Linn. 1518.

O. vulgatum and O. bullosum; Michx. 275-6, Pursh 655.

# CANADIAN NATURALIST.

### SECOND SERIES.

—

## THE DISTRIBUTION OF PLANTS IN CANADA
### IN SOME OF ITS RELATIONS
### TO PHYSICAL AND PAST GEOLOGICAL CONDITIONS.

More than two years ago, in this journal, the writer endeavoured
to indicate and illustrate some of the more obvious features in the
distribution of Canadian plants. It was shown that in taking a
general view of this distribution several distinct floras could be
recognized, viz. :—a general Canadian flora composing species
which range over the whole or greater part of the Province; a
second flora whose species are confined to the districts around the
northern shores of Lakes Superior and Huron; a third to the comparatively narrow district bordering Lakes Erie and St. Clair and
the south-western parts of Lake Ontario; a fourth to the Gulf
and Lower St. Lawrence shores; and a fifth which had an undoubted boreal aspect. Besides these, were a small inland maritime flora, and two other floras whose limits and characteristics
could not then be accurately defined, but which appeared to be
limited—the one to Upper Canada and the other chiefly to Lower
Canada. A number of plants were also indicated which were
apparently confined to the tract of country around the northern
shores of Lakes Huron and Superior and to the more eastern
parts of Lower Canada, whilst several species were named whose
occurrence was quite local. These prefatory references will render
subsequent remarks more intelligible.

In investigating the causes which have influenced the diffusion
of species in Canada, we find that whilst some have in past time
been and are still exerting their influences, others are perhaps
correctly referred to far distant periods. And whilst the operation
of some is confined to narrow limits, others extend their effects

over a wide extent of territory, and many are identical with causes
which produce somewhat similar results in other countries.

There are no long ranges of mountains within the Province to
retard the free interspersion of its different indigenous forms, nor
are the Laurentide hills of such considerable height as to much
impede the admission of the cold boreal winds from around Hudson Bay.  The great breadth of the lakes, however, must, there
is no doubt, preclude a migration from the northern United States
as extensive as under altered circumstances it would be.

To the influences effected by our numerous and extensive lakes
and rivers through their currents, the formation of prairie land,
the evaporation from their surfaces and the necessarily modified
temperature of the land surrounding them, references will, in subsequent parts of this paper, be made.

An eminent writer on botanical as well as geological subjects,
thinks, that many anomalies in the distribution of Canadian vegetation can be explained by considering the chemical constitution of
the soil.  "A little more lime or a little less alkali in the soil renders vast regions uninhabitable by certain species of plants.  For
many of the plants of our Laurentide hills to extend themselves
over the calcareous plains south of them under any imaginable conditions of climate is quite as far beyond the range of possibility as
to extend across the wide ocean."* This view is, in at least a
limited sense, probable.  *Rubus Chamaemorus* Linn. and *Empetrum nigrum* Linn. have been cited as illustrations of the preference maintained by some plants for soils of Laurentian origin.  It
may be more correct to, in part, ascribe the range of these plants
to their known predilections for northern situations.  They are
both in fact sub-arctic plants, and it merely happens to be a coincidence that the Laurentian formations skirt the Lower St. Lawrence
and the northern shores of Lake Superior, on the coasts of the
former of which both of these plants occur, and on those of the
latter *Empetrum nigrum*.  Were their distribution entirely dependent upon the nature of the soil, they should occur in the country
around the Upper Ottawa and elsewhere, but they are not known
to range so far to the southward. *Pinus Banksiana* Lamb.—a less
northern form—and probably *Polygonum cilinode* Michx. would
seem, in our present knowledge of their distribution, to constitute
better illustrations of preference for Laurentian soils and

* Dr. Dawson ; this journal, O. S., vol. vii, p. 342.

strata. It would be interesting, however, to compare the range, in relation to soils, of those plants which are common to Europe and America.

We can conclude from the known distribution in Canada of rocks of the earlier geological formations, and from the direction of the ice-grooves upon them, that soils composed chiefly of Laurentian, or, in some instances, Huronian debris, were spread both over these formations and for at least some distance over the Silurian and Devonian rocks during the epoch of the drift, whilst the strata farther south were carpeted with more calcareous soils. The distribution of these soils was, no doubt, at subsequent periods, somewhat disturbed. Now, the Laurentian strata are composed of such different materials in different localities—some of which lie at but comparatively short distances apart—that knowing the composition of the soil at any given locality, it would be often incorrect to assign a similar composition to soils in the vicinity which we know must have been derived from rocks of the same system. The quartzites have afforded silica in abundance to the soil; the limestones, phosphate and carbonate of lime, and other minerals in variable quantities; the dolomites, carbonates of lime and magnesia; the serpentines, silica and magnesia; and the orthoclase and labradorite, silica, alumina, soda and potash. All of these mineral species, with others, are common in the Laurentian rocks. The Huronian formation also abounds in quartzites and dolomites. Within the limits, then, of a single township there might be met with soils in one case highly calcareous, in another with noticeable quantities of alkalies and but a trace of lime. The very variable proportions in which the same chemical ingredients will frequently occur in soils, at localities not far distant from each other, has been well shown by Dr. T. Sterry Hunt.[*] It is a noticeable circumstance that lime, potash and soda, appeared in all the soils analyzed by him. These facts are mentioned to show that if the composition of soils has such an influence as to affect the presence of plants upon them, conditions must occur in some parts of but limited areas favourable to the existence of many plants which do not in others. Moreover, when we consider the varied compositions of our early formations, it is easy to conceive that over the immense extent of country in which they are developed, whilst many situations afford the requisite conditions for

---

[*] Geology of Canada, 1863, p. 640.

plants requiring much alkali, many other localities must be well suited for species to whose growth lime is more necessary. And again, the different proportions in which lime exists in soils overlying the Silurian and Devonian rocks, make it probable that in many localities the proportion would be so small as to afford suitable habitats for plants preferring non-calcareous soils. However much, then, there may be in the relation existing between plants and the chemical constituents of the soils in which they grow, it seems exceedingly difficult to arrive at any satisfactory conclusions regarding the effect of this relation upon the general distribution of our native plants.

In the above remarks I do not of course include any reference to sea-shore plants, which, without a doubt, derive sustenance from the chloride of sodium, with which both the air and soil, in the vicinity of the coast, are to some extent impregnated. But the very fact that many of these plants are met with in localities far distant from any possible influence of the ocean, clearly shows that this alkali may not be entirely essential to the existence of all maritime species.

Before leaving the subject, a few instances of apparent preferences for particular soils or locations may be cited. The whitewood, *Platanus occidentalis* Linn., is, at London, only met with on the low alluvial flats on either side of the River Thames, and the two or three trees occurring at Toronto exist in a similar situation on the banks of the River Don. At Chatham, and nearer the mouth of the River Thames, this one of the largest of Canadian trees occupies like locations, and is said to attain there a magnificent size. *Pinus rigida* Miller, again, has only been detected among the Thousand Islands—which form the connecting link between the Laurentide hills of Canada and the Adirondacks of New York State—and in the Township of Torbolton on the Upper Ottawa, in the immediate vicinity of which the Laurentian strata are also largely developed. *Corydalis glauca* Pursh, *Kalmia angustifolia* Linn., *Asplenium ebeneum* Aiton, and *Woodsia Ilvensis* R. Brown—for the most part easily recognized plants— are, judging by our present knowledge of their distribution in Canada, limited in range to the area occupied by the Laurentian rocks. The distribution of these and other species is not, however, so definitely established as to warrant any perfectly safe conclusions regarding the effects upon them of particular soils and locations, and other reasons already mentioned would further induce the withholding of any conclusion. Besides, it seems difficult to escape the conviction that very often local circumstances—to some of which reference will hereafter be made—will, more than the general climate or the presence of any particular ingredients in the soil, account for the occurrence of plants in specific localities.

Other features of interest may be also cited. Those who have visited the Thousand Islands in the River St. Lawrence must have been struck with the vast abundance of *Rhus typhina* Linn. and *Pteris aquilina* Linn. Neither of these plants is, however, limited to Laurentian soils, and it is very probable that the profusion here of at least the former is in part due to the rugged, rocky nature of almost all of the islands. It may be also mentioned that the capacity of land for cultivation is often in Canada judged of by the timber trees growing naturally upon it. Eastern farmers look upon the red pine, *Pinus resinosa* Aiton, as characterizing a poor soil, whilst there are many in the Erie district, where the red pine is unknown, who regard the chestnut, *Castanea vesca* Linn. as evidencing some sterility.

Another circumstance affecting distribution is not to be overlooked. The Laurentian rocks, which are very largely developed in Canada, are remarkable for their rugged, corrugated character—in some places forming ranges of high hills, in others, individual elevations of considerable height, and everywhere, to a greater or less extent, displaying the same characteristic rugged surface. The whole breadth of the strata is, besides, dotted with basins of varying sizes and forms, which have been worn out of the softer material of the rock, and are now filled with sheets of water. The surface of the Laurentian rocks is, to a very considerable extent, bare and only tenanted by numerous saxicolous Parmelias, Lecideas, and other lichens, with mosses and some ferns, and a few often stunted phanerogams maintaining an existence in the little soil collected in the frequent cracks and fissures. The very numerous little hollows and depressions in the surface—probably in most instances grooved out by the action of ice—are covered by a layer of soil generally scanty, but which is often very rich and supports a prolific vegetation. On the other hand, the Silurian and Devonian formations have either a level or somewhat undulating surface, and are everywhere covered by clays, sands, gravels and loams, which attain very often a great thickness, especially in the Upper Canada peninsula, where numerous illustrations are afforded

by the oil-well borings.    Between the River Ottawa and the
Georgian Bay and Lake Superior, the Algoma sands form a prominent feature in the surface deposits, whilst over the Upper
Canada peninsula and along Lake Ontario, are chiefly distributed
the Erie clays and Saugeen clays and sands.    This varied nature
of the rock surface, the presence of these very numerous lakes over
the Laurentian strata, and the great diversity in the depth as well
as general characters of the surface deposits, must have a not
inconsiderable influence upon the vegetation of the country,
especially in the multiplication or diminution of the numbers
of many species.

In many localities throughout Western Canada, there are
terraces and ridges of soil extending over, in some cases, considerable surfaces of country—evidences of the much higher levels
attained by the Great Lakes and certain rivers in some recent times
than exist at the present day.    My correspondent, Mr. John
Macoun, of Belleville—other of whose careful observations
obligingly communicated, are elsewhere in the present paper
referred to—has informed me that in his neighbourhood the
ridges (the surface soil of which is generally a fine sand slightly
mixed with clay, with a subsoil of usually limestone gravel or fine
sand) support a vegetation of a southern and western aspect
not met with in localities of a different nature in the same section
of country.    This would appear to be attributable rather to the
general nature and state of aggregation than to any particular
chemical condition of the materials composing the ridges.    When
of such loose materials as the sand, clay and gravel referred to,
these ridges are always well drained, and where exposed to the
action of the sun, absorb the heat with great readiness.    This
heat in radiating again into space, continues to supply the plants
growing upon the ridges with warmth during the intervals of
night.    Now, much less heat is absorbed, and, consequently, less
radiated into the atmosphere by a wet and stiff clay, than by a
loose, gravelly, or somewhat sandy soil, and the oxygen of the air
has much less access to the organic substances in and the roots of
plants growing upon the soil.    These consequences are observable
among all our surface deposits, in a greater or less degree in proportion to the state of aggregation and general character of their component materials, and would be similar, though in a less marked
manner, if the soil were not in ridges.    The rather rare *Ranunculus
rhomboideus* Goldie, *Helianthemum Canadense* Michx., and

*Viola sagittata* Aiton, I have found at London, growing along with other interesting plants, nearly side by side, on a gentle slope, well exposed to the rays of the sun, and composed of a very sandy clay. Mr. Macoun has found the same plants upon the ridges of Northumberland County, growing with *Anemone cylindrica* Gray, *Linum Virginianum* Linn., *Trifolium stoloniferum* Muhl., *Liatris cylindracea* Michx., *Aster ericoides* Linn., *Rudbeckia hirta* Linn., *Artemisia biennis* Willd., and a few others. Both southern and western forms require a higher degree of heat than plants of our eastern districts, even under the same parallel of latitude. As in many parts of Western Canada similar ridges of sand and gravel occur, the circumstances detailed are not of mere local interest.

In connection with the subject of soils, Mr. Macoun points out the fact, that in his neighbourhood, western plants, where not aquatic, always occur in either a sandy soil, or a soil holding much limestone gravel. My own observations at London, and elsewhere, would tend to confirm this in regard to, at least, some plants.

The flora of the Lake Superior districts, in some of its features, is very different from that of other parts of Canada. Many of the familiar trees and herbaceous plants of the more southern parts of the province are absent, whilst there occur—mingling with the very large number of our more abundant species, and the few northern forms—a little assemblage of plants, more characteristic some of the western woody country and plains, and others of the middle and southern States. Additional species are met with upon the American side of the lake. *Ranunculus abortivus* Linn. var. *micranthus* Gray, *Matricaria inodora* Linn., *Tanacetum Huronense* Nutt., *Senecio canus* Hook., and some others, extend as far eastward as the Lake Huron shores, but the majority have only been found in the vicinity of Lake Superior. It is not difficult to account for their presence in these localities, but why do we not find them about Lakes Erie and Ontario, and farther eastward, as well as around the Upper Lakes? Questions of a similar nature will occur to United States botanists. What precludes the eastward range of the characteristic vegetation of the western prairies, and of the central wooded plains of the continent; and to what cause can be ascribed the very peculiar north-westward range of many American plants, by which they occur in Ohio, Michigan, Wisconsin and westward, and about the

Saskatchewan. but are altogether absent from the New England States. and the eastern and central parts of Canada ? Two questions are, in fact, involved in considering, in the present place, the distribution of the vegetation of the country surrounding Lake Superior.

The vegetation of the prairies. like that of the pampas of South America and the steppes of Russia, is of a peculiar type— approached. however. in general characters, by that of the marshes and swamps. Lesquereux, Henry Engelman, and others, have pointed out many of the distinctive features of the prairies and their flora.* Conditions are not suitable for the extension of this flora into the more eastern parts of the United States and Canada. In our Erie district, however, there are a few forms which remind us much of the western prairies. To these some allusion will be hereafter made.

With regard to the vegetation of the central wooded districts of British America and the adjoining American States, doubtless the colder climate of Lake Superior and the rugged nature of the surrounding country preclude the eastward distribution of more of its plants. Climatal and physical conditions would, besides, on principles hereafter explained, encourage a different range.

The north westward diffusion of many American plants has been referred, perhaps correctly in part, to the direction of the valleys in the United States and British America. Other causes must, however, be also taken into account. The principal ranges of North American mountains have a general northern and southern course, with considerable inclinations to either the eastward or westward. The prevalent trends are in fact parallel with the coast lines of the continent. The directions of the large rivers, again, are generally north-east, south-east, or nearly south-west. Here we have furnished to us as the general course of the valleys, along which the southern temperate flora may with facility migrate, two directions—one to the north-east, and the other to the north-west. Still further, the central parts of the continent are comparatively low lying, not exceeding at the headwaters of the Mississippi 1700 feet above the ocean; and the watershed, which separates the rivers which flow into the great lakes and the St. Lawrence from the tributaries and subtributaries of the Mississippi, crosses the northern part of the State of Wisconsin, and

* Amer. Journal of Science [2] xxxvi. 384 ; id., xxxix. 317.

almost skirts the southern and western parts of Lake Michigan.
Now, it is generally known that the north-eastern parts of North
America have a temperature lower than that of the central plains
and wooded countries in similar latitudes, and that the lines of
mean temperature rise very considerably as they cross the continent from the New England States and Canada westward.   The
reason for this lies in the much greater mass of land on the western
half of the continent extending far into the Arctic Sea, the large
areas of polar land on the eastern side separated by extensive
bodies of water from the mainland, and the Great Lakes—all of
which tend, on principles long since stated by Lyell, Humboldt,
Dana, and others, to produce a lower temperature in the north-
eastern sections of the continent.   Other influences, arising from
proximity to the sea, from the Labrador current, and the general
configuration of the coast, also lend their aid.   Now, a plant from
the warmer temperate zone, in migrating northward, would not
range far up those valleys having a north-eastward bearing from
the gradually lower temperature met with there, and yet, favoured
by the course of the valleys and the warmer climate, would be
found in much higher latitudes farther inland.   Further, the Appalachian chain of mountains must form to some extent a barrier
to eastward distribution.   It is also a noteworthy circumstance,
when taken in connection with the lower temperature in proceeding northward, that at least the larger river valleys of eastern New
York and the New England States have a general southern direction.   In this way, it seems to me, the apparently anomalous
north-westward range of many American plants can be fully accounted for.   To some of the causes mentioned, added to the configuration of the coast lines of Lakes Superior, Michigan, Huron,
and Erie, must be also ascribed the presence of the few south
temperate plants which occur around Lake Superior.   The lower
temperature and the broken character of the country must alone
prevent many other species from also finding homes there.

    In the districts which border Lake Erie there is a not unex-
pected intermingling of northern temperate with more southern
forms.   The most casual observer will not fail to account for this.
Separated on the one side by the River Niagara from the western
part of the State of New York, the district extends westwardly
along Lake Erie, widening gradually in its course, consequent on
the form of the lake, until it almost touches upon a not inconsiderable part of Michigan.   We would be quite prepared to meet

within the limits of this district many of the characteristic species of the western portions of the States of New York and of Michigan; and from their relatively lower latitude, and their position near the bend at the head of Lake Erie, we would be as well prepared to find in the townships fronting the Detroit River some of the rarer species of Southern Michigan and Northern Ohio.

The prairie lands around Lake St. Clair, and extending towards Chatham, indicate the considerably greater breadth of surface of that lake at a recent period (geologically considered). These prairie soils are, very probably, the most recent surface deposits of any extent existing in Canada. Their deposition took place after the waters of the Great Lakes had assumed their present level, and, consequently, subsequent to the formation of the ancient lake ridges, terraces and beaches, so frequently observed in Canada West. They do not here, however, as in the Western States, occupy extensive tracts of country. At the present day the formation of prairies is in progress along some of our lake shores. On the American side of Lake Erie, the Bay of Sandusky is—as has been well explained by Leo Lesquereux—in process of transformation into prairie land, and on the Canadian side of the same lake, Point Pelée affords an illustration of more recent commencement.

I am not aware that our Canadian prairies have been explored. There are, however, elsewhere, within the Erie district, some outliers, as it were, of the western prairie flora. Illustrations are found in *Vernonia fasciculata* Michx., *Solidago Ohioensis* Riddell, *S. Riddellii* Frank, *Silphium terebinthinaceum* Linn., *Hieracium longipilum* Torrey, and *Phlox pilosa* Linn.

Mr. Macoun, more than a year ago, pointed out to me the very interesting fact, that on the Lake Ontario beach at Wellington and Presquile, occur a few plants which are not to be met with farther inland, and which have been hitherto thought to be limited in range to the more southern districts of Canada, or to New York, Ohio, and other of the middle States. The more interesting species which he has thus far detected are *Jeffersonia diphylla* Pers., *Lithospermum hirtum* Lehm., *Rhynchospora capillacea* Torrey, *Scleria verticillata* Muhl., *Sporobolus cryptandrus* Gray, *Panicum virgatum* Linn., and *Hypnum trifarium* Web. and Mohr. Upon these beaches the same discerning botanist has obtained *Cladium mariscoides* Torrey, and *Scirpus pauciflorus* Smith, neither of which have been hitherto familiar as Canadian plants, nor has the latter been observed in the Northern States ; and he has also collected *Conopholis Americana* Wallroth, *Physostegia Virginiana* Benth., *Eleocharis tenuis* Schulter, and *Carex Œderi* Ehrh., species which have been observed elsewhere in the central, or in more northern parts of Canada, but which he had never met with in the Counties of Hastings and Northumberland. The occurrence of these species in the localities named was, I conceive, rightly ascribed by Mr. Macoun, to the drift of Lake Ontario. The currents of the lake take a direction from the Niagara River to the entrance to the St. Lawrence, and the Prince Edward peninsula, extending far into the lake would—aided by the prevailing winds—readily intercept the drift.

It is easy to conjecture that a similar cause to that which occasioned the presence of the above-mentioned plants upon the northern shores of Lake Ontario, would lead to the occurrence of forms still more southern upon the Lake Erie shore, at Point Pelée and Long Point, localities, the very formation of which was due, in the first place, to the action of the winds and current. Some plants not at present familiar to us as Canadian, will yet, I suspect, be detected there. The action of the currents of Lake Huron and of the River St. Clair is, I think, exemplified in the occurrence of *Primula farinosa* Linn. and *P. Mistassinica* Michx. upon the shores of that lake and Lake St. Clair.

It has long been a fact familiar to American botanists that a number of strictly maritime plants are diffused along the shores of the Great Lakes, in the immediate vicinity of some smaller lakes, and extensive swamps, situated at a short distance away, and near salt springs in New York State and Wisconsin. The number of these has been, within the last two years, slightly increased. The Rev. Mr. Paine and Judge Clinton, have detected *Naias major* All., *Ruppia maritima* Linn., and *Leptoclon fascicularis* Gray—a perhaps sub-maritime species—near the margin of the Onondago Lake, in New York State and Canadian botanists, although they have not added to this section of their lake shore flora, have yet thrown some further light upon its distribution. The brief catalogue hereunder, probably includes all the maritime plants, with one or more, perhaps, strictly sub-maritime species, now known to have this peculiar range.

| | |
|---|---|
| Ranunculus Cymbalaria, Pursh. | Polygonum articulatum, Linn. |
| Cakile Americana, Nutt. | Rumex maritimus, Linn. |

Hudsonia ericoides, Linn.            Euphorbia polygonifolia, Linn.
H. tomentosa, Nutt.                  Naias major, All.
Hibiscus moscheutos, Linn.           Ruppia maritima, Linn.
Lathyrus maritimus, Bigel.           Triglochin maritimum, Linn.
Atriplex hastata, Linn.              T. palustre, Linn.
Salicornia herbacea, Linn.           Scirpus maritimus, Linn.
Polygonum aviculare, Linn.           Calamagrostis arenaria, Roth.
   var. littorale, Link.             Leptochloa fascicularis, Gray.
              Hordeum jubatum, Linn.

It is to be observed that some of these plants have a very
extended inland range, whilst others are apparently distributed over
very limited areas. *Hudsonia tomentosa, Lathyrus maritimus,*
and *Triglochin maritimum* are, perhaps, the most widely diffused.

It is conceived that this peculiar distribution owes its origin to
successive changes in the physical aspect of the province during
the post-pliocene epoch, and the gradual adaptation of the plants
to the new conditions in which they were, by force of circumstances,
placed; and further, that these plants indicate the probable
existence of a much more extensive maritime flora which flourished
on the ocean shores during this epoch. I have already briefly
detailed my views on the subject in this journal. I may, however,
here explain, that it has not yet been satisfactorily established,
what in post-pliocene times were the conditions of land and water
in what is now known as Western Canada. The precise age,
and the marine or lacustrine origin of the Erie clays, which are
largely developed there, are yet involved in some uncertainty from
the absence of any fossil evidence; nor is it yet known what
relations they bear to the marine sands and clays of Eastern
Canada, although they may have been contemporaneously deposited. If, however, I am correct in referring the origin of the
distribution of the inland maritime flora to the post-pliocene epoch.
it will furnish an argument for the marine character of such
deposits as are coeval with those of the eastern sections of the
province referable to this epoch. If the Great Lakes were in
these distant and yet comparatively recent times, bodies of salt-
water, or if they were united into one vast inland sea, as, judging
from geological evidence, was probably the case, we can readily
account for the migration of the sea-shore species along the coasts.
And if these seas or united seas gradually became fresh-water.
it does not require much stretching of the imagination to picture
the struggle for life which must have taken place among these
wanderers from the ocean coast, in consequence of the gradual

change in at least one of those conditions, hitherto so apparently essential to their very existence. As year followed year, and the lakes became imperceptibly more fresh, successive individuals of some of the species would, as it were insensibly, become more and more reconciled to the new conditions, whilst, perhaps, most of the species would gradually diminish in both numbers and luxuriance, and finally, unable to perform those functions necessary for their reproduction, would die, and thus completely disappear from the lake coasts. As the lakes receded to their present limits, the survivors, lured by the presence of the waters, would follow, leaving, however, some of their number around the saline springs of New York State and elsewhere. These survivors probably constitute a more hardy race than their fellows on the ocean coast. This would seem to be illustrated by the more northern inland range of some, the extended diffusion along the lake margins of others, and the adaptation of all to new conditions.

These inland maritime plants have only as yet been detected on or near the shores of broad lakes, and extensive bays, on the borders of large swamps, or in the immediate vicinity of salt springs and "salt licks," showing the marked preference which these little ramblers still retain for the neighbourhood of saline waters or for homes near the lake or bog margin, in which the saline element alone is wanting to render complete. It is further to be observed that the greatest number of species exist around, or at smaller sheets of water, not far from the shores of lake Ontario, the lake which, of all our inland fresh-water seas, is much the nearest to, in fact, almost adjoins what formed in post-pliocene times, the ocean coast, and to the shores of which the first migration of sea-shore plants was probably effected.

The animal kingdom affords illustrations of a distribution analogous to that indicated by these little inland maritime plants. Dr. Leconte has recognized upon the north shores of Lake Superior, insects of a sea-shore type ; and in fresh-water lakes in Norway have been observed two marine crustaceans whose presence is attributed to a submergence and subsequent rise of the land during the post-tertiary epoch, and a change in the conditions of the waters of the lake from a state of saltness to that of freshness, which these species survived.

There is a probability that many existing species of plants in Canada can date their period of creation as far back as the post-pliocene epoch, and, it may be, to a more distant age. In the Leda clays of Green's Creek, near Ottawa, occur numerous nodules enclosing, among other organic remains, many fragments of plants. Dr. Dawson has, after careful examination, identified *Drosera rotundifolia* Linn., *Acer spicatum* Linn., *Potentilla Canadensis* Linn., *Gaylussacia resinosa* Torrey and Gray, *Populus balsamifera* Linn., *Thuja occidentalis* Linn., *Potamogeton perfoliatus* Linn., *P. pusillus* Linn., and *Equisetum scirpoides* Michx.‡   Now, it will be noticed not only that all of these plants are of still existing species, but also that four, *Drosera rotundifolia*, *Potamogeton perfoliatus*, *P. pusillus*, and *Equisetum scirpoides*, are common to Europe and America. This would appear to establish the fact, irrespective of any evidence which may exist in other countries, that the intermingling of European and American forms, so noticable a feature in our North American vegetation, took place either during this epoch or at an earlier period. Still further evidence of this is afforded by the inland maritime flora. No less than eleven of these have a European as well as an American range. Thus, a part of the temperate floras of both continents can mark the dawn of its existence at a very early period in this epoch, and probably during the antecedent age.

All of our high northern forms occur either in the districts fronting the Gulf and upon the shores of the Lower St. Lawrence, or upon the coasts of Lake Superior. We have no mountains known to us to be capped with little assemblages of arctic and sub-arctic plants, since Mt. Logan and other considerable elevations in the extreme eastern parts of Lower Canada, on which some may be supposed to occur, remain as yet unexplored. The Island of Anticosti, the Mingan Islands, and, it is to be presumed, the neighbouring districts of the mainland on the northern coast, have a nearly arctic aspect, while the north shores of Lake Superior are as nearly sub-arctic in their floral characters. On the former occur a number of characteristic arctic forms, but associated with many plants of more temperate range; and on the latter, whilst there are sub-arctic species present, they are also accompanied by numerous others which have an extensive diffusion to the southward.

It is a circumstance to be somewhat expected, in consequence of the difference of latitude, that the flora of the south shore of Lake Superior, and of the north shore of Lake Huron, is much less boreal in its aspect than that of the northern coasts of the former lake.

It is a fact of considerable interest that far up the River St. Lawrence, upon both sides, even towards Quebec, are found, mingling with sub-arctic forms, some species of truly arctic range. *Rubus Chamæmorus* Linn., *Gentiana acuta* Michx., *Pleurogyne rotata* Linn., *Empetrum nigrum* Linn., and *Woodsia hyperborea* R. Br.,[*] among others, range as far up the river bank as Riviere-du-Loup, where they have been detected by Dr. Thomas; and *Astragalus alpinus* Linn., *A. secundus* Michx., *Vaccinium Vitis Idæa* Linn., *V. uliginosum* Linn., *Euphrasia officinalis* Linn., with one or two other boreal forms, extend to the Island of Orleans and Quebec. In seeking for an explanation of this somewhat peculiar diffusion, it must be borne in mind that arctic plants delight in a low equable temperature, accompanied by a moist atmosphere, and wherever these conditions exist, whether on mountain summits or on northerly ocean coasts, there these little plants can find a home. Now, the coasts of the Lower St. Lawrence amply supply these conditions. They occupy a rather high latitude, and besides frequently rise to considerable elevations, forming extensive cliffs. The broad and deep expanse of water fronting them necessarily has the effect of lowering and equalising the temperature, and the evaporation, which must be very great, continuously taking place, aided by the winds, moistens the surrounding air. Further, a branch of the cold Labrador current flows through the Straits of Bellisle, carrying with it, no doubt, amongst other drift, seeds of arctic and sub-arctic species, and extends its influence far up the St Lawrence. This current would further aid in lowering the temperature of the immediate shores, but its effects, the more marked because the waters are chilled by recent connection with icebergs, would be especially experienced upon the island of Anticosti, which, from its position, would intercept the current, and tend to direct it towards the entrance of the river. To these causes must be ascribed this climate which seems so suited to these little arctic and sub-arctic species of the more eastern sections of the Province.

Upon the northern shores of Lake Superior some of these causes likewise operate. There is the same moist atmosphere and more

---

[*] EDITOR'S NOTE.— *Woodsia hyperborea* R. Br., has been found by Mr. Horace Mann in north-western Vermont; *W. Ilvensis* (Linn.) is abundant on the rocks of the Quebec group south of the St. Lawrence. W.

equalized and lower temperature resulting from the proximity to the widely extended and deep waters of the lake. The higher latitude does not, by any means, alone account for these coasts forming suitable stations for plants of a northern range.

It is a circumstance not without considerable interest that in the alpine and sub-alpine flora of the New England States there is a remarkable paucity of peculiarly American species. With the exception of *Alsine Groenlandica* Fenzl, *Geum radiatum* Michx. var. *Peckii* Gray, *Arnica mollis* Hook., *Solidago thyrsoidea* E. Meyer, *Nabalus nanus* DC., *N. Bootii* DC., *Vaccinium cæspitosum* Michx., *Salix Uva-Ursi* Pursh, *Carex scirpoidea* Michx. and *Calamagrostis Pickeringii* Gray, all of these alpine plants are likewise of European range. This circumstance will, it may be thought, have considerable bearing upon the question with respect to the antiquity of the peculiar flora of Arctic America. The presence of these few species may be thought to be possibly due to the migrations of birds, or to other agencies at work in existing or recent times, and not to causes which, operating in post-pliocene ages, are believed to have given rise to the occurrence of the other members of the flora. In glancing, however, over the arctic plants of Newfoundland, the extreme eastern parts of Canada, and the adjacent coasts of Labrador, it is also somewhat noticeable how comparatively few of these high northern American forms descend, even with the increased facilities afforded now for migration, as far southwards as these districts. In a climate relatively of but little greater severity, we can accordingly conceive the high range which these American arctic plants must have also had in post-pliocene times, and how few could be expected to occur upon the then almost submerged mountain summits of New England.

In the number of this Journal before alluded to, reference was made to an apparent anomaly in the range of *Anemone parviflora* Michx., *Potentilla tridentata* Aiton, *Pinus Banksiana* Lambert, *Allium schœnoprasum* Linn., *Botrychium Lunaria* Swartz, and a number of other species, whose distribution in Canada seems to be confined to the northern coasts of Lakes Superior and Huron, and the Lower St Lawrence, with, at least in some instances, a range between these limits. Without referring to others whose intermediate diffusion is known, I may here mention that the little northern Scrub Pine alluded to has been met with by the Rev. J. K. Macmorine in a few localities in the southern sections of the County of Renfrew.  To the species cited might be added *Saxifraga Aizoon* Jacq., *Viburnum pauciflorum* Pylaie, *Aster graminifolius* Pursh, *Vaccinium Vitis-Idæa* Linn., *Primula farinosa* Linn., *P. Mistassinica* Michx., *Comandra livida* Richards., *Tofieldia palustris* Hudson, *Carex Vahlii* Schk., *Aspidium fragrans* Swartz, and many others.  I have already suggested the probability that the composition of the soil may, to some extent, affect the range of one of these plants, and it is just possible that the distribution of a few others may be modified by the same cause.  It is, however, an observable fact that whilst none of these plants is arctic or perhaps even sub-arctic in its aspect, all have a high northern range.  In the United States their distribution is limited to northern New England and Wisconsin, or to mountain sides and summits.  The vicinity of the lakes and the broad waters of the St. Lawrence, and their equalizing effects upon the temperature, account in part for the presence of the more boreal forms, and their general northern range for that of others.  The little Primulas occur on the American shores of Lakes Huron and St. Clair, but probably the winds, and especially the currents, have brought their seeds from the Manitoulin Islands and the upper shores of the former lake, where both species have been frequently met with by Dr. John Bell.  It may be mentioned that in the St. Clair River, especially where the waters of Lake Huron enter it, the current is very considerable.

Montreal, April, 1867.

---

# ON THE GEOLOGICAL FORMATIONS OF LAKE SUPERIOR.

By THOMAS MACFARLANE.

The crystalline rocks of Lake Superior present many features of interest to the lithologist, and to the student of primary geology; and the sedimentary rocks of that region, being almost destitute of organic remains, have been the subject of much discussion among scientific men, which can, nevertheless, scarcely be said to have settled unequivocally the question of their age.  Having, as I believe, observed certain new facts concerning the composition and association of these rocks, which are calculated to throw some light on their origin and age, I have attempted to describe them in the following paper.

Four different formations are distinguishable on the north, south and east shores of the Lake, where I have had an opportunity of examining their constituent rocks and mutual relations, but the same formations may be observed elsewhere in this region. These formations have been designated as follows: The Laurentian system, the Huronian series, the Upper copper bearing rocks of Lake Superior and the St. Mary sandstones. The two first-named (and older) formations usually occupy those parts of the shores which form high promontories and precipitous cliffs, and they constitute, almost exclusively, the areas which have been explored in the interior. On the other hand, the Upper rocks and St. Mary sandstones are never found far inland, but occur close to the shore in comparatively low-lying land and rocks. They seem to have had, as the theatre of their eruption and deposition, the bottom of the Lake, at a time when its surface was at a higher level than it is at present, although not so high as the general surface of the surrounding Laurentian and Huronian hills.

## I.—THE LAURENTIAN SYSTEM.

Under this name it has become usual, in Canada, to class those rocks which, in other countries, have been regarded as forming part of the primitive gneiss formation, of the primary or azoic rocks, or of certain granitic formations.

The most prevalent rocks of the Laurentian series on Lake Superior present a massive crystalline character, partaking much more of a granitic than of a gneissic nature. Some of these I shall endeavour to describe first. To the north of the east end of Michipicoten Island, on the mainland, there is a very large area of reddish-coloured granite, which exhibits, in a marked degree, the phenomena of divisional planes, and huge detached blocks. The rock is coarsely granular, has a specific gravity of 2·668 to 2·676, and consists of reddish orthoclase, a small quantity of a triclinic felspar, dark green mica (also in small quantity), and greyish white quartz. The mica is accompanied by a little epidote, and an occasional crystal of sphene may be detected. A few miles to the east of Dog River a grey granite occurs extensively, which does not show any divisional planes. The felspar of this variety is yellowish white, with dull fracture, and is fusible without difficulty. It is associated with black, easily fusible mica, in considerable quantity, and with quartz, which is occasionally bluish tinted.   The specific gravity of the rock is 2·750 to 2·763.   Large-grained granite is of very frequent occurrence on Montreal River and on the coast betwixt it and Point-aux⁻ Mines.   It consists principally of orthoclase, in pieces from one to several inches in diameter, a comparatively small quantity of quartz, and a ,still smaller proportion of white mica.   The promontory of Gros Cap, at the entrance of the Lake from River St. Mary's, is composed of coarse-grained and characteristic syenite.   In some places its hornblende is soft, seems decomposed, and is accompanied by epidote.   The rock is seldom free from quartz, and some of it contains so much as to be justly termed syenitic granite.   A chloritic granite appears to occur at a few points on the north side of Bachewahnung Bay, and a small-grained granite, consisting exclusively of felspar and quartz, occurs in large masses at the north-western extremity of the same Bay.   It has not the structure of granulite, and might be properly named aplite or granitelle.

These rocks are all unequivocally granular, without a trace of parallel structure.   They far exceed in frequency and extent those which possess a thoroughly gneissic character; indeed, characteristic gneiss was only observed at Goulais Falls and at Point-aux-Mines.   The rock of the latter locality varied from the closely foliated, resembling mica schist, to that of a granitic character. Granitic gneiss is found on the north shore of Bachewahnung Bay, between Chippewa River and Bachewahnung Village, on the road between the latter and the Bachewahnung Iron Mine, in the neighbourhood of the Begley Copper mine, and at other points on the north shore of Bachewahnung Bay.

Almost equal in frequency to these thoroughly granitic and gneissic rocks, there are found certain aggregates of rocks which present different lithological aspects almost at every step, and which can only be generally described as brecciated and intrusive gneissic, granitic, or syenitic rocks.   There is, however, to be detected a certain uniformity in the manner of their association with each other, which is of the greatest interest, and several instances of which it is now proposed to refer to.   On the north shore of the Lake, about twenty-five miles west of Michipicoten Harbour, one of these rock-aggregates may be observed.   Here fragments of a dark schistose rock, consisting of felspar and horn-blende (the latter largely preponderating), are enclosed in a coarse-grained syenitic granite, and both are cut by veins of another granite containing much less hornblende than the second-mentioned rock.   These veins are, in their turn, intersected by a vein of fine-grained granite, consisting of quartz and felspar, with traces only of mica or hornblende.   The specific gravities of these different rocks were found to be as follows:—

Hornblendic schist........................... 2·836
Syenitic granite................................. 2·787
Granite............................................ 2·608
Fine-grained granite....................... ...... 2·630

That the specific gravity of the last-mentioned rock should be greater than the one preceding, is attributable to its containing more quartz.   Figure 1 gives a representation of the phenomen t here observed.   No chemical analysis of these rocks is required to

Fig. 1.

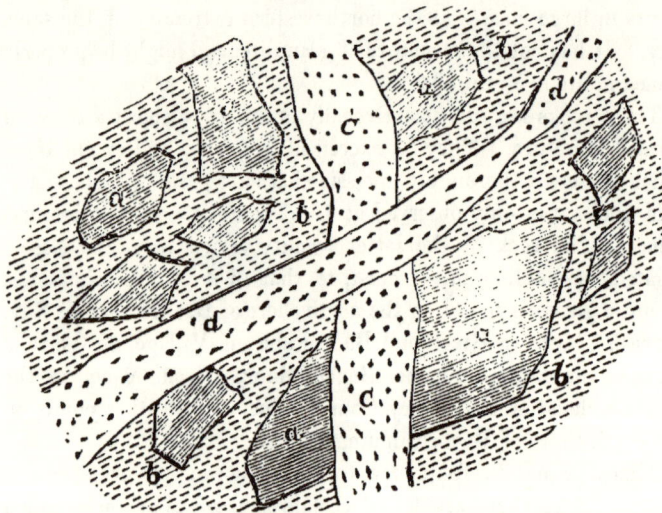

a. Fragments of hornblendic schist. | c. First intersecting granite.
b. Enclosing syenitic granite.       | d. Second intersecting granite.

show that the newer they are the greater are their contents in silica.   This is evident as well from their specific gravities as from their mineralogical composition.   The following relations, similar to these are observable on the north side of the Montreal River, at its mouth.   The prevailing rock here is small-grained granitic gneiss, which contains lighter and darker coloured portions, according as the black mica which it contains is present in smaller or larger quantity.   A triclinic felspar is also noticeable in it. Pieces of this rock are seen to be cut off and enveloped in a finer-grained granite, of a much lighter colour than the gneiss, and comparatively poor in the black mica. The specific gravity of the gneiss is 2·667, and that of the granite, 2·648. Veins of large-grained granite, containing very little mica, traverse both of the rocks just mentioned. The appearance of these rocks is shewn in Figure 2. At the falls of the Chippewa or

Fig. 2.

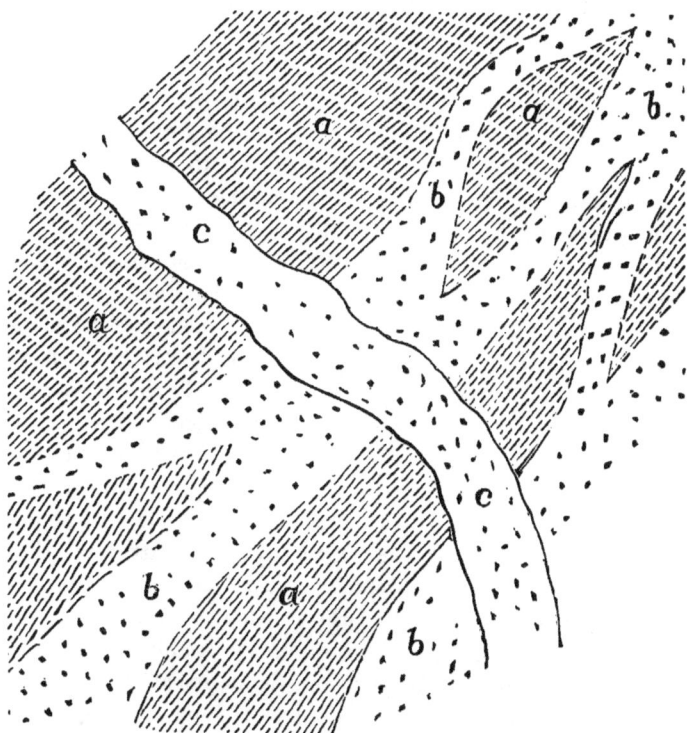

a. Granitic gneiss. | b. Fine-grained granite. | c. Large-grained granite.

Harmony River, which empties into Bachewahnung Bay, the predominating rock is highly granitic gneiss, consisting of reddish orthoclase, quartz and dark-green mica. It is rather small-grained, and, when observed in mass, shows sometimes a schistose appearance, the direction of which ranges from N. 10° W. to N. 57° E. Occasionally, in the more micaceous portions, broad felspathic bands occur, with selvages rich in mica, forming the nearest approach to gneiss. The direction of these bands is altogether irregular. This is also the case with veins of large-grained granite which intersect the rock just described. This granite consists mainly of red orthoclase, with a comparatively small quantity of quartz, with which a still smaller quantity of greenish mica is associated.    The specific gravity of the granitic gneiss is 2·676, and that of the coarse-grained rock of the veins 2·594.    On the north-east shore of the Bay, close to the landing place of the Begley Mine, rocks are observed consisting principally of granitic gneiss, in hand specimens of which, no parallel structure can be detected.    At some places, however, in larger masses, a schistose appearance is observable, with a strike of N. 75° E.    This rock, which is syenitic, contains masses and contorted fragments of gneiss very rich in hornblende.    Both the fragments and enclosing rock are intersected by veins of large-grained granite, containing little or no hornblende or mica.    In the most south-easterly corner of Bachewahnung Bay, rocks occur, which, although they are totally devoid of any approach to gneissic structure, and possess a very different composition, bear some resemblance in the manner of their association to those just described.    A dark-coloured, small-grained mixture of felspar and greenish-black mica, with occasional crystals of reddish orthoclase, and, more rarely, of greenish-white oligoclase, is enclosed in and intersected by another rock consisting of a coarsely granular mixture of orthoclase and soft dark-green mica, enclosing crystal of orthoclase (but no oligoclase) from one-quarter to three-quarters of an inch in diameter.    Both of the rocks might be called micaceous syenites, but as they possess a pdelorphyritic structure, they probably belong to the rock species called minette.    The matrix of the first-mentioned and darkest coloured rock is fusible, but the orthoclase which it encloses is less readily so.    In both rocks, where exposed to the action of the waters of the Bay, the micaceous constituent has been worn away, and the grains and crystals of orthoclase project from the mass of the rock    The specific gravity of the small-grained rock is 2·85, and that of the coarse-grained enclosing rock 2·65.    They are both intersected by narrow veins of granite, consisting of felspar and quartz only, the specific gravity of which is 2.62.    At Goulais Falls, about fifty miles up the Goulais River, gneiss occurs, which is very distinctly schistose, contains a considerable quantity—about one-third—of brownish black mica, interlaminated with quartzo-felspathic layers, in which a transparent triclinic felspar is observable.    The gneiss possesses a specific gravity of 2·74 to 2·76.    Its strike and dip are variable; the former seems, however, to average N. 55° E.,.

and the latter varies from 14° to 26° north-westward. It is interstratified with a small-grained granitic gneiss, containing much less mica than the last—about one-twentieth only,—no triclinic felspar, and having a specific gravity of 2·71 to 2·72. The same granitic gneiss intersects the characteristic gneiss in veins, and both of these rocks are cut by a coarse-grained granite, almost destitute of mica, and completely so of schistose structure. The strata of the gneiss are much contorted in various places. The intersecting granitic gneiss and granite are almost equal in quantity to the gneiss itself; and although they occur as irregular veins, they are, at the point of junction, as firmly united with the gneiss as any two pieces of one and the same rock could well be. Figure 3 is intended to represent the relations observable at Goulais Falls. Between Goulais Falls and the

Fig. 3.

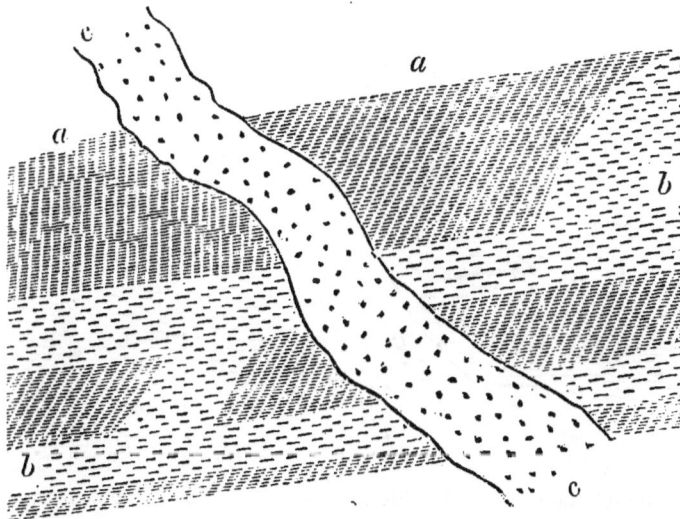

*a*. Gneiss.   |   *b*. Granitic gneiss.   |   *c*. Coarse-grained granite.

point where the line of junction between the Laurentian and Huronian rocks crosses Goulais River, there are numerous exposures of gneissoid rocks, but characteristic gneiss is of rare occurrence among them. At several places hornblende schist, in fragments, is observed enclosed in a gneissoid granite. Some of them are longer than others, and have their longer axes running N. 50° to 60° W. Hand specimens of the enclosing granite show little or no mark of foliation, but when seen in place, a faint parallel structure is observable, the strike of which is N. 50° to 60° W. Both the hornblendic fragments and the gneissoid granite are cut by veins of newer granite. On the south-east shore of Goulais Bay, a beautiful group of syenitic rocks is exposed, the mutual relations of which are similar to those above described. Fragments of hornblende rock or schist, varying from half-an-inch to three feet in diameter, are enclosed in a coarse-grained syenitic granite, in which, occasionally, a rough parallelism of the hornblende individuals is observable, the direction of which is N. 57° E., and coincides with that of the longer axes of the hornblendic fragments. The specific gravity of the hornblendic rock is 2·94 to 3·06, and of the enclosing granite 2·74. Both are intersected by a coarse-grained granite, having a specific gravity of 2·61 only, and containing little or no hornblende or mica. The appearance here described are represented by Fig. 4.

Fig. 4.

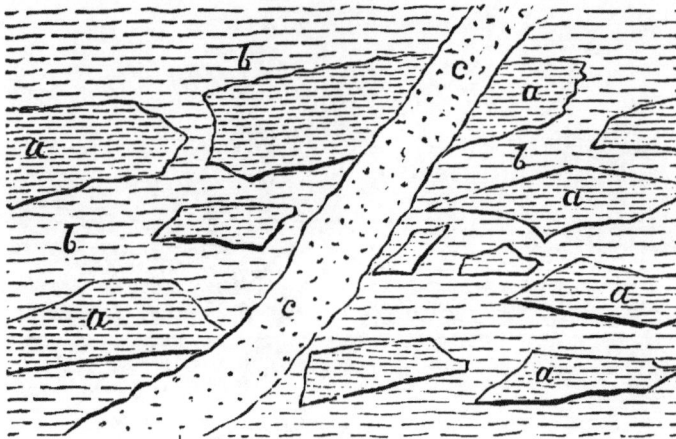

*a*, Hornblende schist. *b*, Syenitic gneiss-granite. *c*, Coarse-grained granite

The mutual relations of these brecciated and intrusive rocks in eight different localities, some of them upwards of one hundred miles apart, have here been described, and it will be observed that, in every one of the instances mentioned, the oldest rock is the most basic in constitution, and this appears to be the case, without regard to the mineralogical composition or structure of the rocks associated together as above described. It matters not whether the older rocks be brecciated or entire, hornblendic or micaceous, granular, schistose or porphyritic, it is always most deficient in silica. It appears, further, that the newer the rock

which encloses or penetrates older ones, the more siliceous it
becomes.  On reference to the specific gravities above given of
the various rocks, it might be supposed that their relations as to
age might be equally well expressed by saying, the older the rock
the heavier ;  the more recent, the lighter it is;  and, in the
majority of instances, this applies.  But, as in the case of the rock-
aggregate occurring to the west of Michipicoten Harbour, when
we come to the very newest granitic veins, consisting only of orthoclase and quartz, those are the heaviest which contain most of the
latter mineral, its mean specfic gravity being 2·65, while that of
orthoclase is only 2·55.  It is to be remembered that these newest
veins are altogether different in appearance from certain veins of
large-grained granite, with distinct side joints, which are occasionally found intersecting these rocks, and the origin of which has
been indicated by Dr. Hunt in his recent valuable report on
mineral veins.  Near Point-aux-Mines a vein of this nature is
found, the rock of which is pegmatite, consisting of orthoclase,
quartz, and greenish white mica, together with occasional grains
of purple copper, copper pyrites, galena, and molybdenite.

It may not be out of place here to advance certain considerations regarding these Laurentian rocks, and especially concerning
the peculiar rock aggregates just described.  The relations of
these rocks to each other we have seen to be as follows :—The
older the rock the more basic is its nature, and the richer it becomes in triclinic felspar, hornblende, and mica.  The newer the
rock the more siliceous it becomes, and the more such minerals as
orthoclase and quartz predominate.  It can scarcely be supposed
that this relation is an accidental one, for it is observable in every
one of the instances above given, the localities of many of which
are very far distant from each other.  It would seem to be the
consequence of an unvarying law which was in operation at the
time when these rocks were first formed.  At first sight, the facts
above described would appear to militate against the idea of the
igneous origin of these rocks, and, in fact, the relation is a similar
one to that which has been observed among the constituent
minerals of granite, and which is one of the chief difficulties in
explaining the origin of that rock on the igneous hypothesis.  In
granite the quartz is frequently found filling up the interstices
between the other minerals, and sometimes it even retains impressions of the shape of the latter.  Nevertheless the felspar and
mica are the most fusible, and the quartz the most infusible of

the constituents of granite. Similarly, the older basic rocks, among the brecciated and intrusive aggregates above described, are the most fusible, while the newer rocks, being most siliceous, are most infusible. At first sight, it is difficult to conceive how a basic and fusible rock could solidify from a melted mass previous to a more siliceous one. But the geological relations of these rocks are such as to afford the fullest proofs of their igneous origin. It may be urged that such an origin for the oldest and more basic fragments does not appear proved, but their similarity in mineralogical composition with the intrusive members of the aggregate is in favour of such a view. Furthermore, these older fragments show, in every instance, such an analogy as regards their relation to the intrusive rocks that they cannot be regarded as accidental fragments of other rocks brought from a distance. If their origin were of this nature, they would not invariably be more basic in composition than the enclosing rock. The fact of their always bearing a certain relation, as regards composition, to the enclosing rock renders it unlikely that their source is similar to that of boulders in a conglomerate or fragments in a breccia. On the contrary, it would appear more reasonable to regard them as the first products of the solidification of the fluid mass from which the granites, and other rocks above described, resulted. In pursuing this subject further, it would appear not unreasonable to base some such theory as the following upon the facts above stated. The area now covered by these rocks must at one time have been occupied by a mass of fused silicates. The temperature of this fluid magma and of the surrounding crust has been intensely high, although perhaps very gradually on the decrease, and the extent of the igneously fluid material must have been such as to render uniformity in its chemical composition an impossibility. Variations in its composition, as well as in the manner of its solidification, may therefore be supposed to have obtained in different parts of the fluid area. According to the proportion of silica and bases present where crystallisation commenced and progressed, hornblendic rock, mica syenite, or comparatively basic granite, first assumed the solid form, leaving a part of the fluid or magma beneath or on the outside of it still in a plastic state, but changed in its chemical composition, and rendered more siliceous than the original magma. If the solidification commenced at a point where the fluid mass was comparatively undisturbed, the granular varieties of the rocks above described may have been produced.   If, on the other hand, the solidification took place while the fluid mass was in motion, the hornblendic and micaceous schists and gneisses were most probably the results of this process, and the strike of these would indicate the direction of the current at the time of their formation.   The rarity or indistinctness of parallelism in the Laurentian rocks of Lake Superior shews, however, that no very constant and persistent motion in one direction took place in the fluid mass which produced them.   This first solidification of part of the fluid magma most likely continued for a long period, and spread over a large surface; but there seems at last to have arrived a time when, from some cause or other, these first rocks became rent or broken up, and the crevices or interstices became filled with the still fluid and more siliceous material which existed beneath them.   Gradually, this material solidified in the cracks, or in the spaces surrounding the fragments, and the whole became again a consolidated crust above a fluid mass of still more siliceous material.   Further solidification of this latter material doubtless then took place, and continued until a second general movement of the solidified crust opened other and newer crevices, which became filled with the most siliceous material which we see constituting the newer veins among the rocks above described.

Although the theory here given as to the origin of these rock aggregates is in thorough harmony with the facts related concerning them, it is doubtless possible to urge objections against it founded upon the relative fusibility of their constituent rocks. There is no doubt that the point of temperature at which these various rocks become fluid under the influence of heat is higher with the newer than with the older rocks, but it does not follow that in cooling they solidify, that is, become quite hard and solid at the same point of temperature at which they fuse.   Bischof describes an experiment which proves that the temperature at which certain substances solidify does not at all correspond with their fusing point.   He prepared a flux, consisting of common glass and carbonate of potash, which fused at a temperature of 800° R., and melted it along with some metallic bismuth in a crucible.   This metal fuses at 200°, and solidifies with a very uneven surface, on account of its tendency to crystallize.   Although the difference between the fusing point of the bismuth and of the flux amounted to 600°, nevertheless, when the crucible cooled, all the irregularities of the surface of the metal were found to have imprinted themselves upon the lower surface of the solidified flux, a very plain proof being thus furnished that at a temperature of 200° R., the flux was still soft enough to receive the impression of the solidifying metal. If we further observe the various fused slags which flow from different furnaces, we shall obtain some idea of the manner in which the rocks above described may have behaved during their solidification. The scoriæ of iron furnaces are usually very acid, containing as much as 60 per cent. of silica. They generally fuse at a temperature of 1450° C. As they flow out of the breast of the furnace, they may be observed to do so very leisurely, to be sluggish and viscid, but nevertheless to continue fluid a long time, and even in some cases to flow out of the building in which they have been produced, before solidifying. On the other hand, slags from certain copper furnaces, or from those used for puddling iron, are more or less basic, containing from 30 to 45 per cent. silica. As they flow out they are seen to be very fluid, and to run quickly, but they solidify much more rapidly than iron slags. Yet these basic slags fuse at about 1300° C., or about 150° less than the more acid slags. Those who have been accustomed to observe metallurgical processes will not find it difficult to conceive how a very siliceous slag might continue fluid at a temperature at which a more basic one might become solid. We conceive, however, that the rocks which we have described must have solidified under circumstances altogether different from those under which furnace slags cool. We suppose that these rocks must have solidified at temperatures not very far below their fusing points; that the temperature of the atmosphere, and of the fluid mass itself, had sunk somewhat beneath the fusing point of the more basic rocks before solidification began, and that at this point it was possible for the basic rocks to crystallize, while a more siliceous magma still remained plastic. This latter supposition does not appear unreasonable when the experiment above referred to, and the behavior of furnace slags above described, is taken into consideration.

It becomes a question of much interest as to whether these rocks are to be regarded as constituting one and the same, or several and distinct, geological formations. There cannot be a doubt as to the fact that some of them are of more recent origin than others; but, on the other hand, many of the veins above described do not present such distinct joints as are visible where trap or basalt dykes traverse sedimentary strata. Although the cementing material

of the brecciated rocks above described differs in composition from
the fragments which it encloses, we nevertheless find that the two
are usually so intimately combined with each other as to behave
under the hammer like one and the same rock.   There is, in the
majority of cases, no joint to be found at their junction with each
other; and in fracturing them, they very often break just as
readily across as along the line which separates them.   It would
appear, therefore, that, although these rocks solidified at different
times, the dates of their formation were not sufficiently far
distant from each other to enable the previously existing rock to
cool thoroughly before it became penetrated by or enclosed in the
newer one; that consequently the older rock, being in an intensely
heated condition, readily amalgamated at its edges with the next
erupted and fused mass, and formed with it a solid compact whole.
Apart from the difficulties which would doubtless attend any
attempt to distinguish separate geological groups among these
rocks, it would appear just as unreasonable so to separate them, as
to regard each distinct stratum of sedimentary rock as distinct
geological formations.   According to Naumann, a geological
formation consists of a series of widely extended or very numerous
rocks or rock-members (*Gebirgs-glieder*), which form an independent whole, and are by their lithological and palæontological
characters, as well as by their structure and stratigraphical succession (*Lagerungs folge*), recognisable as contemporaneous (geologically speaking) products of similar natural processes.   According
even to this definition, it would appear just to class all the rocks
above described, in spite of the distinctly intrusive character of
some of them, as belonging to one and the same geological formation,—in short, to the Laurentian series of Sir W. E. Logan,
or the Primitive Gneiss formation of Naumann.   The last named
geologist certainly distinguishes a separate granite formation, but
the rocks included in it are generally more recent than the primitive gneiss or primitive schists.   Where, as in Silesia, in Podolia on
the Dnieper, in the central plateau of France, in Finland, in Scandinavia, and in the Western Islands of Scotland, granite occurs
in similar intimate association with gneissoid rocks as on Lake
Superior, Naumann always regards it as part and portion of the
primitive gneiss.   As early as 1826, Hisinger, in his work on
Swedish mineralogy, shewed that the granite which occurs in
intimate combination, by lithological transition and otherwise, with
the primitive gneiss of Scandinavia, was of contemporaneous origin

with it; and in the Pyrenees, La Vendee, Auvergne, the Black Forest and Hungary, according to Coquand, Riviere, Rozet, Rengger, and Beudant respectively, the gneiss and granite of these countries cannot be separated into distinct formations, but form one and the same mass of primitive rock.

## II.—THE HURONIAN SERIES.

The rocks of this system, as developed on Lake Superior, present at first sight rather a monotonous and uninteresting aspect to the student of lithology. Large areas are occupied by schistose and fine-grained rocks, the mineralogical composition of which is, in the most of cases, exceedingly indistinct. These rocks are, to a very large extent, pyroxenic greenstones and slates related to them. On closer examination, they are found to exhibit many interesting features, and it is possible to distinguish among them the following typical rocks:—

*Diabase.*—The granular varieties among these greenstones belong to this species. It is developed at several points on Goulais River, at some distance to the west of the Laurentian rocks already referred to. It is usually fine-grained, pyroxene is the preponderating constituent, and chlorite is present in considerable quantity in finely disseminated particles. The felspar is in minute grains, and, in many instances, it is only on the weathered surface of the rock that its presence can be recognized. One variety of this rock from the Goulais River has a specific gravity of 3·001. Its colour is dark green, and that of its powder light green. The latter, on ignition, lost 2·29 per cent. of its weight, and changed to a brown colour. On digestion with sulphuric acid, 22·99 per cent. of bases were dissolved from it, which circumstances would seem to indicate that the felspathic constituent is decomposable by acids, and is therefore, in all likelihood, labradorite. This rock is underlaid to the south-west by greenstone schist, striking N. 65° W., and dipping 75° north-eastward, and is overlaid by amygdaloidal diabase and greenstone slates, striking N. 66° W., and dipping 49° north-eastward. Granular diabase is also met with a few miles higher up the river from the rocks just mentioned, associated with porphyritic diabase and diabase schist, the latter striking N. 55° to 65° W., and dipping 60° north-eastward. Similar rocks were observed on the hills between Bachewahnung and Goulais Bay, and at several points on the north shore of the lake between Michipicoten

Harbour and Island. In the neighbourhood of, and on the road to, the Bachewahnung Iron Mine, they are also plentiful. Not unfrequently the pyroxene in them assumes the appearance of diallage.

*Augitéporphyry.*—The porphyritic diabase above referred to is a small-grained diabase, in which are disseminated crystals of pyroxene, about three-eighths of an inch in diameter. The specific gravity of the rock is 2·906. Its fine powder has a light greenish grey colour, which changes on ignition to dark brown, 2.01 per cent. of loss being at the same time sustained. Hydrochloric acid dissolves from it 23·48 per cent. of bases.

*Calcareous Diabase.*—The amygdaloidal diabase above mentioned is the same rock as is termed by Naumann *Kalkdiabase.* It is a fine-grained diabase, somewhat schistose, in which oval-shaped concretions of granular calcspar occur. The latter are not, however, always sharply separated from the mass of rock, which is slightly calcareous. The amygdules, if such they can be called, have their longer axis invariably parallel with each other, and with the schistose structure of the rock.

*Diabase Schist.*—This rock occurs much more frequently than either of those just described. It is, indeed, difficult to find a diabase among these Huronian rocks which does not exhibit a tendency to parallel structure, or which does not graduate into diabase schist. But the latter rock occupies considerable areas by itself, not only on Goulais River, but also on that part of the north shore referred to in this paper. The higher hills to the north-east of Goulais Bay consist, to a large extent, of this rock. Apart from its schistose structure, it possesses the characters of diabase. For example, a specimen of the rock from the north shore has a specific gravity of 2·985. Its powder, which is light grey, changes on ignition to light brown, losing 1·43 per cent. of its weight. On digestion with hydrochloric acid, it loses 14·24 per cent. of bases; and with sulphuric acid, 16·12 per cent. It is fusible before the blow-pipe. Many of these schists are pyritiferous and calcareous, and these graduate frequently into greenstone slate.

*Greenstone and Greenstone Slate.*—The rocks above mentioned, being small-grained, are recognizable without much difficulty; but, besides these, and occupying much more extensive areas, there occurs finely granular and schistose rocks, many of them doubtless of similar composition to the above mentioned diabase and diabase schist. Where the transition is traceable from the latter rocks to those of a finer grain, the same names are perhaps applicable. But since this is not always the case, it would seem advisable to make use of other terms for them until their composition is more accurately determined. The names aphanite and aphanite slate have been applied to rocks such as these, but since the former term has been applied by Cotta to compact melaphyre, it would seem better for the present to continue the use of the other terms, compact greenstone and greenstone slate, especially since the signification of the first of these has been so limited by Naumann as to denote pyroxenic greenstones only, thus distinguishing them from the hornblendic greenstones or Diorites. These pyroxenic greenstones, or fine-grained diabases, frequently contain more chlorite than the coarser-grained varieties. They are very frequent on the Goulais River, in the district between it and Bachewahnung Bay, and in the neighbourhood of the Bachewahnung Iron Mine. One specimen from a point four miles north-east of Goulais Bay yields 21·44 per cent. of bases to sulphuric acid. Its powder is dark green, changing on ignition to dark brown, and losing 1·72 per cent. of its weight. These greenstones are seldom destitute of iron pyrites. Quartz never occurs in them as a distinct constituent, and even in veins it is rare; but there are a few occurrences of greenstones which are lighter in colour, more siliceous, and harder than others, and which have possibly become so by contact with quartzose rocks. On the other hand, they are frequently found impregnated with calcareous matter. By assuming a schistose structure, these greenstones often graduate into greenstone slate, an apparently homogeneous rock, generally of a dark greenish grey colour and slaty texture. The latter character is sometimes so marked, that it becomes difficult to distinguish it from clay slate. The greenstone slates however, would seem to differ from the latter rock in the small quantity of water which they contain, their generally higher specific gravity, and in their yielding nothing which would form a good roofing slate. On the other hand, they are related to the greenstones and diabase schists not only by gradual transition, but in some of their physical characters. For instance, a greenstone slate from Dog River, on the north shore, of a dark grey colour, has a specific gravity of 2·738, and loses 1·62 per cent. of its weight on ignition, in which operation the colour of its powder changes from a greenish white to a decided brown. It yields to hydrochloric acid 16·44, and to sulphuric acid 10·29 of bases.

*Siliceous Slate.*—In many places bands of such dark coloured slate as that just described are interbedded with others which are lighter coloured and more siliceous. Such banded slates may, for instance, be observed on the north-east shore of Goulais Bay. Here the darker slate is very evenly foliated, of a dark greenish-grey colour, and has a specific gravity of 2·685. Its powder is light green, changing on ignition to light brown, and losing 2·02 per cent. of its weight. It yields to sulphuric acid 16·75 of bases. The rock of the lighter bands is highly siliceous, and in fusibility equal to orthoclase. The powder has a reddish grey colour, which changes on ignition to brownish grey, 0·54 per cent. of loss being at the same time sustained. Hot sulphuric acid removes only 3·79 per cent. of bases. A similar association of slates is found at a point bearing 41° 30' E. from the east end of Michipicoten Island. Here, a series of lighter and darker coloured bands of very decided slate occur, striking N. 78° to 86° W., and dipping 50° to 52° northward. They are overlaid by a band of dark green slate, which contains granitic pebbles, and this band is again overlaid by light coloured slates. Small bands may be observed to leave the dark green slates and to join with those of a lighter colour. The latter are not only lighter in colour, but harder and less dense, and occasionally show on their cleavage planes a silky lustre. A specimen gave a specific gravity of 2·681, and its powder, which was almost quite white, lost 1·12 per cent. on ignition, becoming slightly brown. It fuses only in fine splinters, and, generally, the fusibility of these slates is the greater the darker their colour.

*Chlorite Schist.*—Some of the greenstone slates occasionally contain an unusually large quantity of chlorite, and sometimes so much as to form chlorite schist. This schist forms the side rock of the Palmer Mine on Goulais Bay.

*Quartzite.*—This rock is of less frequent occurrence than I had anticipated. It is most frequent on the west and south-west side of the hills between Bachewahnung and Goulais Bay, and in the district north-eastwards from Sault Ste. Marie.

*Hematite.*—This mineral often occurs in such quantity as to constitute rock masses. It will however be referred to under the economic minerals of the series.

*Greenstone Breccia.*—The occurrence of angular fragments of other rocks in the greenstones above described is by no means rare, and the resulting breccias are common between Bachewahnung and Goulais Bays. In the majority of instances where the matrix is granular, the fragments are angular; on the other hand, where the matrix becomes schistose, the fragments are generally rounded, and there results the slate conglomerate so characteristic of the Huronian series.

*Slate Conglomerate.*—This rock is extensively developed at the mouth of the Dore River, some distance to the west of Michipicoten Harbour. Its matrix is the greenstone slate above described. The boulders and pebbles which it encloses seem, for the most part, to be granite, and are rarely quite round in form. The most of them are oval or lenticular shaped, and then their outlines are scarcely so distinct as in the case of those which approach more closely to the round form. Very frequently those of a lenticular form are drawn or flattened out to such an extent that their thickness decreases to a quarter or half-an-inch, and they are sometimes scarcely distinguishable from the slate, except by their lighter colour. Part of the rock exhibits merely a succession of lighter and darker coloured bands, the former of which sometimes resemble in form the flattened pebbles above-mentioned. On account of the presence of these lighter bands, it is often impossible to select a piece which may be regarded as the real matrix of the rock. As in the case of some of the rocks above described, the light bands are more siliceous and less dense than the darker ones. The latter are, not unfrequently, calcareous. A specimen of this character had a density of 2·768 to 2·802. Its powder was light green, which changed on ignition to light brown, with a loss of 2·75 per cent. On treatment with sulphuric acid, it effervesced strongly, and experienced a loss of 36·85 per cent. Iron pyrites impregnates the matrix quite as frequently as calcareous matter. The direction of the lamination in the matrix is parallel with the longer axis of the lenticular pebbles, and where the boulders are large (they seldom exceed twelve inches in diameter) and round, the lamination of the slate winds round them, and resumes its normal direction after passing them. Occasionally a flattened pebble is seen bent half round another, and, among the very thin pebbles, twisted forms are not uncommon. The nature of the pebbles, especially of those which have been flattened, is sometimes very indistinct. The quartz is generally easily recognized in the larger boulders, but the felspar has lost its crystalline character, and the mica is changed into dark green indistinct grains, where it has not altogether disappeared. Besides the granitic pebbles, there are others which seem to consist of quartzite. An idea of the structure of this rock is attempted to be given in figure 5.

Fig. 5.

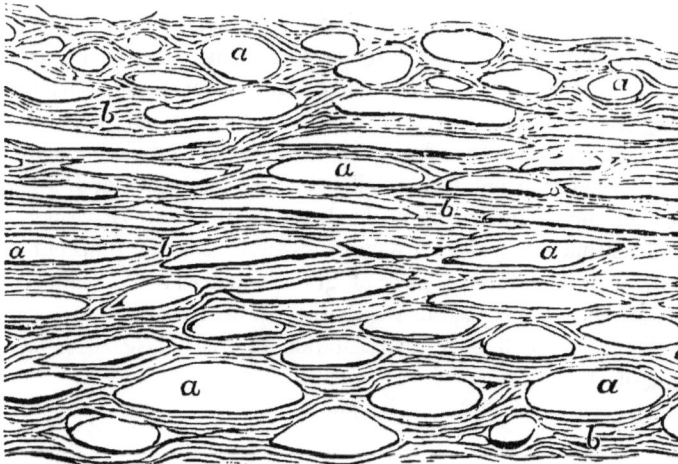

*a.* Granite boulders, and long drawn masses. *b.* Schistose matrix.

The manner in which these rocks are occasionally associated with each other is calculated, as in the case of the Laurentian rocks, to suggest to the observer some definite ideas regarding their origin. Equally instructive is the manner in which they adjoin the Laurentian areas at several points on the north shore, between Michipicoten Harbour and Island. I paid some attention to that point of junction which lies to the west of Eagle River, the precipitous cliffs to the east of which consist principally of diabase schist and greenstone slate. A few miles to the west of these cliffs, and at a point bearing N. 29 ° E. from the east end of Michipicoten Island, the Laurentian granite is penetrated by enormous dykes of dense basaltic greenstone (having the peculiar doleritic glitter when fractured), which contain fragments of granite. This greenstone is also seen in large masses, which can scarcely be called dykes, overlying the granite and enclosing huge masses of that rock, one of which I observed to be cut by a small vein of the greenstone. From this point to Eagle River those two rocks alternately occupy the space along the shore, seldom in such a manner as to show any regular superposition of the greenstone on the granite, but almost always more or less in conflict with each other. The greenstone, however, becomes more frequent towards the east, and at Eagle River it has almost wholly replaced the granite, and assumed a lighter colour and an irregular schistose

structure.   The strike of these schists is, at places, quite incon-
stant; they wind in all directions, and what appear, at first sight,
to be quartz veins, accompany their contortions.   On closer
inspection, however, of the largest of these, they are seen to be of
granite, but whether twisted fragments of that rock or really veins
of it, is, at first glance, very uncertain.   Observed superficially,
they have the appearance of veins, but they do not preserve a
straight course, and bend with the windings of the enclosing schist.
They often thin out to a small point and disappear, and, a few
feet or inches further on in the direction of the strike, reappear
and continue for a short distance.   Sometimes a vein thins out at
both ends and forms a piece of granitic material of a lenticular
shape, always lying parallel with the lamination of the enclosing
slate.   Figure 6 is a representation of the phenomena here
described.

Fig. 6.

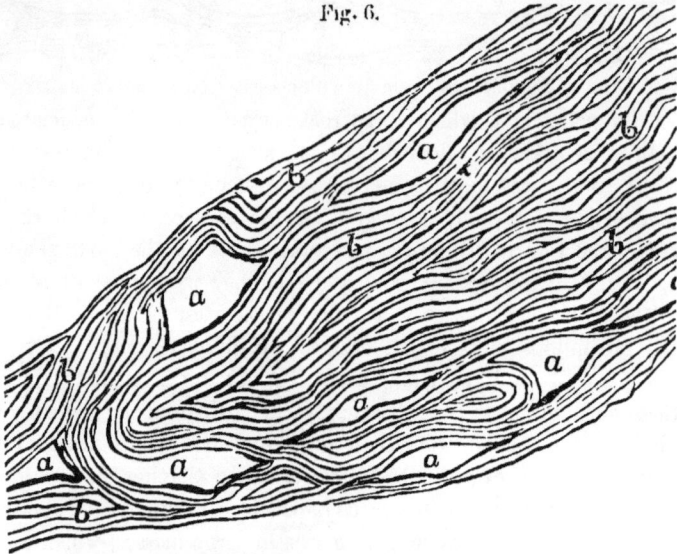

*a*. Fragments and contorted pieces of granite.
*b*. Slates enclosing same.

At another point of junction, on the north shore, to the east of
that above described, there is a large development of similar
basaltic greenstone.   Its constituents, with the exception of iron
pyrites, are indistinguishable; it has a greenish black colour, and
a specific gravity of 3.   Its powder has a dark green colour, which
changes on ignition to dark brown, with a loss of 1·79 per cent. of
its weight.   It yields to sulphuric acid 18·41 per cent. of bases.

It exhibits numerous divisional planes and a tendency to slaty structure, the direction of which is not, however, parallel with that of the divisional planes. It contains numerous fragments and long drawn contorted masses of granite, which are best discernible on the worn surface of the rock, and not readily so where it is freshly fractured. To the eastward it changes to a much harder light grey siliceous rock, having a specific gravity of 2·709 only. In fine powder this rock is white, but on ignition becomes brownish, and loses 0·55 per cent. of its weight. It yields only 4·62 per cent. of bases to sulphuric acid. At one place it seems to contain fragments and twisted pieces of the dark greenstone, and further eastward it assumes the character of a breccia, granite fragments being enclosed in the slaty rock, which is at some points darker, at others lighter, coloured. The fragments are sometimes quite angular, and sometimes rounded off, and not sharply separated from the matrix. Their longer dimensions are invariably parallel with the lamination of the matrix. The distance over which the transition extends renders it impossible to give any accurate sketch of the phenomena described.

Similar relations are observable at the junction of the two formations in the north-east corner of Bachewahnung Bay. Here the greenstone is compact, but still possesses the glittering basaltic fracture. The Laurentian rock is a highly granitic gneiss, and pieces of it are enclosed in the dark greenstone, which at one place seems to underlie the granite. A reddish grey felsitic rock, with conchoidal fracture, is observed at the point of junction. Eastward from it banded traps occur, striking N. 55° W., together with greenstone—breccia, and conglomerate. On ascending the hills behind this point another breccia is observed, of which the matrix is greenstone and the fragments granite.

With regard to the succession of these rocks, it will doubtless be found a matter of very great difficulty to establish any such, even if any order of superposition of a tolerably regular character should exist among them. That this is not very likely to be the case, will appear from the considerations yet to be advanced regarding the origin of these rocks. As to their general strike, it is scarcely possible to give any such, but within certain limits a tolerably constant strike may be observed. In the Huronian area, betwixt Goulais River and Bachewahnung Bay, although there are occasional north-easterly directions, the strike generally ranges from N. 40° to N. 80° W. On the north shore it is generally east and west, seldom deviating more than 20° to the north or south of these points. The following observations were made in the neighbourhood of Eagle River, at points where the slates appeared most regular: N. 83° E., dip 45° northward; N. 80° W., dip 46° northward; N. 45° E., dip 34° north-westward.

In the foregoing description an attempt has been made to delineate with fidelity the most important features of the Huronian formation as developed on Lake Superior. It is now proposed to give a fair unstrained interpretation of the characters stamped upon the rocks of that series. The fact of the Laurentian granite being pierced, as above described, by Huronian rocks, and the fact of their enclosing fragments of such granite, proves incontestably that some of them are of eruptive origin, and of later age than the Laurentian series. The enclosure of the huge sharply angular fragments of granite in the very basic greenstone, above described, stands in intimate connection with the enclosure of smaller and contorted granite fragments in a matrix of similar chemical composition, but different (slaty) structure. The appearances visible near Eagle River, of which figure 6 is an illustration, prove that enclosed granitic fragments sometimes undergo modifications of form through contact with certain Huronian rocks. In Foster and Whitney's Lake Superior Report (Part II., pp. 44 and 45), analogous phenomena are described, but the exactly opposite conclusion is arrived at, viz., that the granite is in the form of veins, and is the newest rock. There would seem to be only the two methods of explaining the facts described : either the granite forms veins penetrating the schistose greenstones, in which case the latter are the oldest rocks, or it is in the form of contorted fragments, in which case the enclosing rocks must be of eruptive origin. The fact that the granitic fragments do not cut but run parallel with the slates which enclose them, is the strongest argument against considering them to be veins. The supposition that they are long drawn and contorted fragments seems to be most in harmony with the facts stated, and with what is known as to the relative ages of the Laurentian and Huronian rocks. The true explanation most likely is, that the basic greenstone, after enveloping the granitic fragments, continued for some time in motion, and, previous to solidification, softened and rendered plastic the fragments, which then became drawn out in the direction of the flow of the igneous mass, and forced to accompany its sinuosities, and that the motion of the fluid mass previous to and during solidification developed in the greenstone its schistose structure.  The other facts, described above as observable at a considerable distance east of Eagle River, shew that something more than a mere modification of form is caused by the action of basic greenstone upon granite fragments. Not only are the latter there observed to be enclosed in, softened by, and twisted around with the greenstone, but the phenomena observed fully justify the supposition that they have been dissolved in it, that is to say, actually fused in and incorporated with its material.  The fragments are seen to be firmly joined together with the enclosing rock, especially where the latter becomes more siliceous.  Furthermore, their sharp angles are often rounded off, indicating plainly that these parts were first melted away by the fluid greenstone.  Moreover, the product of the union of the latter with the dissolved parts of the granite is plainly visible.  It is the siliceous slate rock described above as forming in places the matrix of the breccia.  This siliceous rock, the specific gravity of which is much lower than that of the greenstone, is further seen to be twisted about with the latter in such a manner as, in its turn, to envelope parts of the greenstone, thus shewing that motion assisted the incorporation of the two.  The reddish grey felsitic rock, mentioned as occurring at the junction of the two formations in the north-east corner of Bachewahnung Bay, has doubtless had a similar origin to t'at of this siliceous rock, and it is not unlikely that the banded traps and slates, so frequently found among Huronian rocks, are attributable to a similar mode of formation. Closely connected with the breccias just alluded to, so far as regards the cause of its peculiar structure, is the Huronian slate conglomerate.  It is impossible to examine closely this rock without being impelled to the conclusion that its origin is not very different from that of the breccias; that its matrix has been a fused mass, flowing slowly but constantly in the one direction; and that its boulders are merely fragments which have been half melted and rounded off by contact with the igneous rock.  The oval, twisted, lenticular and long drawn forms of the boulders are such as could never have been produced by ordinary attrition, and they frequently furnish examples of such intimate amalgamation with the matrix as are never found in aqueous conglomerates. Further, the fact of the boulders being frequently drawn out into what are simply bands of light coloured slate, not only disproves the sedimentary origin of the conglomerate, but indicates the manner in which the association of greenstone slate and siliceous slate above described have been formed. They have simply been produced where no tumultuous motion was at hand thoroughly to incorporate the material of the greenstone with that derived from the softened fragments, but where a steady continuous motion, always in the one direction, drew out the materials of the different slates into long bands side by side with each other. It thus seems to us reasonable, and quite compatible with a scientific interpretation of the facts above given, to explain the origin of by far the greater number of the above enumerated Huronian rocks upon a purely igneous theory; and it has occurred to us that many of the instances of local metamorphism, recorded by geologists, in which the contact of an igneous rock caused the silicification or lamination of another, might be capable of thorough explanation in a manner similar to that in which we have tried to account for the origin of the breccias, conglomerates, siliceous greenstones and banded slates, which constitute such a large part of the Huronian series.

The Huronian series, whatever its mode of origin may have been, must undoubtedly be regarded as an independent geological formation. It has been represented as being " a mixture of the St. Alban's group of the upper Taconic with the Triassic rocks of Lake Superior, the trap native-copper bearing rocks of Point Keeweenaw, and the dioritic dyke containing the copper pyrites of Bruce mine on Lake Huron " * but surely such a description is based upon a misconception of Sir W. E. Logan's views on the subject. Until its discovery by Sir William, the Huronian formation was unknown to geologists as a separate and independent system, and even now it is only in comparatively few countries besides Canada that it has been shown to exist. On a former occasion, in the columns of the *Naturalist* † I endeavoured to shew that the Azoic schists of Tellemarken, in Norway, were almost identical in lithological characters with the Huronian rocks, and Dr. J. J. Bigsby ‡ shortly afterwards insisted upon the fact of their being the same formations. Dr. Bigsby is of opinion that the Huronian also occurs on the Upper Loire, in France, and that it is a totally distinct formation from the Cambrian, with which it has hitherto been customary to associate it. The Huronian forms part of what Naumann calls the primitive slate formation.

---

* Marcou; The Taconic and Lower Silurian Rocks of Vermont and Canada.

† Vol. vii, p. 113.

‡ Quart. Journ. Geol. Soc. Vol. xix, p. 49.

Besides the black and greenish black dykes which occur in the neighbourhood of, and stand in connection with, Huronian rocks, there are others which occur at a distance from Huronian areas, and whose rocks differ somewhat from those of that formation. This is the case, for instance, with a set of dykes which occur on the south-east shore of Goulais Bay, cutting Laurentian rocks. They are there separated from the gneissoid rocks by very distinct joints. They vary in thickness from nine to seventy feet, and strike N. 72° to 75° W. In the widest veins the rock is fine grained at the side and small grained in the centre, so that even there it is difficult to determine its constituents. They seem, however, to be dark green pyroxene and greyish felspar, with magnetic and minute grains of iron pyrites. The rock has a specific gravity of 2·974. Its powder, from which a magnet extracts magnetite, has a grey colour, which changes on ignition to a dirty brown, with a loss in weight of 1·67 per cent. Hydrochloric acid produces no effervescence, but removes 21·74 per cent. of bases. Sulphuric acid removes 20·83 per cent. The presence of magnetite and absence of chlorite would seem to indicate that the rock inclines more to the nature of dolerite than diabase. A similar vein of fine grained rock penetrates the syenite of Gros Cap, on the summit of that hill, striking N. 40 ° W. A very large mass of small grained doleritic rock likewise occurs at the mouth of the Montreal River, on its south bank. It probably forms a dyke of very large dimensions in the granitoid gneiss there. It consists, seemingly, of black augite, white or greyish white felspar (on some of the cleavage planes of which parallel striæ are distinctly observable), and magnetite. Its specific gravity is 3·090. Its powder yields magnetite to the magnet, and does not effervesce on treatment with sulphuric acid, which removes 11·15 per cent. of bases. Other dykes of this nature cut the reddish granite of the north shore opposite Michipicoten Island, and, nearer to Michipicoten Harbour, a sixty feet dyke of diorite cuts the grey granite. It is fine grained at the sides, but granular and even porphyritic in the centre. Its direction is N. 63 ° E. About a mile further east another dyke occurs, which seems to contain fragments of granite. Close to the landing place of the Begley Mine, in Bachewahnung Bay, a dioritic dyke, bearing N. 80 ° E., cuts gneissoid rocks Further investigation is necessary to determine what relation, if any, these dykes bear to the Huronian series.

(To be continued.)

# ON SOME REMAINS OF PALÆOZOIC INSECTS

### RECENTLY DISCOVERED IN

## NOVA SCOTIA AND NEW BRUNSWICK.

### By J. W. DAWSON, LL.D., F.R.S., F.G.S.

In connection with the preparation of the second edition of "Acadian Geology," I have obtained, from friends who have been engaged in geological investigations in Nova Scotia and New Brunswick, some interesting illustrations of the entomology of the Carboniferous and Devonian Periods, which I have thought it might be useful to publish in advance of the appearance of my work.

## 1. CARBONIFEROUS INSECTS.

The existence of insects in the Carboniferous period has long been known. The coal formations of England and of Westphalia afforded the earliest specimens; and, more recently, some interesting species have been found in the Western States.[*] They belong to the order of the Neuroptera (shad-flies, etc.), the Orthoptera (grasshoppers, crickets, etc.), and Coleoptera (beetles, etc.)

In the coal-field of Nova Scotia, notwithstanding its great richness in fossil remains of plants, insects had not occured up to last year, except in a single instance—the head and some other fragments of a large insect, probably Neuropterous, found by me in the Coprolite or fossil excrement of a reptile enclosed in the trunk of an erect Sigillaria at the Joggins, along with other animal remains. This specimen was interesting, chiefly as proving that the small reptiles of the coal period were insectivorous, and it was noticed in this connection in my "Airbreathers of the coal period." Last year, however, Mr. Jas. Barnes, of Halifax, was so fortunate as to find the beautiful wing represented in Fig. 1, in a bed of shale, at Little Glace Bay, Cape Breton. The engraving is taken from a photograph kindly sent to me by Rev. D. Honeyman, F.G.S. It will be observed that in consequence, probably, of the mutual attraction of loose objects floating about in water, a fragment of a frond of a fern, *Alethopteris lonchitica*, lies partly over the wing, obscuring its outline, but bearing testimony to its carboniferous date. The wing has been examined by Mr. S. H. Scudder, of Boston, who has made such specimens his special study, and who

* See Lyell's Elements, and Dana's Manual for references.

refers it to the group of Ephemerina (day-flies, shad-flies) among the Neuroptera, and has named it *Haplophlebium Barnesii*. It must have been a very large insect—seven inches in expanse of wing—and, therefore, much exceeding any living species of its group. When we consider that the larvæ of such creatures inhabit the water, and delight in muddy bottoms rich in vegetable matter, we can easily understand that the swamps and creeks of carboniferous Acadia, with its probably mild and equable climate, must have been especially favorable to such creatures, and we can imagine the larvæ of these gigantic ephemeras swarming in the deep black mud of the ponds in these swamps, and furnishing a great part of the food of the fishes inhabiting them, while the perfect insects emerging from the waters to enjoy their brief space of aerial life, would flit in millions over the quiet waters and through the dense thickets of the coal swamps.

Mr. Scudder describes the species as follows:—

Fig. 1.

(*a*) Profile of base of wing.

"HAPLOPHLEBIUM BARNESII Scudder; (Fig. 1.)—This is probably one of the ephemerina, though it differs very much from any with which I am acquainted. The neuration is exceedingly simple, and the intercostal spaces appear to be completely filled with minute reticulations without any cross-veins. The narrowness of the wing is very peculiar for an Ephemeron. The form of the wing and its reticulation remind me of the Odonata, but the mode of venation is very different; yet there is apparently a cross-vein between the first and second veins in the photograph (not rendered in the cut) which, extending down to the third vein, occurs just where the "nodus" is found in Odonata, and if present would, unquestionably, remove this insect to a new synthetic family between Odonata and Ephemerina. I cannot judge satisfactorily whether it is an upper or an under wing. The insect measured fully seven inches in expanse of wings —much larger than any living species of Ephemerina."

## 2. DEVONIAN INSECTS.

The only known remains of insects of this age are the wings of four species found by Mr. C. F. Hartt, in the plant-bearing Devonian Shales of St. John, New Brunswick. The figures now given of these remains, taken from drawings made by Mr. Scudder, though they represent fragmentary specimens only, are of the highest interest, as the most ancient remains of insects known to us, and contemporary with the oldest known land flora; their age being probably about that of the Hamilton or Chemung formations of New York.

Their geological date is unquestionable, since they are found in beds richly stored with species of Devonian plants, and unconformably underlying the oldest portion of the carboniferous series. The containing beds are fully described in a paper by Mr. Matthew, in the *Journal of the Geological Society of London*, and also in Prof. Bailey's *Report on the Geology of Southern New Brunswick* —Appendix A, on the Devonian Plant locality of Lancaster, by Mr. C. F. Hartt.

These insects, it will be observed, are of older date than the carboniferous species previously noticed, and they bore the same relations to the land and the water of the Devonian which the former did to those of the carboniferous period. They were all Neuropterous insects, and allied to the Ephemeras. It is interesting, however, to observe that, like many other ancient animals, they show a remarkable union of characters now found in distinct orders of insects; or constitute synthetic types, as they have been named. Nothing of this kind is more curious than the apparent existence of a stridulating or musical apparatus like that of the cricket, in an insect otherwise allied to the Neuroptera. This structure also, if rightly interpreted by Mr. Scudder, introduces us to the sounds of the Devonian woods, bringing before our imagination the trill and hum of insect life that enlivened the solitudes of these strange old forests.

Mr. Scudder has kindly furnished descriptions of these insects as follows:—

Fig. 2.

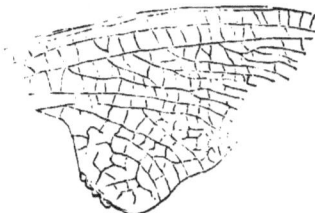

" PLATEPHEMERA ANTIQUA Scudder; (Fig. 2.)—The direction of the principal nervures in this insect convinces me that it belongs to the Ephemerina, though I have never seen in living Ephemerina so much reticulation in the anal area as exists here—so, too, the mode in which the intercalary nervules arise is somewhat peculiar. It is a gigantic species, for it must have measured five inches in expanse of wings—the fragment is a portion of an upper wing.

Fig. 3.

" HOMOTHETUS FOSSILIS Scudder; (Fig. 3.)—At first sight the neuration of the wings seems to agree sufficiently with the Sialina to warrant our placing it in that family; but it is very interesting to find, in addition to minor peculiarities that near the base of the wing, between the two middle veins, there is a heavy cross-vein from which new prominent veins take their rise; this is characteristic of the Odonata, and of that family only. We have, therefore, a new family representing a synthetic type which combines the features of structure now found in the Odonata and Sialina, very distant members of the Neuroptera. The fragment is sufficiently preserved to shew the direction, extent and mode of branching of nearly every principal nervure. It is evidently a portion of an upper wing; the insect measured not far from three one-half inches in expanse of wings.

Fig. 4.

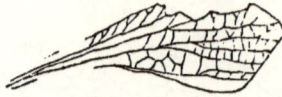

"LITHENTOMUM HARTTII Scudder; (Fig. 4.)—This was the first specimen discovered by Mr. C. F. Hartt. I have therefore named it after him :—apparently, it does not belong to any family of Neuroptera represented among living forms. It agrees more closely with the family Hemeristina, which I founded upon a fossil insect discovered in Illinois, than it does with any other ; but is quite distinct from that, both in the mode of division of the nervures and in the peculiar cross-veining. The fragment which Mr. Hartt discovered is very imperfect; but, fortunately, preserves the most important parts of the wing. I am inclined to think that it was a lower wing. The insect probably measured three one-half inches in expanse of wing.

Fig. 5.

"XENONEURA ANTIQUORUM Scudder; (Fig. 5.)—Although in this fragment we see only the basal half or third of a wing, the peculiar mode of venation shows that the insect cannot belong to any known family of Neuroptera living or fossil; yet it is evidently a neuropterous insect. In addition to its other peculiarities, there is one of striking importance, viz.:—the development of veinlets, at the base of the wing, forming portions of concentric rings. I have endeavored in vain to explain these away as something foreign to the wings, accidentally introduced upon the stone ; and I know of nothing to which it can be compared but to the stridulating organ of some male Orthoptera ! It is difficult to tell whether the fragment belongs to an upper or an under wing. Its expanse of wings was probably from two to two one-half inches."

# ON THE RELATION

BETWEEN THE

## GLACIAL DEPOSITS OF SCOTLAND AND ,THOSE OF CANADA.

By the Rev. HENRY W. CROSSKEY.

Principal Dawson, of Montreal, among his other great services to Geology, has very carefully investigated the Canadian glacial beds, and the following notes are suggested by a study of his writings:—

I. The difference between the glacial fossil fauna of Canada and that now existing in the Gulf of St. Lawrence is far less marked than the difference between the glacial fauna of the Clyde beds and that now existing in the Firth. The fossil fauna of Canada, in its general aspect, and in the proportions and characteristic varieties of its species, is slightly more arctic than that of the Gulf, but does not present that broad contrast with which we are familiar between the fossil contents of our local clays and·the living inhabitants of our waters. There are only two species in Canada which can be regarded as locally extinct, viz., *Leda Portlandica* (Gould), and *Astarte Laurentiana* (Lyell); while in Scotland there is a very remarkable list of species fossil in the clay, but extinct through the whole range of the neighbouring seas. Upon the west, we find :

| | |
|---|---|
| Tellina calcarea *(proxima.)* | Mangelia pyramidalis. |
| Saxicava (*Panopæa*) Norvegica. | Natica affinis (*clausa*). |
| Astarte borealis. | Trophon clathratus (*scalariformis*). |
| Leda pernula. | Velutina undata. |
| Pecten Islandicus. | Cyclostrema costulatum. |
| Modiolaria discors. | Balanus cariosus (*Darwin*). |
| Littorina limata (*Loven*). | |

The eastern clays comprise extinct species even more artic in character, viz. :—

| | |
|---|---|
| Leda arctica (*Portlandica, Gould*). | Thracia myopsis. |
| „ lucida. | Cardium Grœnlandicum. |
| „ thraciæformis. | Scalaria Grœnlandica. |
| Pecten Grœnlandicus. | |

It is evident, therefore, from this very marked contrast, that the change of climate in Scotland has been more complete than in Canada. From this fact important physical consequences ensue : the glacial epoch cannot have been caused by any of those cataclysmal agencies to which it has been attributed. Any heaping up of the land at the North Pole; or passage of the earth

through colder regions of space; or shiftings of the earth's axis;
or alteration in the heat-conducting power of the atmosphere,
would leave, I apprehend, a more uniform distribution of climatic
results, and obliterate those delicate proportions of species, varying
in different beds of the same epoch, in exact analogy to those
variations produced by the causes now at work. To account for
the fact we are examining, there must have been a deflection of
the Gulf Stream from our coasts. The effect of the Gulf Stream
is shown by the lingering of a species like *Saxicava* (Panopæa)
*Norvegica* upon the Dogger bank, which is protected from its
influence, and subject to an arctic current, while it is extinct on
the west of Scotland. Moreover, the existence of *Pecten Islandicus*
in its natural position over large beds in the glacial clay, combined
with the fact of its total absence, not only from our present sea,
but from any intermediate bed, renders its comparatively sudden
extinction by warmer currents taking the place of the more arctic,
the most probable hypothesis. The cause of extinction must have
been quiet, or its position would not have been so natural, and at
the same time sufficiently marked to permit little lingering. The
deflexion of the Gulf Stream must be considered in connection
with those movements of the land which we know to have been
going on in Scotland during the whole epoch. The subsidence
indicated by the shell beds at Airdrie and elsewhere was followed
by an elevating movement, which, judging from the peculiarly
undisturbed arrangement of different clays in various uplifted
beds, must have been very gradual. This elevating movement
itself also, is proved by the sections given by Mr. Jamieson[*] to
have been broken by a second, although slighter subsidence. The
shifting arrangements of the boundaries of land and water,
occasioned by these undulations of the earth's crust, would
materially affect climate, distributing variously the points of
insular and more continental temperatures, and in connection with
the deflection of the Gulf Stream, would (I am at present disposed
to think) sufficiently account for the cold of the glacial epoch.
Upon this point, however, Mr. Croll's most able and remarkable
papers give him a right to be heard, and I would venture to
suggest to him the consideration of the variable eccentricity of the
earth's orbit (as claimed by his theory) upon the climate of
Canada, so as to account for the fact that its temperature was,

---

[*] Journal of Geological Society, Vol. xxi.

during the glacial epoch, so little different from that now prevailing, while in Scotland the contrast has been so extreme.

II. Another most important point connected with the Canadian glacial beds, as compared with those of Scotland, is that they occur in a distinct order, whereas in the Clyde district, their order is only a matter of inference.

Dr. Dawson gives some instructive sections. In the lower beds are the deep water fossils, while littoral species occur in ascending order, manifesting the gradual alteration of the old sea bottom.

In collections of Clyde fossil shells we have a mixture of deep-sea coralline, laminarian, and littoral species ; but while we have superimposed beaches, we have no orderly succession in any exposed section, equivalent *e.g.* to that of Logan's farm, Montreal.

By carefully collecting the fossils from each separate pit in Scotland, and comparing them together, it may be proved, I think, that we have beds equivalent to those of Montreal, although our local sections are physically more obscure. Taking our glacial beds as a whole, it cannot be said that they co-existed at one depth, or were even synchronous. The Canadian beds justify the conviction I have long entertained and endeavoured to work out in the field, that our clay beds can be classified, and that there exists a definite order to reward patient research. They also support the proofs we have accumulated in this district of the theory that the rise of land was gradual, and that the passage from the ice epoch to the present was accomplished by forces extending over that vast period of time, necessarily demanded for those very delicate changes, involved in the distribution and redistribution of a specific fauna. It is not simply that a few mollusca disappear from their accustomed haunts—a great deal more is involved in a change of climate as it affects a fauna. Zoophytes, Foraminifera, Entomóstraca, must gradually alter their proportions and their specific representatives, as well as mollusca, so that between any two marked points of contrast, must stretch vast periods of geologic time.

III. All our Clyde shells occur in beds, resting upon the oldest boulder clay. The absolute absence of fossils, and the superposition of the shell-bearing clays, are facts which prove that the old boulder clays of the west of Scotland are the produce of land ice. The boulder clay appears the base of the section quoted from Logan's farm, just as it is of our Clyde series.

Undoubtedly, however, it is possible to have a boulder clay with marine remains. This may happen in two ways—(1) a glacier may lap over the sea, and melting, deposit the striated stones and mud which it has gathered on its course; or (2) striated boulders may be dropped from floating ice upon the mud beneath, and when the sea-bottom is uplifted, there will be a boulder clay of marine origin.

Patches of boulder clay containing shells may thus occur along the seaboard, as, for example, at Caithness, and on the east coast of England; but these patches of marine boulder clay will be *newer* than the clay at the base of the Clyde sections. Upon this point I hope soon to submit a detailed argument to the Society. Meanwhile, I remark, as a curious coincidence, that Dr. Dawson pronounces the shells collected from an "indubitable instance of a marine boulder clay" at Rivière-du-Loup, to be, on the whole, a more modern assemblage than those of the Leda clay of Montreal, which rests UPON the boulder clay.

Dr. Dawson gives one or two localities for fossils in "stony clays of the nature of true till;" but in the greater part of his sections, the fossiliferous beds are superimposed on the boulder clay, exactly as in the Clyde sections.

IV. Very curiously, a bed is noted beneath the boulder clay, for which we have a Scottish equivalent. A peat deposit, with fir roots, is found beneath boulder clay at Cape Breton, while at Chapelhall, Airdrie, we have vegetable remains in the same position—indicating the existence in both countries of land in parts afterwards depressed beneath the sea and again uplifted. The exact climate when this land existed, is believed by Dr. Dawson to have been, at Cape Breton, that of Labrador—in this country I believe it to have been such as to support the *Elephas primigenius*, whose remains have been found beneath boulder clay (certainly) at Kilmaurs, and (probably) at Airdrie.

V. The researches of the last few years have brought the Clyde list of fossils into nearer relation to the Canadian list than has hitherto been supposed. The *Leda arctica* from Errol is undoubtedly the *L. Portlandica* of the Canadian beds. This species occurs in such large quantities at Errol as to be characteristic of that clay. The *Astarte compressa* of the Clyde beds is not identical with *A. Laurentiana*, but often approaches exceedingly near to it. *Menestho albula* has been found at Paisley. It is doubtful whether the *Menestho albula* of the Canadian beds is Möller's species. Mr.

J. Gwyn Jeffreys considers a specimen from Quebec to which that name has been affixed to be *Scalaria borealis*. Taking the contents of one section, as collected by Dr. Dawson (this journal, April, 1865), out of twenty species of Lamelli-branchiata, fifteen occur fossil in Scotland, and seventeen out of twenty-seven species of Gasteropoda.

Speaking generally, about two-thirds of the Scottish fossils at present collected are also fossil in Canada, while the differences are no greater than those which geographical position might easily cause. At the period, therefore, when our glacial fossils lived in the Scottish seas, the climate was nearly the same as that prevailing in Canada during the same epoch—that is, slightly colder than in the present Gulf of St. Lawrence. The fossils, however, can not be considered as marking the extreme point of cold reached during the epoch, but rather as indicating the commencement of slightly milder climatic conditions than had hitherto prevailed. When the deposition of the oldest boulder clay commenced (which it must always be remembered is beneath the shell beds in the Clyde sections), the land must have stood higher than at present, and the temperature would be more intense than during its subsidence.

The question of climate as indicated by the fauna, thus resolves itself into this—what conditions would produce in the Clyde a temperature slightly colder than that of the Gulf of St. Lawrence?

The existence of an arctic current, the wide expanse of land in the American Arctic regions, exercising its chilling influence, and other circumstances connected with the directions of the mountain ranges and heights of the watershed, well known to the physical geographer, sufficiently account for the climate of Canada. A corresponding series of circumstances, therefore, would adequately explain the existence of a more arctic climate in Scotland. There is no necessity to introduce causes for the production of cold which do not now exist. Those alterations of level, for which there is ample evidence, would involve re-arrangements of the relative proportions of land and water, and vital changes in the directions of the arctic currents. For the solution of the problems involved in the great history indicated by the fossil fauna of Canada and Scotland, we must first consult those great principles of physical geography, which may now be studied in hourly action over the surface of the globe. From Transactions of the Geol. Society of Glssgow.

# ON A SUBDIVISION
# OF THE ACADIAN CARBONIFEROUS LIMESTONES,

### WITH A DESCRIPTION OF A SECTION ACROSS THESE ROCKS AT WINDSOR, N.S.

### By C. Fred. Hartt, A.M.

During several excursions made to Nova Scotia, previous to the year 1864, I visited Windsor, Brookfield, Shubenacadie, and Stewiacke, making extensive collections of the fossils of the carboniferous limestone, so abundant at these localities. Taking care to keep all the species obtained from any one bed or set of beds separate from those from any other, I soon found that certain groups of fossils were limited in their occurrence to certain beds, and that by means of these the whole series might be subdivided somewhat after the manner of the sub-carboniferous limestones of the west. In the summer of 1864, I spent some time in examining the same ground, and in working out a section exposed on the river Avon, at Windsor. The collection made at that time I had an opportunity, through the kindness of Prof. Agassiz, of examining at the Museum of Comparative Zoology ; but before my studies had been brought to completion, they were interrupted by my Brazilian journey, and as I have in this city no facilities for resuming them, I have sent, for determination, a considerable number of these fossils to Dr. Dawson and Mr. Billings, so that ample material will be afforded for the establishment of the faunal differences of the subdivisions of the Acadian carboniferous limestones, which I shall attempt to point out in this paper.

On the right bank of the river Avon, at Windsor, a few rods below the bridge, there begins a bluff, which, attaining in some places a height of fifty or sixty feet, skirts the shore for the distance of about half-a-mile above the bridge, when it gradually descends into a tract of marsh, which occupies the shore for nearly three-quarters of a mile further up, where there is a good exposure of a heavy bed of limestone seen in a bluff, called the Otis King rock. The bluff below the toll-house of the bridge is composed of drift, a great part of the mass being derived from the under-lying dark red, soft, friable, calcareous, marl-like sandstone. At the toll-house the first rocks *in situ* appear buried deeply under the drift deposit, thence southward, for about half the length of the bluff above the bridge, the beds of carboniferous limestone, clayey sandstone, etc., crop out under the drift. Some of the harder beds extend from top to bottom of the cliff, but owing to the softness and friable nature of the marly beds, and the way in which the beds of limestone are broken up by the action of the weather and hidden by drift and *debris*, the section is not easy to work out. Fortunately, the line of strike of the beds is such as to carry them out on the sloping shore, and though they are much hidden by shingle and mud deposited by the turbid Avon, we are able to gather material for the piecing together of our section, and occasionally to gain a clue as to the arrangement of the beds which is not given on the cliff.

Beginning at the beds of the toll-house, and going thence southerly along the shore, we find the following succession of beds :—

The first rocks seen at the toll-house are beds of limestone, having a strike of E. 15 ° S., and a dip of 65 ° to the north-ward, and of which a thickness of about twelve feet is visible. In the upper part these limestones are, in their weathered state, cream coloured, earthy, soft, and highly laminated, but with some compact bands. They afford fucoids of a slender, flattened cylindrical kind, without carbonaceous coating, a Productus of the Cora type, exactly like that so common in a reef just south of the bridge ; and a Bakevellia-like shell. In the middle portion is a band of soft, earthy, light lead-colored limestone, apparently full of fucoids, and with a few fragments of shells. In the lower part there is a not very compact, light brown, weathered limestone of a beautiful oölitic structure. Then follows, in descending order, a bed of very friable, fine-grained, greenish sandstone, cemented by carbonate of lime, which is succeeded by a bed of the same character, but of a deep red color from the presence of iron ; but this has several greenish layers. This bed occupies the shore for a distance of about seventy-five feet. In the lower part it is much obscured by rubbish. The cliff is then occupied for a distance of about thirty feet (horizontal) by a limestone of a loose texture and a light blueish color mottled with white, and probably altered by the action of the weather. The bed is much fractured and hidden by *debris*. Then succeeds an irregular mass of breccia, composed of angular fragments of limestone, and this rests on beds of light lead-colored, highly laminated calcareous shales, and lime-stone bands : thickness, six feet ; strike, E. 15 ° S. ; dip, 25 ° northward ; fucoids. Underlying this is a highly vesicular limestone, the cavities being lined with minute crystals of calc-spar : thickness, five feet.  Then come fifteen feet of light, lead-colored fissile, often highly laminated limestone, which, from its hardness, forms the most prominent part of the cliff, and extends in a reef down to low-water mark.  These beds are very rich in fossils.

The most characteristic fossil of this bed is a Productus of the true Cora type, but differing from *P. Lyelli* De Verneuil, in its smaller size, its long perpendicular posterior marginal prolongation, its more prominent and less numerous surface lines, which increase by a more regular and frequent implantation or bifurcation.*  This Productus is exceedingly common in certain layers of the shelly limestone.  Among the few other forms associated with it is a Bakevellia, usually indifferently preserved, and a slender branching fucoid, often preserved as a carbonaceous film ; minute stems of crinoids occasionally occur.  It is worthy of note that crinoidal remains are exceedingly rare in the limestone of the carboniferous about the Basin of Minas, and I have observed only the stems, which are always minute.  The dip of these beds varies from 35 ° to 50 ° northward ; strike, same as last observed.

Succeeding these are beds of a very dark, blackish limestone, very hard, cracking into small irregular pieces, and wearing nodular: thickness five to six feet.  This is full of fossils; the most characteristic is a Spirifer, which appears to differ from *Spirifer glaber* Martin, only in its smaller dimensions; a small Rhynconella, with large plaits (*R. Ida*, nob.) ; a Spirifer like *S. Octoplicatus*, but larger.  I have found here a single specimen of a Phillipsia, which differs from *P. Howi* in wanting the tubercles on the axial rings and pleuræ of the side lobes, in the shape of the pygidium, which is more rounded in outline, and in which the grooves are distinctly marked on the six anterior pleuræ.  For this species, which appears to be new, I have proposed the name of *P. Vindobonensis*.  Dr. Dawson has, in his description of this section, in his Acadian Geology, inadvertently placed this bed on the southern side of the gully about to be mentioned.  There is also a minute plaited Aviculopecten which occasionally occurs in this bed.  For this series of beds, characterized by *P. Cora* Var. *Nova-Scotica*, and *Spirifer Glaber*, I propose the name of Avon Limestone.

---

* Mr. Billings regards this as a variety of *P. Cora*.  It may be designated as Var. *Nova-Scotica*, this name being proposed by Mr. Hartt.

Underlying these beds are seven and one-half feet of calcareous sandstone, of a light lead color, and decomposing into a soft, incoherent mass; then nine feet of compact, flaggy, light brown limestone, with shaly partings, apparently without fossils; and very friable shales of a blueish tint, much decomposed at the surface, and hidden by rubbish. Here we have a fault, a dislocation of about six feet. Then comes a bed of red, very friable, marly, calcareous sandstone, of which a thickness of about thirty feet is exposed. Here the surface water has excavated a considerable gully through the soft sandstone. There can be no doubt, as Dr. Dawson has stated, that there is a fault here, for the beds on the other side of the gully are seen dipping southward, and there is no repetition of the strata.

Continuing the section, the first bed seen on the opposite side of the gully is exactly like that last described, and occupies the shore for some sixty feet. This is overlaid by a bed of limestone, flaggy, with more compact bands. In the cliff these beds have a dip southward of 50°, but at its foot they become more nearly vertical, and run out some twenty feet on the beach, with a strike of E. 10° S., and an almost vertical dip, inclining, however, to the south about 96° to 95°. Crossing a belt of mud on the shore at low tide, we find the same beds appearing, with the same strike, near the bed of the river, but their dip is reversed, and they are inclined to the northward at an angle of 25° to 30°. The thickness of beds just described is twenty to twenty-five feet.

A bed of the red, marly sandstone, about thirty-five to forty feet thick, next follows. It seems to be irregularly stratified, and there are several green layers. This same bed, in ascending order, succeeds at low-water mark to that last mentioned. Beds of limestone, with a strong southerly dip, next come, occupying the cliff for a distance of sixty to seventy feet along its base, whence they extend out on the shore for some twenty feet, with an easterly strike and an almost vertical dip. In their line of strike across a belt of mud and shingle, a few yards down the beach, the same beds appear again, describing a slight curve to the north on the inclined beach. Tracing them towards low-water mark, they gradually change their dip towards the north, until, at the bed of the river, it is about 60° N. Examined at the base of the cliff, the limestone of these beds is of a blueish color, weathering light brown, concretionary in the lower part, and with a band in the middle of a beautiful oölitic structure. This limestone appears to be quite unproductive of fossils, except in one or two thin bands, which are closely packed full of minute gasteropods, and the joints of slender stemmed crinoids. Associated with these are occasionally found a fossil resembling a large Dentalium, but Mr. Meek writes me that it does not belong to that genus.* The fossils which characterize this bed seem to me to be quite distinct from those found in the other beds. I have not observed this limestone elsewhere.

A bed of the red, marly sandstone overlies the limestone, appearing also at the foot of the beach, and this is overlaid in turn by a bed of limestone, fifteen feet in thickness, having a southward dip of 45°. This last bed is seen to be overlaid by a bed of the red marly sandstone, having a layer of a green tint about a foot thick at its base. The face of the cliff is here not very clear, but the limestone is seen to be broken abruptly off by a fault, and the marly sandstone to occupy the face of the bluff from top to bottom. This fault I developed by cutting away the face of the bluff.

The limestone last described is very compact, and of a light, clear, leaden blue color, weathering, however, to a brown. It seems to be made up of alternate layers of a very hard and concretionary limestone, and of a softer kind, so that they wear unequally, which gives to their upturned edges, exposed on the sea shore, a rubbly appearance. This bed has usually been supposed to be non-fossiliferous, and it is not mentioned by Dr. Dawson in Acadian Geology. Struck with the resemblance the highly tinted limestone bore to that which at Kennetcook affords the *Phillipsia Howi* of Billings, I was led to examine it with care, and was rewarded by finding a specimen of that trilobite, together with a Zaphrentis, common in the Kennetcook and Cockmegun limestones, and a number of other fossils. Among these was a Spirifer over two inches long, a valve of what Mr. Meek refers doubtfully to *Athyris lamellosa* L'Eveille, a Productus quite undistinguishable from the ordinary form of *P. semi-reticulatus* and another species like *P. costatus*, with very long spines. There are also several species of Myoid Lamellibranchs, and occasionally one finds a minute fish tooth. An Athyris, somewhat like *A. subtilita*, but distinct, occurs in this bed, both at Windsor and Kennetcook, together with a Stenopora and a Fenestella

---

* It is apparently a *Serpulites.*—J. W. D.

(or Retepora), both of which are not found in the other beds. In Dr. Dawson's collection there is a large Orthoceras and a Bellerophon from Kennetcook. The Kennetcook limestone is quarried for building purposes, and the library of King's College at Windsor is partially built of it. From this limestone Professor How has collected many of these fossils. A fucoid occurs quite abundantly in some of the layers at Kennetcook, but I have never detected it at Windsor.

These same beds appear low down on the shore, but badly exposed, owing to the loose material encumbering the surface. The same limestones, bearing the same fossils, are exposed at Lower Stewiacke, on the Stewiacke River, near the house of Mr. Jacob Stevens, where it has a strike of N. 50 ° E., and a dip of 45 ° S.W. This bed is so well characterized, both faunally and lithologically, and has an extension over so large an area, that it seems to merit a special name, and I would propose for it the name of Kennetcook Limestone.

Continuing our examination of the bluff still farther southward from the fault last described, we find the rocks so disintegrated and stratification so obscured by the falling of rubbish over its sloping face, that little else can be ascertained except the presence of beds of marly sandstone and limestone from the oblique lines seen on the face of the bluff. About one hundred yards beyond the fault occurs a bed of snowy white gypsum, containing stellar crystals of Selenite disseminated through it, which, being of a brownish tinge, are very conspicuous on the weathered surfaces. This gypsum was formerly quarried at this point for exportation. If we cross the hill in the line of strike of the bed, we reach, at a short distance from the river, the principal quarry of this vicinity excavated in this same bed, which is here about thirty feet in thickness, with a strike of E. 35 ° N., and a dip of 15 ° to 30 ° to the southward. The excavation made in quarrying the gypsum is some thirty feet deep, one hundred feet wide, and five hundred feet long. The bed does not seem to be very regular, and it appears to be considerably contorted.

Returning to the river side, we find the section fails from the gypsum bed, and it is not until we reach a fence, where the shore bends eastward, that we meet with any exposure of rock of any interest. Here there is an irregular mass of limestone of a brownish color, exceedingly rich in fossils, being almost wholly made up of shells. These are often empty, so as to give the rocks an open texture. Following the higher land of the shore eastward along a marsh for a few rods, we find it making a bend southward once more along a low bluff of the same limestone, and here, as well as at the first named exposure of this limestone, beautiful specimens of its characteristic fossils may be obtained in great quantity. The bed is so badly exposed that its thickness cannot be determined. It has a slight southward dip.

This bed, which I shall call the Windsor Limestone, has afforded me a large number of very interesting species, among which the following may be named as the most characteristic:—

Of Radiates, a few crinoid joints, very minute, have been detected, but they are by no means common. A Stenopora (*Ceriopora spongites* of Acadian Geology) is exceedingly common, and very characteristic of this bed. The fauna of this bed is not rich in Articulates, but it has afforded a Leperditia, a Serpula (?), and part of the cephalo-thorax of another crustacean (a Decapod?), which is in the hands of Dr. Dawson and Mr. Billings for study.*

It is in Mollusks that this bed is especially rich, and of these the following may be named :—

BRYOZOANS.—A species of Fenestella, different from the species occuring elsewhere; very rare.

BRACHIOPODS.—*Rhynchonella Evangelina, nob†*, very common. This has the characteristic oral supports of Rhynconella, which are easily examined, a large proportion of the specimens being hollow. A small Productus of the Cora type is very abundant. It is very different from the other Producti of Nova Scotia, and it differs from *P. Lyelli* De Verneuil, in being constantly smaller, more globose, and wanting in the large marginal prolongations. A Terebratula (*T. sacculus* Mart.) is a common fossil in this bed. I have examined large numbers of specimens of this form, and have compared them, not only with the *T. sacculus* of Davidson's paper, from the overlying bed, but also with specimens of that species from de Koninck's collections in the Museum of

---

* The specimen is too imperfect for determination.—J. W. D.

† This is probably the shell which Davidson has referred to in his paper on Acadian Carboniferous Brachiopods as *Rh. pugnus,* but it bears a striking resemblance to the form which he has figured as *Camarophoria globulina ?* This is certainly a Rhynconella, tor it has the characteristic oral supports of the genus. It is quite distinct from *Rh. Pugnus.*

Comparative Zoology, but I cannot satisfy myself that they are specifically identical.   There is a not uncommon Terebratula-like shell, which shows, finely preserved, the characteristic loop of Centronella (*C. Anna* Hartt).   This is the first evidence we have of the existence of this genus above the Devonian.

LAMILLIBRANCHS. — Several species of Aviculopecten are especially abundant.   Of one of these, *A. simplex* Daws., Mr. Meek writes me as follows: " There are among the Windsor collection several good specimens of a little shell, exceedingly like the so-called *Pecten pusillus* (not a true Pecten), from the European permian rocks.   They are very similar, and, indeed, almost the only differences observable on direct comparison with good European specimens now before me, are the slightly more ventricose form of the valves, and the rather more prominent anterior ear of the left valve of the Windsor shell.   Perhaps this ear, in its left valve, is also a little more defined from the swell of the umbo in some of the large specimens from Windsor, but on comparing examples of the same size as the German specimens here (which are not near so large as some figured in foreign works), it is difficult to see characters by which they can be distinguished.   They are, in fact, more nearly alike than the figures given of *P. pusillus* by different European authorities, or, in some cases, by the same author, as varieties of that species.   In short, if found associated in the same rock at the same locality with *P. pusillus*, few would suspect them to be distinct species." *Aviculopecten fallax* McCoy ?   Windsor and de Bert River, Dr. Dawson ;   *A. Nova-Scotica* Daws., Schubenacadie, Dr. Dawson ;   *Pteronites Gayensis* Daws., Gay's River, Dr. Dawson ; *Macrodon elegans* De Koninck ?   Windsor, Dr. Dawson and Mr. Hartt;   *Modiola Pooli* Daws., Windsor, Poole and Hartt. Besides the above, there are several other Lamillibranchs not yet determined.

GASTEROPODS.—*Naticopsis Howi, nob.*, one of the commonest fossils of the Avon beds.   I have detected only a single fragment of Conularia in these beds, and this appears to be different from the species of the overlying beds.

CEPHALOPODS.—A single Orthoceras has been collected at Windsor.

The Windsor limestone is well developed at Brookfield and Stewiacke, and Gay's River, where it holds the same fossils as at Windsor.   I have not had an opportunity of examining extensive

collections from the other Acadian localities, so that I am unable
to report its existence elsewhere.

At the eastern end of the little bluff last described, there is an
accumulation of broken masses of a limestone, similar to that of
the Windsor limestone, but it is lighter in color, more compact, of
a light brownish tint, and composed almost entirely of fossil
remains, the species are, with rare exceptions, distinct from those
which are found in the Windsor limestone. Among the masses of
rock here found there is not a single piece from the Avon beds, so
that it is evident that here there is a bed of limestone. which
overlies the Avon beds. Three quarters of a mile farther up the
river, across a wide marsh, is the Otis King rock, which is composed
of the same limestone and furnishes the same fossils. Here,
however, the beds are seen with a slight *northward* dip. The beds
in their lower part are less compact than in the upper, where they
pass into a very hard fine-grained limestone, capable of taking a
high polish. Fossils occur all through the bed, but they are
especially abundant in the upper part. This bed which I would
call the Stewiacke limestone, appears to be overlaid by a bed of
gypsum, seen between the two localities, at the head of the marsh,
which appears to occupy a synclinal valley. The Stewiacke
limestone is very rich in beautifully preserved fossils.

RADIATES.—Of Radiates there is a great paucity of species, as
elsewhere in Nova Scotia; minute crinoid stems are occasionally
found, and there is a pretty Stenopora (*S. exilis* Daws.) which
is very common, and is one of the most characteristic fossils of this
limestone.

ARTICULATES.—Of Articulates there are very few species, a
Serpula (?) tube occurs rarely, together with a Leperditia and a
Spirorbis.

MOLLUSKS are the reigning type. Bryozoans are represented
by a Fenestella, *F. Lyelli* Daws. This is exceedingly abundant
and eminently characteristic of this limestone, wherever it occurs.
Of Brachiopods there are many representatives. *Productus
Lyelli* De Verneuil, (*P. Cora,*) is one of the commonest fossils
both at Windsor and elsewhere, and this is associated with an
abundance of *P. semi-reticulatus*, and the Terebratula referred
by the last mentioned author to *T. Sacculus* Martin, and the forms
referred by him to *Athyris subtilita*, *Spirifer acuticostata* De
Koninck, and *Spiriferina cristata*. Besides there are a number
of Rhynconellæ and other Brachiopods, which appear to be

confined to this bed.  Lamellibranchs are abundant, and among the most characteristic may be named the following :—

*Aviculopecten reticulata,* Daws., Windsor and Gay's River; *A. Nova-Scotica* Daws., (*A. plicata* of Acadian Geology); *Macrodon Hardingii* Daws., very characteristic; *Conocardium Acadicum, nob.,* rare.  The Gasteropods are all minute and as yet undetermined.  A Conularia is occasionally met with at Windsor and Stewiacke.

Of the Cephalopods, we have a large Nautiloid shell, *Nautilus (Cryptoceras) Avonensis* Daws., not uncommon at Windsor and Stewiacke; a Trematodiscus (?), and also two or more species of Orthoceras.  I cannot report a single fragment of a vertebrate for the Stewiacke limestone. *

The question naturally arises as to the relative position of these beds, but this is one which it seems impossible to settle from the Windsor section, and I have seen no localities elsewhere, where their relations to one another were distinctly exhibited.  I think that there can be no doubt that the Stewiacke limestone is the highest, the Windsor limestone coming next below, the Kennetcook limestone appears to come next in order, and the oölitic fossiliferous band, to which I give no name, underlies this again, but the Avon limestone at Windsor, is separated from the rest by a fault, and although I believe it to be the lowest of the four limestones, it may be that subsequent observations made elsewhere, may not confirm that belief.  These carboniferous limestones whenever they occur, are much disturbed and broken up, while the disintegration of the intercalated soft marly strata and gypsum beds, adds to the obscurity of the exposures.

The resemblance borne by the faunæ of the Acadian carboniferous limestone to the permian of Europe, has been ably discussed by Lyell, Dawson and Davidson ; but these gentlemen have united in expressing the opinion that they are really members of the carboniferous system.  In studying the Windsor fossils at the Museum of Comparative Zoology, I failed to find any marked resemblance between them and those of the sub-carboniferous of the West, while I was exceedingly struck with the greater similarity borne by these in their *facies* to the fauna of the Kansas permo-carboniferous ; and in a list of New Brunswick fossils, which I contributed to Professor Bailey's Report on the Geology of the

---

* The whole of the fossils referred to in this paper, will be described in the forthcoming edition of Dr. Dawson's Acadian Geology.—EDS.

Southern Counties of New Brunswick, I ventured to express a
doubt as to the precise age of the Acadian carboniferous limestones,
for a few species collected in the vicinity of the Albert mines had
the same permo-carboniferous look as those at Windsor. Dr. J.
S. Newberry, in looking over my collection, was also impressed
with their permo-carboniferous *facies.* At his suggestion, I sent
a small collection of these fossils to Mr. Meek, who writes me as
follows :—" A small collection of these same fossils from Windsor
was presented to the Smithsonian Institution, by Dr. E. Foreman,
some three or four years since, and they have remained a puzzle
to me ever since. If they had been brought in from some unexplored region of the Rocky Mountains, for instance, I confess I
should have referred them to the horizon of the upper coal
measures, or to that of a series of rocks known in Kansas as the
permo-carboniferous, from the remarkable mingling in them of
coal measure and permian types there; but in reading over the
able publications of Dr. Dawson, Sir Charles Lyell, and Mr.
Davidson, on the age of these Nova Scotian beds, I was led to the
conclusion that this must be one of those very rare cases where
physical structure shows palæontology to be at fault. Although
I am not positively sure that any of the species are absolutely
identical with those of the higher horizon, these fossils certainly
present a remarkable permo-carboniferous look, and, when viewed
collectively, they are unlike the western sub-carboniferous fauna.
For instance, there are here from Windsor several good specimens,
showing both valves, with the surface markings of an Aviculopecten undistinguishable by any characters yet observed from *A.
Occidentalis* of Shumard (*Pecten Cleavelandicus* Swallow), one
of our most common and characteristic coal measure, permo-
carboniferous and permian species in the west, which, so far as
yet known, has never been found below the upper coal measures,
at any rate in the western localities. Another shell represented
in the collections from Windsor by casts, is very similar to
varieties of the so-called *Mytilus squamosus* from the English
permian. It has almost precisely the form, and agrees in size, as
well as in showing between the beaks the cast of a little depression
on a shelf or septum within the beaks, such as we often see in
species of Myalina, to which these shells doubtless belong.
Another little shell, from Windsor, is quite or nearly like a little
permo-carboniferous species in the west, known as *Sedgwickia ?
concava,* M. and H.; while you have from the same casts of an

Edmundia very like a western coal measure form. . . Taking the whole group of Windsor Mollusca, including the Lamillibranchs, any one familiar with the fossils of the western coal measure and permo-carboniferous beds, would, upon palæontological grounds alone, be very strongly inclined to refer the Windsor rocks at least to the upper coal measures." This conclusion Mr. Meek hardly feels that we ought to accept, seeing that so many able geologists have united in placing the beds in the sub-carboniferous, but expresses his opinion that "it *may* be an example of what Barrande would call an upper coal measure, or even permo-carboniferous fauna, ' colonized ' far back in the sub-carboniferous period."

The carboniferous limestones and marls of Windsor certainly overlie the plant bearing shales and sandstones of the lower coal measures, which are seen exposed at Windsor Brook, Horton Bluff, Gaspereaux, and Wolfville, skirting the edge of the carboniferous basin ; and Dr. Dawson has described these marine limestones, marls and gypsums as occupying a synclinal trough in these lower coal measure strata, extending from Windsor to Stewiache, a distance of some fifty miles.*    Over this region the middle coal measures do not occur, so that of these limestones there is no stratigraphical evidence to contradict the evidence afforded by palæontology as to their permo-carboniferous age, and in this region Dr. Dawson has suggested that the upper limestones may represent the coal measures.    I have not had any opportunity of studying these limestones except about the Basin of Minas, neither have I been able to examine sufficient suites of fossils to enable me to determine whether the above divisions I have marked-out obtain elsewhere.    From a careful study of the evidence brought forward by Dr. Dawson, it certainly seems proven that the limestones, with their fossils, underlie the true coal measures in other parts of Nova Scotia.

This whole subject is one of great interest, and needs the most careful investigation.    It will now be of much importance to have the limestones of north-eastern Nova Scotia and of Cape Breton compared with those of the Basin of Minas, in order to ascertain whether the same divisions obtain there as at Windsor. Another interesting point to be studied is the extension of the marly sandstones and gypsums, the conditions of their deposition, and the influence which they may have had in the extinction of

---

* Proceedings of Geological Society, Vol. xv., Part I., pp. 64-65.

life over the regions they occupy. Might not some material be gathered from this new and rich field bearing on that vexed question of descent with modification ?

NEW YORK, May 28th, 1867.

---

NOTE BY DR. DAWSON.—Much credit is due to Mr. Hartt for the careful manner in which he has worked up the succession of fossils in the limestones of the Avon estuary. I have endeavoured, in the new edition of Acadian Geology, to apply his results to other parts of Nova Scotia. In regard to the resemblance of the Windsor fauna to the permo-carboniferous of the west, it is to be observed—(1) That no such distinction as sub-carboniferous and carboniferous can hold in Nova Scotia. The Windsor fauna is simply the marine fauna of the carboniferous, and some of the beds may be coeval with the coal measures, as I suggested many years ago (Acad. Geol. 1st. ED.). (2) The lithological character of these beds is like that of the permian, and similar sea bottoms of different periods often present resemblances of fauna. (3) That the fauna in question actually lived in the lower carboniferous period, is proved by the sections in Cumberland, Pictou and Cape Breton, which show the limestones with these shells lying below the productive coal measures. (4) It is to be observed that the supposed premo-carboniferous *facies* applies to the upper members of the Windsor limestones more especially. I have fully illustrated these points in the new edition of Acadian Geology.

# ON THE CHEMISTRY OF THE PRIMEVAL EARTH.

## By T. STERRY HUNT, LL.D.. F.R.S.[*]

The natural history of our planet, to which we give the name of geology, is, necessarily, a very complex science. including, as it does, the concrete sciences of mineralogy, of botany and zoology, and the abstract sciences chemistry and physics.   These latter sustain a necessary and very important relation to the whole process of development of our earth, from its earliest ages, and we find that the same chemical laws which have presided over its changes, apply also to those of extra-terrestrial matter.   Recent investigations show the presence in the sun, and even in the fixed stars—suns of other systems—the same chemical elements as in our own planet.   The spectroscope, that marvellous instrument, has, in the hands of modern investigators, thrown new light upon the composition of the farthest bodies of the universe, and has made clear many points which the telescope was impotent to resolve.   The results of extra-terrestrial spectroscopic research have lately been set forth in an admirable manner by one of its most successful students, Mr. Huggins.   We see, by its aid, matter in all its stages, and trace the process of condensation and the formation of worlds.   It is long since Herschel, the first of his illustrious name, conceived the nebulæ, which his telescope could not resolve, to be the uncondensed matter from which worlds are made.   Subsequent astronomers, with more powerful glasses, were able to show that many of these nebulæ are really groups of stars, and thus a doubt was thrown over the existence in space of nebulous luminous matter; but the spectroscope has now placed the matter beyond doubt.   By its aid. we find in the heavens, planets, bodies like our earth, shining only by reflected light; suns, self luminous, radiating light from solid matter; and, moreover, true nebulæ, or masses of luminous gaseous matter.   These three forms represent three distinct phases in the condensation of the primeval matter, from which our own and other planetary systems have been formed.

This nebulous matter is conceived to be so intensely heated as to be in the state of true gas or vapour, and, for this reason, feebly

---

[*] Report of a lecture delivered before the Royal Institution of Great Britain, London, May 31st, 1867, and reprinted from the Proceeding of the Royal Institution.

luminous when compared with the sun. It would be out of place, on the present occasion, to discuss the detailed results of spectroscopic investigation, or the beautiful and ingenious methods by which modern science has shown the existence in the sun, and in many other luminous bodies in space, of the same chemical elements that are met with in our earth, and even in our own bodies.

Calculations based on the amount of light and heat radiated from the sun show that the temperature which reigns at its surface is so great that we can hardly form an adequate idea of it. Of the chemical relations of such intensely heated matter, modern chemistry has made known to us some curious facts, which help to throw light on the constitution and luminosity of the sun. Heat, under ordinary conditions, is favourable to chemical combination, but a higher temperature reverses all affinities. Thus, the so-called noble metals, gold, silver, mercury, etc., unite with oxygen and other elements; but these compounds are decomposed by heat, and the pure metals are regenerated. A similar reaction was many years since shown by Mr. Grove with regard to water, whose elements—oxygen and hydrogen—when mingled and kindled by flame, or by the electric spark, unite to form water, which, however, at a much higher temperature, is again resolved into its component gases. Hence, if we had these two gases existing in admixture at a very high temperature, cold would actually effect their combination precisely as heat would do if the mixed gases were at the ordinary temperature, and literally it would be found that " frost performs the effect of fire." The recent researches of Henry Ste.-Claire Deville and others go far to show that this breaking up of compounds, or dissociation of elements by intense heat, is a principle of universal application; so that we may suppose that all the elements which make up the sun or our planet, would, when so intensely heated as to be in that gaseous condition which all matter is capable of assuming, remain uncombined— that is to say, would exist together in the condition of what we call chemical elements, whose further dissociation in stellar or nebulous masses may even give us evidence of matter still more elemental than that revealed by the experiments of the laboratory, where we can only conjecture the compound nature of many of the so-called elementary substances.

The sun, then, is to be conceived as an immense mass of intensely heated, gaseous and dissociated matter, so condensed,

however, that notwithstanding its excessive temperature, it has a
specific gravity not much below that of water; probably offering
a condition analogous to that which Cagniard de la Tour observed
for volatile bodies when submitted to great pressure at temperatures much above their boiling point. The radiation of heat,
going on from the surface of such an intensely heated mass of
uncombined gases, will produce a superficial cooling, which will
permit the combination of certain elements and the production of
solid or liquid particles, which, suspended in the still dissociated
vapours, become intensely luminous and form the solar photosphere. The condensed particles, carried down into the intensely
heated mass, again meet with a heat of dissociation; so that the
process of combination at the surface is incessantly renewed, while
the heat of the sun may be supposed to be maintained by the slow
condensation of its mass; a diminution by $\frac{1}{10000}$th of its present
diameter being sufficient, according to Helmholtz, to maintain the
present supply of heat for 21,000 years.

   This hypothesis of the nature of the sun and of the luminous
process going on at its surface is the one lately put forward by
Faye, and although it has met with opposition, appears to be that
which accords best with our present knowledge of the chemical
and physical conditions of matter, such as we must suppose it to
exist in the condensing gaseous mass, which according to the
nebular hypothesis, should form the centre of our solar system.
Taking this, as we have already done, for granted, it matters little
whether we imagine the different planets to have been successively
detached as rings during the rotation of the primal mass, as is
generally conceived, or whether we admit with Chacornac a process
of aggregation or concretion, operating within the primal nebular
mass, resulting in the production of sun and planets. In either
case we come to the conclusion that our earth must at one time
have been in an intensely heated gaseous condition, such as the
sun now presents, self-luminous, and with a process of condensation
going on at first at the surface only, until by cooling it must have
reached the point were the gaseous centre was exchanged for one
of combined and liquefied matter.

   Here commences the chemistry of the earth, to the discussion
of which the foregoing considerations have been only preliminary.
So long as the gaseous condition of the earth lasted, we may
suppose the whole mass to have been homogeneous; but when the
temperature became so reduced that the existence of chemical

compounds at the centre became possible, those which were most stable at the elevated temperature then prevailing, would be first formed. Thus, for example, while compounds of oxygen with mercury or even with hydrogen could not exist, oxides of silicon, aluminium, calcium, magnesium, and iron might be formed and condense in a liquid form at the centre of the globe. By progressive cooling, still other elements would be removed from the gaseous mass, which would form the atmosphere of the non-gaseous nucleus. We may suppose an arrangement of the condensed matters at the centre according to their respective specific gravities, and thus the fact that the density of the earth as a whole is about twice the mean density of the matters which form its solid surface may be explained. Metallic or metalloidal compounds of elements, grouped differently from any compounds known to us, and far more dense, may exist in the centre of the earth.

The process of combination and cooling having gone on until those elements which are not volatile in the heat of our ordinary furnaces were condensed into a liquid form, we may here inquire what would be the result, upon the mass, of a further reduction of temperature. It is generally assumed that in the cooling of a liquid globe of mineral matter, congelation would commence at the surface, as in the case of water; but water offers an exception to most other liquids, inasmuch as it is denser in the liquid than in the solid form. Hence, ice floats on water, and freezing water becomes covered with a layer of ice, which protects the liquid below. With most other matters, however, and notably with the various mineral and earthy compounds analogous to those which may be supposed to have formed the fiery-fluid earth, numerous and careful experiments show that the products of solidification are much denser than the liquid mass; so that solidification would have commenced at the centre, whose temperature would thus be the congealing point of these liquid compounds. The important researches of Hopkins and Fairbairn on the influence of pressure in augmenting the melting point of such compounds as contract in solidifying, are to be considered in this connection.

It is with the superficial portions of the fused mineral mass of the globe that we have now to do; since there is no good reason for supposing that the deeply seated portions have intervened in any direct manner in the production of the rocks which form the superficial crust. This, at the time of its first solidification, presented probably an irregular, diversified surface from the result of contraction of the congealing mass, which at last formed a liquid bath of no great depth, surrounding the solid nucleus. It is to the composition of this crust that we must direct our attention, since therein would be found all the elements (with the exception of such as were still in the gaseous form) now met with in the known rocks of the earth. This crust is now everywhere buried beneath its own ruins, and we can only from chemical considerations attempt to reconstruct it. If we consider the conditions through which it has passed, and the chemical affinities which must have come into play, we shall see that there are just what would now result if the solid land, sea, and air were made to react upon each other under the influence of intense heat. To the chemist it is at once evident that from this would result the conversion of all carbonates, chlorides and sulphates into silicates, and the separation of the carbon, chlorine, and sulphur in the form of acid gases, which, with nitrogen, watery vapour, and a probable excess of oxygen, would form the dense primeval atmosphere. The resulting fused mass would contain all the bases as silicates, and must have much resembled in composition certain furnace-slags or volcanic glasses. The atmosphere, charged with acid gases which surrounded this primitive rock must have been of immense density. Under the pressure of such a high barometric column, condensation would take place at a temperature much above the present boiling point of water, and the depressed portions of the half-cooled crust would be flooded with a highly heated solution of hydrochloric acid, whose action in decomposing the silicates is easily intelligible to the chemist. The formation of chlorides of the various basis, and the separation of silica, would go on until the affinities of the acid were satisfied, and there would be a separation of silica, taking the form of quartz, and the production of a sea-water holding in solution, besides the chlorides of sodium, calcium, and magnesium, salts of aluminium and other metallic basis. The atmosphere, being thus deprived of its volatile chlorine and sulphur compounds, would approximate to that of our own time, but differ in its greater amount of carbonic acid.

We next enter into the second phase in the action of the atmosphere upon the earth's crust. This, unlike the first, which was subaqueous, or operative only on the portion covered with the precipitated water, is sub-aerial, and consists in the decomposition of the exposed parts of the primitive crust under the influence of the carbonic acid and moisture of the air, which convert the complex silicates of the crust into a silicate of alumina, or clay, while the separated lime, magnesia, and alkalies, being converted into carbonates, are carried down into the sea in a state of solution.

The first effect of these dissolved carbonates would be to precipitate the dissolved alumina and the heavy metals, after which would result a decomposition of the chloride of calcium of the sea-water, resulting in the production of carbonate of lime or limestone, and chloride of sodium or common salt. This process is one still going on at the earth's surface, slowly breaking down and destroying the hardest rocks, and, aided by mechanical processes, transforming them into clays; although the action, from the comparative rarity of carbonic acid in the atmosphere, is less energetic than in earlier times, when the abundance of this gas, and a higher temperature, favoured the chemical decomposition of the rocks. But now, as then, every clod of clay formed from the decay of a crystalline rock corresponded to an equivalent of carbonic acid abstracted from the atmosphere, and equivalents of carbonate of lime and common salt formed from the chloride of calcium of the sea-water.

It is very instructive, in this connection, to compare the composition of the waters of the modern ocean with that of the sea in ancient times, whose composition we learn from the fossil sea-waters which are still to be found in certain regions, imprisoned in the pores of the older stratified rocks. These are vastly richer in salts of lime and magnesia than those of the present sea, from which have been separated, by chemical processes, all the carbonate of lime of our limestones, with the exception of that derived from the sub-aerial decay of calcareous and magnesian silicates belonging to the primitive crust.

The gradual removal, in the form of carbonate of lime, of the carbonic acid from the primeval atmosphere, has been connected with great changes in the organic life of the globe. The air was doubtless at first unfit for the respiration of warm-blooded animals, and we find the higher forms of life coming gradually into existence as we approach the present period of a purer air. Calculations lead us to conclude that the amount of carbon thus removed in the form of carbonic acid has been so enormous, that we must suppose the earlier forms of air-breathing animals to have been peculiarly adapted to live in an atmosphere which would probably be too impure to support modern reptilian life. The agency of plants in purifying the primitive atmosphere was long since pointed out by Brongniart, and our great stores of fossil fuel have been derived from the decomposition, by the ancient vegetation, of the excess of carbonic acid of the early atmosphere, which through this agency was exchanged for oxygen gas. In this connection the vegetation of former periods presents the curious phenomenon of plants allied to those now growing beneath the tropics, flourishing within the polar circles. Many ingenious hypotheses have been proposed to account for the warmer climate of earlier times, but are at best unsatisfactory, and it appears to me that the true solution of the problem may be found in the constitution of the early atmosphere, when considered in the light of Dr. Tyndall's beautiful researches on radiant heat. He has found that the presence of a few hundredths of carbonic acid gas in the atmosphere, while offering almost no obstacle to the passage of the solar rays, would suffice to prevent almost entirely the loss by radiation of obscure heat, so that the surface of the land beneath such an atmosphere would become like a vast orchard-house, in which the conditions of climate necessary to a luxuriant vegetation would be extended even to the polar regions. This peculiar condition of the early atmosphere cannot fail to have influenced in many other ways the processes going on at the earth's surface. To take a single example: one of the processes by which gypsum may be produced at the earth's surface involves the simultaneous production of carbonate of magnesia. This, being more soluble than the gypsum, is not always now found associated with it; but we have indirect evidence that it was formed and subsequently carried away, in the case of many gypsum deposits, whose thickness indicates a long continuance of the process under conditions much more perfect and complete than we can attain under our present atmosphere. While studying this reaction I was led to inquire whether the carbonic acid of the earlier periods might not have favoured the formation of gypsum; and I found, by repeating the experiments in an artificial atmosphere impregnated with carbonic acid, that such was really the case. We may thence conclude that the peculiar composition of the primeval atmosphere was the essential condition under which the great deposits of gypsum, generally associated with magnesian limestones, were formed.

The reactions of the atmosphere which we have considered, would have the effect of breaking down and disintegrating the surface of the primeval globe, covering it everywhere with beds of stratified rock of mechanical or of chemical origin.  These would now so deeply cover the partially cooled surface that the amount of heat escaping from below is inconsiderable, although in earlier times it was very much greater, and the increase of temperature met with in descending into the earth must have been many times more rapid than now.  The effect of this heat upon the buried sediments would be to soften them, producing new chemical reactions between their elements, and converting them into what are known as crystalline or metamorphic rocks, such as gneiss, greenstone, granite, etc.  We are often told that granite is the primitive rock or substratum of the earth, but this is not only unproved, but extremely improbable.  As I endeavoured to show in the early part of this discourse, the composition of this primitive rock, now everywhere hidden, must have been very much like that of a slag or lava ; and there are excellent chemical reasons for maintaining that granite is in every case a rock of sedimentary origin—that is to say, it is made up of materials which were deposited from water, like beds of modern sand and gravel, and includes in its composition quartz, which, so far as we know, can only be generated by aqueous agencies, and at comparatively low temperatures.

The action of heat upon many buried sedimentary rocks, however, not only softens or melts them, but gives rise to a great disengagement of gases, such as carbonic and hydrochloric acids, and sulphur compounds, all results of the reaction of the elements of sedimentary rocks, heated in presence of the water which everywhere filled their pores.  In the products thus generated we have a rational explanation of the chemical phenomena of volcanoes, which are vents through which these fused rocks and confined gases find their way to the surface of the earth.  In some cases, as where there is no disengagement of gases, the fused or half-fused rocks solidify *in situ*, or in rents or fissures in the overlying strata, and constitute eruptive or plutonic rocks like granite and basalt.

This theory of volcanic phenomena was put forward in germ by Sir John F. W. Herschel thirty years since, and, as I have during the past few years endeavoured to show, it is the one most in accordance with what we know both of the chemistry and the physics of the earth.  That all volcanic and plutonic phenomena have their seat in the deeply buried and softened zone of sedimentary deposits of the earth, and not in its primitive nucleus,

accords with the conclusions already arrived at relative to the
solidity of that nucleus, with the geological relations of these
phenomena as I have elsewhere shown; and also with the remarkable mathematical and astronomical deductions of the late Mr.
Hopkins, of Cambridge, based upon the phenomena of precession
and nutation; those of Archdeacon Pratt; and those of Professor Thompson on the theory of the tides; all of which lead to
the same conclusion—namely, that the earth, if not solid to the
centre, must have a crust several hundred miles in thickness,
which would practically exclude it from any participation in the
plutonic phenomena of the earth's surface, except such as would
result from its high temperature communicated by conduction to
the sedimentary strata reposing upon it.

The old question between the plutonists and the neptunists,
which divided the scientific world in the last generation, was, in
brief, this—whether fire or water had been the great agent in giving origin and form to the rocks of the earth's crust. While
some maintained the direct igneous origin of such rocks as gneiss,
mica-schist, and serpentine, and ascribed to fire the filling of
metallic veins, others—the nuptunial school—were disposed to
shut their eyes to the evidences of igneous action on the earth,
and even sought to derive all rocks from a primal aqueous magma.
In the light of the exposition which I have laid before you this
evening, we can, I think, render justice to both of these opposing
schools. We have seen how actions dependent on water and acid
solutions have operated on the primitive plutonic mass, and how
the resulting aqueous sediments, when deeply buried, come again
within the domain of fire, to be transformed into crystalline and
so-called plutonic or volcanic rocks.

The scheme which I have endeavored to put before you in the
short time alloted, is, as I have endeavoured to show, in strict
conformity with known chemical laws and the facts of physical
and geological science. Did time permit, I would gladly have
attempted to demonstrate at greater length its adaptation to the
explanation of the origin of the various classes of rocks, of metallic
veins and deposits, of mineral springs, and of gaseous exhalations.
I shall not, however, have failed in my object, if, in the hour
which we have spent together, I shall have succeeded in showing
that chemistry is able to throw a great light upon the history of the
formation of our globe, and to explain in a satisfactory manner
some of the most difficult problems of geology; and I feel that

there is a peculiar fitness in bringing such an exposition before
the members of this Royal Institution, which has been for so
many years devoted to the study of pure science, and whose glory
it is, through the illustrious men who have filled, and those who
now fill, its professorial chairs, to have contributed more than any
other school in the world to the progress of modern chemistry and
physics.

———

# REVIEW.

## " MANUAL OF THE BOTANY OF THE NORTHERN UNITED STATES."

By ASA GRAY, Fisher Professor of Natural History in
Harvard University. New York, 1867.

A fifth edition of this very useful manual has been recently
issued. The author has, to a great extent, re-written the work,
and, in the elaboration of some parts, has received active co-operation from some other American botanists, prominent among whom
are Dr. George Engelmann and Prof. D. C. Eaton. Important
changes in the arrangement of some of the orders and genera have
been embodied in this edition ; the geographical range of very
many species has been extended ; naturalized as well as indigenous
plants—some familiar as Canadian—not previously known to
occur within the Northern United States (as limited by the author)
have been recently discovered and are now included; and to the
work have been added not a few new species.

In the present edition there are many points of considerable
interest to Canadian botanists.

Among the orders several noticeable changes occur. The
Cabombeæ are treated by the author as constituting a sub-order
of Nymphæaceæ, and the Limnantheæ, Balsamineæ and
Oxalideæ as sub-orders of Geraniaceæ. This comprehensive view
of Geraniaceæ is that originally entertained by Jussieu, the
founder of the order, but regarding which difference of opinion
has existed among later botanists. The irregular, unsymmetrical
flowers, the usually fewer sepals, petals, and stamens, the spur or
sac on the posterior sepal, the simple leaves, as well as other distinguishing characters of the Balsams, seem to entitle, at least,

them to rank as a distinct order. The Parnassiæ, which in many respects approach the Hypericaceæ, but the flowers of which are, as indicated in former editions of the Manual, sometimes clearly perigynous, and the Grossulariæ are removed by the author to Saxifragaceæ. The Haloragæ, formerly regarded as a sub-order of Onagraceæ, he now considers to have characters sufficient to constitute an order. Under Liliaceæ, as here extended and re-arranged, are included the Trillideæ, Melanthiceæ, and Uvularieæ.

Among the genera there are not many changes to note. Atragine, distinguished by the presence of petals which gradually merge into stamens, is included in Clematis. Iodanthus and Turritis are referred to Arabis, and Alsine, Mœhringia, and Honkenya, also considered by some authors, as well as in former editions, as genera, and of which the last named has considerable claims to generic distinction, are comprised in Arenaria. Further, among endogens, the older genus Habenaria, distinguished from Orchis by its naked and exposed separate glands or viscid disks. is revived in this edition, and the Gymnadenia and Platanthera of former editions referred to it. Whilst on the subject of genera, it may be added that it admits of grave doubt whether an author when changing a species from one genus to another should wholly suppress the name of the original describer.

Mr. Paine's new Water Lily, *Nymphæa tuberosa*, from Oneida Lake and other parts of the Union, and which has been recently observed near Belleville by Mr. Macoun, is fully described. *Arabis petræa* Lam., which occurs on the Canadian side of Lake Superior, appears now as a United States plant, having been found on Willoughby Mountain by Mr. Horace Mann. *Oxytro-pis campestris* DC., it will interest Quebec and New Brunswick botanists, is to be looked for about the Maine boundary line. The other species, *O. Lamberti* Pursh, an interesting local plant of the Province of Quebec, is another noticeable addition to Dr. Gray's work. Such are also *Parnassia parviflora* DC., one of our Anticosti plants, which has been observed on the north-west shore of Lake Michigan, and *Sedum Rhodiola* DC., a rather boreal plant of Anticosti, Labrador, and Newfoundland, which has been met with in Maine and, curiously enough, in Pennsylvania on cliffs of the Delaware River above Easton. Among other recent additions of interest to the flora of the Northern United States there may be mentioned, as species previously known to occur in either Quebec or Ontario, *Matricaria inodora* Linn., a native of the far-west. introduced from Europe into Maine; *Senecio pseudo-arnica*, Less., a plant of Anticosti and northward, detected on Grand Manan Island, off the coast of Maine; *Polemonium cæruleum* Linn., and *Corispermum hyssopifolium* Linn., both western plants, the latter apparently extending eastward; *Rumex patientia* Linn., a stray introduction into both countries from Europe, and *Sagittaria calycina* Engl., also a recent addition to our flora from Grand Manitoulin Island, where it has been collected by Dr. John Bell.

A hasty enumeration of the number of genera and species shows that, numerically, considerable additions have been made. Of Exogenous plants there are 627 genera and 1842 species, and of Endogenous plants 174 genera and 716 species. The increase has chiefly taken place in the orders Leguminosæ, Compositæ, Naidaceæ, Cyperaceæ, and Graminæ.

Six lithographic plates, illustrative of the genera of the Cyperaceæ, have been added to the fourteen illustrating the Graminæ and Filices. These will prove useful aids to the young botanist.

A. T. D.

Mr. Eaton's elaboration of the ferns is painstaking, able and thorough. Four sub-orders are represented within the limits, Polypodiaceæ, Schizæaceæ, Osmundaceæ and Ophioglossaceæ. The second of these contains the genera Schizæa and Lygodium which have not yet been detected in Canada. Some changes have been made in the arrangement of the genera composing the Polypodiaceæ. Phegopteris has been seperated from Polypodium and put next to Aspidium, its proper place, as was long ago indicated by Roth (who included it in his genus Polystichum, the equivalent of Swartz's Aspidium) and by Fée, who founded the genus. Struthiopteris has been removed from Pterideæ to Aspidieæ and placed next to Onoclea, but not included in that genus, chiefly because of its different venation. Pellæa has been' seperated from Allosorus, the only species which retains the latter name being *Allo. acrostichoides*, an inapt section, inasmuch as Bernhardi's name is not appropriate to any other genus than Cheilanthes of Swartz, and moreover, Robert Brown's well named and well defined genus, Cryptogamme, was constituted expressly for this species. Sir William Hooker held (probably correctly) that our North American plant was identical with the European *C. crispus;* Mr. Eaton appears to consider them distinct. Mr.

Eaton has here cleared up the confusion which existed among our
species of the genus Cheilanthes. He has confirmed the Abbé
Brunet's observations that *Neph. lanosum* of Michaux is the
"*Ch. vestita* Willdenow" of former editions and of American
botanists generally, — the "*Ch. vestita* Swartz" of the present
edition of this work. It is matter for regret that Michaux's
name has not been respected, but, having been continued through
so many editions and now confirmed in this one, and being probably
applied to the plant which the founder of the genus had in view,
the name *vestita* must now stand; it is, however, noteworthy that
Swartz misunderstood Michaux's plant—he believed it to be an
Aspidium, and that Sir William Hooker and other European
botanists have applied *vestita* to the plant here named *Ch. lanuginosa*. *Ch. vestita* has been found by Mr. Denslow, as far north
as the island of New York, and his specimens appear to be as
vigorous as those of more sourthern latitudes. *Ch. tomentosa* of Link
(Lindheimer No. 743; *Ch. Bradburii*, Hook. Sp. Fil.) is not stated
to be rare, and yet specimens of it appear to be very scarce in the
herbaria of American botanists. The third species of Cheilanthes
which occurs within the geographical limits is here named *Ch.
lanuginosa*, a MS. name given by Nuttall; this must give place to
Riehl's earlier name *Ch. gracilis* which has been adopted by Fée
and by Mettenius: this plant is the *Ch. vestita* of Hooker's
Species Filicum. In Asplenium one new species has been admitted,
*A. ebenoides* R. R. Scott, which is only an abnormal form of *A.
ebeneum*. In Woodsia, Mr. Eaton has receded from the position
assumed in his paper on the genus contributed to this journal,
having readmitted *W. glabella* to the rank not merely of a good
species but of a purely American species. The truth would appear
to lie between his two extremes; those glaberous Lapland plants,
named *W. hyperborea* by Scandinavian botanists, are certainly
identical with our *W. glabella*, and are possibly what Liljeblad had
before him when describing his *Acros. hyperboreum*, and also what
Wahlenberg named *Polypodium hyperboreum* in "Flora Lapponica." Our plants are, however, certainly distinct from the
*Acros. alpinum* of Bolton (*W. hyperborea*, R. Br., Hook. etc.)
which is very near to *W. Ilvensis* if, indeed, it be separable from
it. In Botrychum, *B. simplex* is admitted as a species, as is also
*B. lanceolatum*, neither of which have much claim to that rank.
With his views of generic limits, Mr. Eaton might very fairly have
seperated Dryopteris from Polystichum; he has, however, following

Swartz and Mettenius, combined them in Aspidium. Leaving
out *Aspl. ebenoides*, there are described fifty-six good species and
six well-marked ("black-letter") varieties. Of these the following
were not recognized in the second edition :—*Ch. lanuginosa, Allo.
acrostichoides, Aspd. Filix-mas, Woodsia Oregana, Botrychium
Lunaria, B. simplex* and *B. lanceolatum.* Twenty species are
marked with the contraction "Eu." as being common to Europe
and America, and from three it has been accidentally omitted (*B.
simplex, B. lanceolatum* and *B. virginianum*); to these I would
add for the reasons above stated *Allo. acrostichoides, Woodsia
glabella,* and also probably *Aspd. fragrans* (which appears to extend
all round the Arctic circle), which would increase this number to
twenty-six. But the remaining thirty species are not all of them
confined to America; *Adiantum pedatum, Pellæa gracilis, Aspl.
thelypteroides* and *Onoclea sensibilis* are also Asiatic and *Aspl.
ebeneum* has been collected in Africa. Of the sixty-two species
and varieties forty-nine are known to me as Canadian, in addition
to which the following may be looked for within our boundaries
with good prospect of success,—*Ch. vestita, Woodwardia areolata,
Aspl. Ruta-muraria, Aspd. Filix-mas* (about Lake Superior),
*Woodsia obtusa, W. Oregana* (about Lake Superior), and
*Lygodium palmatum ;* on the other hand, *Polypodium incanum,
Ch. tomentosa, Ch. gracilis, Aspl. montanum, A. pinnatifidum*
and *Schizæa pusilla* are pretty surely beyond our reach. We
have left to us but two Canadian ferns not noticed by Mr. Eaton,
*W. hyperborea* R. Brown, and *Aspl. viride* Hudson, but as the
first-named has been found on Willoughby Mountain by Mr.
Horace Mann, and the latter is most probably a native of the
northern parts of Maine, etc., Mr. Eaton might as well have
included them and thus had the opportunity of fully revising his
former views on the genus Woodsia. *Aspidium spinulosum* has
been split up into four varietal forms—dilatatum, intermedium,
verum, and Boottii, the var. intermedium (*Aspd. intermedium*
Willd.) being our common narrow form; this would seem to be a
somewhat too minute subdivision. The large broad form of *A.
cristatum* which we have been calling var. *majus* is here named
var. *Clintonianum,* in compliment to Judge Clinton of Buffalo.
The var. *Braunii* of *A. aculeatum* is hardly entitled to "black
letters," it being merely a form of the var. *angulare*—the *A.
angulare* of Willd. etc.—unless on the supposition that this latter
is a good species. The plates being unaltered, the genera Allosorus

and Phegopteris are not illustrated; the figures of *Ch. vestita*
and *W. glabella* are very indifferent, the latter particularly so.

The following extracts from the preface are of interest :—

" This work is designed as a compendious Flora of the Northern
portion of the United States, for the use of students and of
practical botanists.

" The first edition (published in 1848) was hastily prepared to
supply a pressing want. Its plan, having been generally approved,
has not been altered, although the work has been to a great extent
twice rewritten, and the geographical range extended. The second
edition, much altered, appeared in 1856. The third and fourth
were merely revised upon the stereotype plates, and some pages
added, especially to the latter.

" The *Garden Botany*, an Introduction to a Knowledge of the
Common Cultivated Plants, which was prefixed to this fourth
edition in 1863, is excluded from the present edition, and is to
be incorporated into a simpler and more elementary work, but of
wider scope, designed especially for school instruction, and for
those interested in cultivation, — entitled *Field, Forest, and
Garden Botany*.

" In the present edition, it has been found also expedient to
remand to a supplementary volume the *Mosses* and *Liverworts*,
so carefully and generously elaborated for the previous editions
of this work by my friend, Wm. S. Sullivant, Esq. It is hoped
that the *Lichenes*, if not all the other orders of the Lower
Cryptogamia, may be added to this supplementary volume, so
that our students may extend their studies into these more
recondite and difficult departments of Botany.

\*　　\*　　\*　　\*　　\*　　\*　　\*　　\*

" There is abundant reason, I doubt not, for me to renew the
request that those who use this book will kindly furnish information of all corrections or additions that may appear to be necessary,
so that it may be more accurate and complete hereafter, and
maintain the high character which it has earned.

" Geographical Limitation, Distribution, etc. As is stated on the
title-page, this work is intended to comprise the plants which grow
spontaneously in the United States, north of North Carolina and
Tennessee, and east of the Mississippi. A Flora of the whole
national domain, upon a similar plan (the issue of which I may

now hope will not be delayed many years longer), would be much
too bulky and expensive for the main purpose which this Manual
fulfils.   For its purpose, the present geographical limitation is, on
the whole, the best,— especially since the botany of the states
south of our district has been so well provided for by my friend,
Dr. Chapman's *Flora of the Southern States*, issued by the same
publishers.   The southern boundary here adopted coincides better
than any other geographical line with the natural division between
the cooler-temperate and the warm-temperate vegetation of the
United States; very few characteristically Southern plants occurring north of it, and those only on the low coast of Virginia, in the
Dismal Swamp, etc.   Our Western limit, also, while it includes a
considerable prairie vegetation, excludes nearly all the plants
peculiar to the great Western woodless plains, which approach
our borders in Iowa and Missouri.   Our northern boundary, being
that of the United States, varies through about five degrees of
latitude, and nearly embraces Canada proper on the east and on
the west, so that nearly all the plants of Canada East on this side
of the St. Lawrence, as well as those of the deep peninsula of
Canada West, will be found in this volume.

\*        \*        \*        \*        \*        \*        \*        \*

" Distinction of Grade of Varities.   Vain is the attempt to draw
an absolute line between varieties and species.   Yet in systematic
works the distinction has to be made absolute, and each particular
form to be regarded as a species or a variety, according to the
botanist's best judgment.   Varieties, too, exhibit all degrees of
distinctness.   Such as are marked and definite enough to require
names are distinguished here into two sorts, according to their
grades : 1. Those which, I think, cannot be doubted to be
varieties of the species they are referred to, have the name printed
in small capitals.   These varieties make part of the common
prragraph.   2. Those so distinct and peculiar that they have
been, or readily may be, taken for species, and are some of them
not unlikely to establish the claim : of these the name is printed
in the same [black letter] type as that of the species ; and they
are allowed the distinction of a seperate paragraph, except where
the variety itself is the only form in the country."

The whole work is a model of accurate description, correct
orthography and typographical excellence.                         W.


Published, Montreal, 1st January, 1868.

# CANADIAN NATURALIST

SECOND SERIES.

## ON THE GEOLOGICAL FORMATIONS OF LAKE SUPERIOR.*

By THOMAS MACFARLANE.

### III. UPPER COPPER-BEARING SERIES.

The name of the Upper Copper-bearing Rocks of Lake Superior was given to this series by Sir W. E. Logan, to distinguish it from the Huronian or Lower Copper-bearing Rocks. The geographical and geological position, lower altitude, regular bedding, and peculiar lithological character of these Upper Rocks cause them to be easily recognised and readily distinguished from the Huronian. They have been separated into an upper and lower group, the latter of which seems, however, to be confined to the north-west parts of the lake. Along its eastern shore, between Sault St. Marie and Michipicoten, there are frequently found, betwixt the water and the high Huronian or Laurentian hills, narrow strips or patches of the rocks of the upper group, which often jut out as small islands into the lake, and doubtless extend out great distances beneath its waters. Such limited strips of these rocks are found, for instance, skirting the base of Gros Cap, along the south shore of Bachewahnung Bay and at Cape Gargantua. But besides these and much more important for the study of the upper group of the Upper Copper-bearing series, there are occasional extensive developments of its rocks, many thousand feet in thickness, such as at Cape Mamainse, Michipicoten Island, and Point Keweenaw

---

* Continued from page 201.

on the south shore. These rocks have been generally described
in the Geology of Canada as sandstones, conglomerates, stratified
traps and amygdaloids. In referring to them more minutely, the
following rock-varieties may be distinguished as belonging to the
upper group of the series :—

*Granular Melaphyre.*—A large number of the rocks of this
series which have hitherto been described as traps and greenstones,
belong to this species. The simplest variety of it is seen at the
north-west end of Michipicoten Island, and consists of two
minerals only, a felspar and a greenish black mineral. The
felspar is the principal constituent, possesses a red, almost pink,
colour, which it loses on ignition, and being readily fusible and
but slightly decomposed by acids, is most probably oligoclase, or
closely allied to that species in composition. The dark coloured
mineral is easily fusible and has the appearance of augite. Some
of it appears soft and decomposed, and has most probably been
converted into delessite. These two minerals are combined into a
small grained, distinctly compound rock, which does not effervesce
with acids, and whose red colour is visible at a considerable
distance. It is very seldom however that this rock is observed
with such a bright colour, or with constituents so distinctly
separated. Much more frequently the felspar has a dark reddish-
brown colour, and the grains of augite or delessite have a very
indistinct contour. This is the case with some of the melaphyres
of Mamainse and Gros Cap. When the brown coloured felspar
predominates, and the augitic or chloritic constituent becomes
scarcer and even more indistinct, rock-varieties are developed
belonging to the species Porphyrite, hereafter to be described.
When, on the other hand, the dark greenish constituent gains the
upper hand, and is recognisable as consisting almost exclusively
of delessite, it gives rise to the variety of melaphyre next
described.

*Delessitic Melaphyre.*—This rock has a greenish-gray colour,
and consists of a granular mixture of felspar and delessite, with
small portions of magnetite and undecomposed augite. In some
instances mica is also found as a constituent. The delessite,
besides occurring in small grains, often forms larger rounded
particles and amygdules, without however imparting to the rock
a very marked amygdaloidal structure. The rocks enclosing the
cupriferous beds of the Pewabic and Quincy Mines, and that from
the Quincy adit are examples of this variety, and have already been

described by me in this journal.*   The delessite which enters
so largely into their composition can scarcely have been one
of the original constituents, and has probably resulted from
the gradual alteration of augite, since authenticated instances
are on record of the conversion of that mineral into delessite and
green-earth.   The specific gravity of these rocks varies from 2.83
to 2.89.  When ignited they lose 1.32 to 3.09 per cent. of their
weight, the powder changing from light greenish-grey to a light
brown colour.   Digested with hydrochloric acid from 32.44 to
35.72 per cent. of bases are removed from them, the greater part
of which belongs to the chloritic constituent.   While the variety
of melaphyre first above described is seldom found with
amygdaloidal structure, the delessitic melaphyres are exceedingly
prone to be developed as amygdaloids.   In this case the rock
contains amygdules of small size but very numerous, and they are
either filled with delessite alone, or are lined with a coating or
rind of that mineral, in which latter case calcspar generally
fills out the centre of the cavity.   Quartz or agate is
comparatively rare in amygdaloids the matrix of which is
delessitic melaphyre.

*Compact Melaphyre.*—When the small grained melaphyres
above described become so fine-grained as to render the recognition
of their constituents impossible, there results the fine-grained traps
which are so numerous on the south-west coast of Mamainse and
on Michipicoten Island.   These rocks vary from reddish, bluish,
greenish, or greyish black, to decided black in colour, and possess
not unfrequently conchoidal fracture and resinous lustre.   Their
specific gravities vary from 2.67 to 2.898, and they fuse before
the blowpipe to glasses of black or brownish black colour.
Occasionally their material becomes less homogeneous, and presents
the appearance of an intimate mixture of reddish grey and green
coloured specks, which may perhaps represent partially developed
constituents.   They exhibit various phenomena as regards
divisional joints.   Some possess a rudely columnar structure,
others have planes of separation forming various angles with the
plane of bedding, several shew a tendency to separate into flags,
while a few instances are observable of curved shaly separation,
(*Krummschaalige Absonderung*).   Transitions can frequently be
traced from these compact melaphyres to others approaching in

---

* Vol. iii., Second Series, p. 2.

character to porphyrite. For instance, to the west of the entrance to the harbour on the south side of Michipicoten Island, there is found, forming part of a bed of undoubted compact melaphyre, a rock of a greenish-grey colour, with conchoidal fracture. It had a specific gravity of 2.589, and could only be glazed at the edges before the blowpipe. To the east of the same harbour entrance, another rock occurs intermediate in character betwixt compact melaphyre and porphyrite. It is black, impalpable, with imperfectly con-choidal fracture. It bears some resemblance to pitchstone, but differs from that rock in its specific gravity, which is 2.774, and in being readibly fusible to a black glass. It possesses a slightly resinous lustre, and contains an occasional crystal of colourless triclinic felspar. It exhibits planes of separation at right angles, or nearly so with the inclination of the bed, and agate veins are observable, which seem to accompany the divisional joints. This latter phenomenon is also seen in some of the beds of compact melaphyre, and in one of these, curved joints are visible, standing at right angles to the plane of bedding and filled out with calcspar. Brecciated quartz veins occasionally permeate these rocks, and agatic geodes are very frequent among them. The latter are sometimes so frequent as to form amygdaloids, but they are much larger, and never so numerous as are the cavities in the amygdaloids of which delessitic melaphyre is the matrix. There is further this peculiarity with the amygdules of the compact melaphyres, that they contain little or no delessite, agate occupying its place, with occasionally calcspar filling the centre of the geode.

*Tufaceous Melaphyre.*—Interstratified with the rocks above described, and much more frequently associating with, and gradu-ating into the delessitic melaphyres than the other varieties, there are occasionally found beds of comparatively soft, dark brown, porous rock, with almost earthy fracture and seldom destitute of amydaloidal structure. These frequently carry metallic copper, and constitute the 'ash beds' so extensively worked in the mines of the south shore. Although they are generally of a dark brown or chocolate colour, as in the case of the 'Pewabic lode,' there are rocks of this species which are bluish-brown and green coloured. The matrix is generally fusible, and in places impregnated with grains of metallic copper, sometimes of a very minute size. The larger grains of the metal are frequently found in the amygdules, either alone or accompanied by green-earth, calcspar, quartz, delessite, laumontite, and prehnite. Besides the rounded grains or 'shot copper' of the amygdules, these rocks often contain huge masses of metallic copper, with which small quantities of native silver are associated. Large irregular patches and veins of calcspar, and smaller masses of epidote are frequently met with in these tufaceous melaphyres.

*Porphyrite.* - The transitions, which are frequently observable on the south side of Michipicoten Island, from compact melaphyre to porphyrite have been referred to above. Undoubted porphyrite is to be found at the south-west corner of the Island. It possesses a fine-grained greenish red matrix, containing small flesh-coloured crystals of felspar, some of which have striated cleavage planes. The specific gravity of the rock is 2.619, and the matrix is fusible at the edges. In the upper part of the bed the matrix of the rock becomes coarser grained, shewing distinctly felspar and a darker coloured mineral as constituents, with the small felspathic crystals still scattered through it. The felspar predominates in the matrix and determines the colour of the rock, which is dark red. Its specific gravity is 2.626, and it is fusible, although not readily, before the blow-pipe. It separates into blocks, with very decided divisional planes, but of no regular form. Similar rocks are found at the south-east corner of the Island, where also rocks resembling pitchstone and pitchstone porphyry are extensively developed. The black shining impalpable trap, which has the appearance of pitchstone, has a specific gravity of 2.573. It is fusible to a brown glass, and sometimes contains small colourless felspar crystals. Where these accumulate, there results the rock resembling pitchstone porphyry. The crystals in this rock are frequently recognisable as triclinic. The matrix is fusible to a brown blebby glass, and the specific gravity of the rock as a whole is 2.631 to 2.678. Since the specific gravity of the rock in which no crystals occur is lower than that usually ascribed to melaphyre, and since it is greater than that of true pitchstone, it would appear reasonable to class both these rocks with the porphyrites, or with these porphyries which contain no quartz, to which they probably bear the same relation as true pitchstones bear to felsitic or quartzose porphyries.

*Melaphyre Breccia.*—Among the newest of the beds of compact melaphyre, developed on Michipicoten Island, there are sometimes observable beds of a breccia consisting of fragments of dark brown melaphyre, cemented together by a reddish-brown trappean sand. Occasionally the fragments appear rounded, and present more of the character of a conglomerate. Similar rocks are seen in the Point Keweenaw district.

*Porphyritic Conglomerate.*--At the south-west corner of Michipi-coten Island there is visible a conglomerate bed, the boulders of which consist principally of porphyrite, in which a few minute felspar crystals are discernible. Some of the boulders are granitic, and occasionally pebbles occur consisting of or containing agate. These are enclosed in a matrix consisting of coarse-grained and red-coloured porphyritic or trappean debris. In the upper part of the Mamainse group similar conglomerates are found, but in one instance the matrix seems to consist of the same crystalline materi l as the boulders and fragments, and is very firmly cemented to these. The most interesting example of this rock is that of the Albany and Boston mine, near Portage Lake. Here the matrix of coarse-grained porphyritic sand is accompanied by calcspar, and in some places fine metallic copper.* Other porphyritic conglo-merates occur to the south of Portage Lake, some of the boulders of which consist of quartzose porphyry, and the matrix of some of which contains quartz as well as calcspar.

*Felsite-tuff.*—Overlying the Albany and Boston conglomerate a bed of so-called 'fluckan' occurs, which is a fine-grained, dark-red shaly rock, in which pieces of a greenish blue colour are sometimes seen. Both substances are fusible before the blow-pipe, and contain occasionally small grains and flakes of copper.

*Polygenous Conglomerate.*—This name is applied by Naumann and Zirkel to those fragmentary rocks whose boulders consist of two or more different rocks. Conglomerates of this nature are especially frequent among the inferior rocks of the Mamainse group, and among those of Keweenaw Point. The boulders of these Mamainse conglomerates are chiefly of granite, gneiss, quartzite, greenstone, and slate, and some of the newer beds con-tain boulders of melaphyre and amygdaloid in abundance. The matrix is generally a dark red sandstone.

*Sandstone.* — Among the melaphyres and conglomerates of Mamainse and Point Keweenaw an occasional stratum of sand-stone is found of the same character as that which forms the matrix of the polygenous conglomerates.

The manner in which the rocks above described are associated with each other, is much more regular than the architecture of

---

* This Journal, Vol. iii., Second Series, p. 9.

the Laurentian and Huronian rocks. They are regularly interstratified with each other, and even among the melaphyres and
porphyrites distinct bedding is observable. They do not
seem to have been disturbed to such a degree as to occasion the
formation of anticlinal and synclinal folds, and in each of the
principal areas of distribution a tolerably persistent strike and dip
can be observed.

The general strike of the rocks of the Mamainse group is N.
20° to 50° W., and the dip 20° to 45° south-westward. They
are beautifully exposed along the west coast of Mamainse, and
the highest strata of the group form the south-west extremity of
the cape. The lower part of the group consists of granular and
delessitic melaphyres, polygenous conglomerates and sandstone. In
the upper part compact melaphyres and porphyritic conglomerates
predominate. The total thickness of the group, according to an
approximative measurement, is 16,208 feet, of which the conglomerates occupy 2,138 feet. The succession of the beds along the
coast is quite regular; but on attempting to follow them inland,
they are found to thin out and disappear, while others take their
places. This is especially the case with the conglomerates. Were
the beds continuous throughout, the section above given ought to
be repeated on the south coast, and round to Anse-aux-Crêpes.
But there, although some of the melaphyre beds have the same
strike and dip as on the west coast, there is not the same regularity,
nor the same plentiful development of conglomerates. There
are moreover evidences of great disturbances and of a conflict
between the rock of some of the igneous beds and a sandstone,
which here appears in highly contorted and sometimes vertical
strata. On coming round the south coast of Mamainse, from
Anse-aux-Crêpes, strata of sandstone are observed very much
disturbed and dipping inland. As near as it can be ascertained,
their strike is about N. 85° W., dip 25° to 40° northward. The
sandstone is red coloured, and contains streaks and spots of a
cream coloured felspathic substance, which also forms bands crossing
the stratification. Many thin cracks filled with calcspar also
traverse the beds. The same sandstone continues for about a
hundred and forty yards further to the west, becoming still more
disturbed, and containing between its layers the felspathic
substance. The strike, where the beds are at all regular, is N. 10°
W., and dip 52° eastward. Further west it changes to N. 52°
E., with dip vertical, and in places 75° S. W. Here the sandstone

becomes utterly broken up into a breccia, which has pieces from one inch to a foot in diameter invariably angular, and a matrix consisting of the white felspathic substance above mentioned, with occasionally calcspar. Further westward the measures are concealed for two hundred yards; then strata of bluish-grey calcareous sandstone are exposed, striking N. 40° E., and dipping 75° S. E. From this point for three hundred yards further north-westward, disturbed sandstone occupies the coast where the measures are not concealed. It is followed by a breccia similar to that already mentioned, with angular fragments of sandstone, and then by beds of trappean rocks, striking N. 75° W., and dipping 40° S. W. Rocks of this nature occupy the coast, where not concealed, for one and a half miles further north-westward. Here sandstone again becomes visible, in strata almost vertical, but nevertheless much bent. It is covered by a breccia consisting of sandstone fragments with a trappean matrix, and this again is surmounted by regular trap. In many places there would seem to be the clearest evidence that the trap lies unconformably upon the upturned and contorted edges of the sandstone. Besides the breccia above mentioned, other rocks of a peculiar nature are found at the junction of the sandstone and trap. One of these is indistinguishable from quartzose perphyry, and another seems to consist of fragments of trap bound together by this same quartzose perphyry. There are good grounds for supposing that the latter rock is the product of the action of the more basic trap upon the sandstone, and results from the igneous amalgamation of the two rocks last named. These confused rocks occupy about a quarter of a mile of the coast. To the north-westward, although the sandstones occasionally protrude, they become much less frequent, while the overlying melaphyres become much more regular, and gradually assume the same strike and dip as the strata on the west coast. The hills to the north of Anse-aux-Crêpes consist of the same beds of melaphyre and conglomerate as were observed on the west coast, with similar strike and dip.

The eruptive origin of the melaphyres and traps of this group is evidenced not only by their crystalline character, and by some of their relations in contact with undoubted sedimentary rocks, but also by their occurring as intrusive masses in the gneiss of Point-aux-Mines, and in the granitoid gneiss of Chippewa Falls. At the latter place the melaphyre is in the form of a dyke, and at Point-aux-Mines it is seen to form a dome-shaped mass, completely

surrounded by gneissied rocks.  Furthermore, the lower members
of the Mamainse series are intersected by numerous dykes, consisting of compact melaphyre.  In some of them, the constituents
of that rock are distinguishable, but most of them are almost
impalpable, vary from a reddish-brown to a dark green colour, and
frequently exhibit at their sides bands of slightly different colours,
which run parallel with the side-walls of the dyke.

The average strike of the Upper Copper-bearing rocks of Michipicoten Island is N. 68° E., and the dip 25° south-eastward.  An
approximative estimate of their thickness is as follows:—

Granular, delessitic and compact melaphyres,
    and conglomerates.................10,000 feet.
Compact melaphyres with agate amygdules. 4.500  "
Resinous traps, porphyrites and breccias... 4,000  "

                            18,500 feet.

If we compare the rocks of Michipicoten Island with those of
Mamainse, it would appear that the inferior rocks of the latter
group do not come to the surface at Michipicoten Island, and that
the higher rocks of the Michipicoten group have not been developed at Mamainse, or lie beneath the waters of the lake to the
south-west of the promontory.  It would therefore appear just, in
estimating the thickness of the Upper Copper-bearing rocks of the
eastern part of Lake Superior, to add to the Mamainse series the
above mentioned 4000 feet of resinous traps or porphyrites, which
would make the whole thickness at least 20,000 feet.  The rocks
of the west and south shores of Michipicoten Island present the
most regular appearance, and it might be expected that those of
the south shore would, from their strike and dip, repeat themselves on the east side.  But, as in the case of Mamainse, such an
expectation is disappointed.  On examining the rocks of the east
shore, the upper beds, consisting of the porphyrites above mentioned, seem regular enough, but beneath these come brecciated
melaphyre, delessitic melaphyre cut by a porphyritic rock, and
others in which the evidences of bedding are very indistinct.
Among these rocks the two following may be particularised as
occuring in large masses.  The first has an impalpable flesh-red
or reddish-grey matrix, wherein occur numerous grains of dark
grey quartz, and also light-coloured soft particles, which
seem liable to removal by atmospheric agencies, giving
the rock where this has taken place a porous appearance.

It also contains light red and ash-grey crystalline grains of felspar, and others which appear earthy and decomposed. The matrix is fusible, in fine splinters only, to a white enamel. The rock has an uneven fracture, a specific gravity of 2.493, and is probably a porphyritic quartz-trachyte. The other rock, which occupies a very considerable area, partakes more of the character of felsitic porphyry, although the felspar crystals are very often indistinct. It contains, besides these, numerous grains of greyish quartz, sometimes one-eighth of an inch in diameter, and a fine-grained, dark red, difficultly fusible, matrix. The specific gravities of three different specimens were found to be 2.548, 2.579, and 2.583. The bedding of the rock, if it possesses any, is very obscure; but it shews in places a tendency to separate into flags. It has a very rough uneven fracture, and is probably also quartzose trachyte. At the north-east corner of the Island it seems to overlie, unconformably, beds of trap, which here assume something like the ordinary strike and dip, viz., N. 72° E., dip 25° S. E.

The islands which lie opposite the mouth of the harbour on the south shore are composed of a peculiar rock, which is nowhere visible on the main island. It consists of a reddish-brown impalpable matrix, with a hardness but slightly inferior to that of orthoclase, in which minute spots of a soft yellowish-white material are discernible. There are also lighter flesh-coloured grains observable, which seem to be incipient felspar crystals. The matrix is difficultly fusible to a colourless blebby glass, and the specific gravity of the whole rock, where freshly broken, is 2.469. A piece slightly bleached to a greyish-white, from its adjoining a crack in the rock, gave a specific gravity of 2.477. Some parts of it exhibit a slightly porous structure, but this was not the case with either of the pieces whose specific gravity were determined. The rock has a very uneven fracture, and is probably trachytic phonolite. The occurrence of these trachytic rocks on Michipicoten Island is very interesting, for they are the only ones of the region which have in other countries been found in connection with undoubted volcanoes.

The general strike of the strata of the rocks of Point Keweenaw, at least in the neighbourhood of Portage Lake is N. 30° to 40° E., and the dip 55° to 70° north-westward. The melaphyres predominate, although polygenous and porphyritic conglomerates are also frequent. The copper-bearing tufaceous melaphyres seem to be more plentiful here than in the other areas, or at least the mines to which they give rise are more extensively worked.

At the other points in the east shore of the lake, where rocks of the character of melaphyre have been observed, the area occupied by them is very limited, and confined to narrow strips of beach and rocky ground, between the lake and the much more elevated Laurentian or Huronian rocks. In the most westerly cove on the south shore of Bachewahnung Bay, red sandstone is observed striking N. 12° W., and dipping 15° south-westward. It is interstratified with conglomerate, the boulders of which are principally of quartzite, dark green slate and red-jasper conglo_ merate, which have doubtless been derived from the Huronian hills in the rear. They range in diameter from one to twelve and even eighteen inches. The matrix is generally red sandstone, but the interstices are sometimes filled out with quartz. A short distance along the shore to the north-east exposures occur of a reddish-brown melaphyre tuff, containing amygdules of calcspar and quartz, the matrix of which is very soft and decomposed. The beds appear to strike N. 8° E., and dip 25° to 29° westward. They would therefore seem to be conformable with the sandstone and conglomerate. Further north-eastward the rock becomes more compact, of a reddish-green colour, and exhibits curves of igneous flow. The geodes become much less frequent and consist almost exclusively of agate. The next rock to the north-east is a light red sandstone, striking N. 65° W., and dipping 35° to 40° N. E. Its contact with the trap is not visible, but its dip is such as to lead to the supposition that it has been disturbed by that rock. There is a great thickness of this sandstone exposed here, in strata frequently vertical, striking generally east and west, or to the north of west, and exhibiting dips varying from 35° N. to 57° S., and at least two anticlinal axes. From what has been stated here and also concerning the south shore of Mamainse, it would appear that there is evidence of the existence of a sandstone of greater age than the bedded melaphyres and conglomerates, and it would appear not unreasonable to suppose that it belongs to what has been called the Lower group of the Upper Copper-bearing series.

The trap rocks which surround the south-west base of Gros Cap, although comparatively seldom amygdaloidal, are readily distinguished as melaphyres. They are sometimes coarse-grained, consisting of reddish-grey felspar, soft dark-green iron-chlorite (delessite), and occasional spots of yellowish-green epidote. From this they graduate into finer-grained varieties, but they very seldom become impalpable, or their constituents altogether indistinguishable. Sandstone was not observed in contact with the traps, but a large mass of quartzose porphyry is seen at a short distance from the shore.

Another large development of traps and sandstones occurs to the north of Pointe-aux-Mines, where an occasional bed of tufaceous melaphyre is also found.

Besides the rocks above described, there are found on the low ground betwixt Goulais and Bachewahnung Bays, betwixt the latter and Pancake Bay, and on many of the islands of the east shore, large areas of red sandstone, almost horizontal, which are supposed to be the continuation of that occurring at Sault St. Marie, and usually called the St. Peter Sandstone. The true relations of this rock to those of the upper group of the Upper Copper-bearing series have not yet been made out. It closely resembles, in lithological character, the sandstone described above as occurring in almost vertical strata on the south shore of Bachewahnung Bay. The disturbance of the latter is reasonably attributable to the neighbouring melaphyres, in which case the sandstone would be the earlier rock. On the other hand, as Sir W. E. Logan observes, " the contrast between the general moderate dips of these sand- " stones and the higher inclination of the igneous strata at " Gargantua, Mamainse, and Gros Cap, combined with the fact " that the sandstones always keep to the lake side of these, while " none of the many dykes which cut the trappean strata, it is " believed, are known to intersect the sandstones (at any rate on " the Canadian side of the lake), seem to support the suspicion " that the sandstones may overlie unconformably those rocks " which, associated with the trap, constitute the copper-bearing " series."* The following facts are confirmatory of this view. In the bay immediately south of Point-aux-Mines, where the Mamainse series adjoins the Laurentian rocks, the lowest member of the former is unconformably overlaid by thin bedded bluish and yellowish-grey sandstones, striking N. 50° E., and dipping 18° north-westward. The lowest layer is a conglomerate, with granitic and trappean boulders, and a bluish, fine-grained and slaty matrix.

---

* Geology of Canada, p. 85.

It is about six feet thick, and is followed by thirty feet of the thin bedded sandstones, some parts of which might yield good flag-stones. Some of the surfaces of these are very distinctly ripple-marked. Above these come thin, shaly, rapidly disintegrating layers, in which are found spheroidal concretions from five to ten inches in diameter. It is not possible to ascertain the total thickness of these sandstones, since they descend beneath the level of the lake. They are similar in lithological character to the sandstones which occur on the north side of Point-aux-Mines. Although there is no doubt that these sandstones unconformably overlie the melaphyre series, still their lithological characters are very different from those of the horizontal red sandstone above referred to. The latter is evenly small-grained, is coloured red by iron oxide, and contains here and there small pieces of red shale, which have evidently furnished the colouring matter. It frequently consists of evenly bedded red and yellowish-grey layers, and exhibits sometimes the phenomenon named by Naumann, discordant parallel-structure, and by Lyell, diagonal or cross stratification.

In enquiring next as to what geological formation in Europe most closely resembles the Upper Copper-bearing series of Lake Superior, the opinion expressed by Delesse ought not to be lost sight of, viz., that the constituent minerals have the same meaning and importance for eruptive rocks which organic remains have for those of sedimentary origin. Therefore, where the palæonto-logical evidence does not entirely contradict it, that derived from lithological resemblance ought to be allowed its full weight. The melaphyres of the upper rocks being interbedded with conglomer-ates and sandstones, the age of the latter may be ascertained approximatively by enquiring under what circumstances and during what period the melaphyres of Europe were developed. Upon this point Naumann thus expresses himself : " With regard " to the eruption-epochs of the melaphyres, there appears, indeed, " to have been many of them, but the most occur in the period " of the Rothliegende, or in the first half of the Permian forma-" tion, and all are probably more recent than the Carboniferous " system. This applies at least to the melaphyres on the south " side of the Hundsrück, to those of the Thuringian Forest, of the " neighbourhood of the Hartz, of Lower Silesia, Bohemia, and " Saxony. Many of these melaphyres were deposited soon " after the commencement, others towards the end, of the

" Rothliegende period, and generally the latter, in many coun-
" tries, shews a decided coincidence, both as regards time
" and space, with the formation of the melaphyres." Zirkel,
in his recent work on " Petrographie," gives a description
of the melaphyre deposits of Germany, of which the following is a translation: "In districts which are older than the
" Carboniferous formation melaphyre rocks are but seldom found.
" The melaphyres of the southern Hundsrück and of the Pfalz,
" whose stratigraphical relations are better known than their
" mineralogical composition, appear in the Carboniferous system
" or the lower Rothliegende.  This melaphyre region extends
" from Düppenweiler to Kreuznach, a distance of twelve miles,
" with a breadth between St. Wendel, Birkenfeld, Kirn, and
" Grumbach of several miles. Very few irregular masses are known,
" but, on the other hand, numerous veins have been observed with
" thicknesses varying from four to sixty feet.  They possess
" mostly a vertical dip, cut sharply the Carboniferous strata, and
" often extend on their strike considerable distances.  The mass
" of the vein frequently encloses fragments of the side rock, slate-
" clay or sandstone.  But most frequently in this region, the
" melaphyres present themselves in the form of beds, which are of
" very variable dimensions, (often only five to ten feet, sometimes
" two hundred feet thick,) and lie, for the most part, evenly
" inserted between the strata of the Carboniferous system.  Some
" of these can be traced for a distance of two miles.  Besides
" these a melaphyre layer appears in this region, extending over
" many square miles.  It is superimposed upon the upper strata
" of the Carboniferous system, and upon it rest the Conglomerates,
" sandstones and slate-clays of the Rothliegende.  This great
" covering of melaphyre is at its edges accompanied by melaphyre-
" tuffs, which are in many places developed as melaphyre-amygda-
" loids.  In very few instances only has it been observed that
" these melaphyres have exerted altering influences upon the side-
" rock. Within the limits of the Rothliegende melaphyres are very
" frequent.  According to Naumann the melaphyre of Ilfeld in
" the Hartz, must be regarded as a thick layer bedded into the
" Rothliegende.  It nevertheless in places lies immediately over
" the Carboniferous system, on account of its extending beyond
" the edges of the lower strata of the Rothliegende.  Naumann
" also mentions a mass of melaphyre which in Tyrathal covers the
" junction of the Greywacke with the Rothliegende, and in its

" further extension overlies also the latter formation.   The
" melaphyre-amygdaloid of Planitz, near Zwickau in Saxony,
" forms also a covering regularly inserted into the Rothliegende,
" above its inferior strata.   On the western declivity of the
" Oberhohndorfer Hill, near Zwickau, the melaphyre which here
" contains numerous green-earth and calcspar amygdules, shews an
" interesting intercalation with the brownish-red slate-clays of the
" Rothliegende, irregular lumps and patches of which being as it
" were kneaded into the mass of the melaphyre.   The melaphyric
" rock of the Johann-Friedrich and Zabenstadter Adit, in Mansfeld,
" is evenly interstratified in the Rothliegende.   G. Leonhard
" mentions that in the Rothliegende of the neighbourhood of
" Darmstadt, at Gœtzenhain and Urberach, the melaphyre forms
" distinct outbursts of considerable size in the form of domes
" (Kuppen,) which consist in the centre of solid melaphyre, and
" towards the periphery of amygdaloidal rocks, and shews in
" places both flagstone-like and columnar separation.   In Silesia
" the melaphyres appear in two places : in the country between
" Lœwenberg and Læhn, where they, according to the investi-
" gations of Beyrich, occur in several courses, striking from
" north-west to south-east, intersecting the Rothliegende, and
" in still more extended measure at the edge of the great
" bay opening towards south-east in the Grauwacke at Landeshut,
" in which the carboniferous formation and the Rothliegende
" have been deposited, and in which they form; according to Zobel
" and Von Carnal, a range extending from Schatzlar to Neurode.
" In north-eastern Bohemia, according to Emil Porth, and
" Jokély, malaphyres are found as numerous, and sometimes very
" thick layers, in the Rothliegende.   Jokély describes, in the
" district of Jicin, five beds of melaphyre in various parts of the
" Rothliegende, which exhibit very distinctly observable strati-
" graphical relations.   They prove to be, for the most part, true
" melaphyre streams, which have flown like lavas, and in visible
" connection with undoubted vein-like outbursts.   According to
" Porth, the neighbourhood of the melaphyre veins is frequently,
" for great distances round, a field of melaphyric ash and
" scoriæ."*

From these quotations it is plain that, in Europe, melaphyres
only made their appearance during the Carboniferous and Permian

---

* Zirkel ; Petrographie.   Vol. ii., p. 71.

periods, and especially characterised the latter.  The occurrence
of porphyritic conglomerates in Germany is similarly limited.  On
this point Zirkel says: "As porphyritic eruptions principally fall
" in the period of the Rothliegende, so the whole of the clastic
" rocks of the porphyry family stand in close connection with the
" deposition of its strata, to which they have also contributed a
" considerable amount of material.  For instance, coarse porphy-
" ritic conglomerates form members of the Upper Rothliegende
" in the Oschatz-Frohburg basin, in the Döhlen basin, at Wieser-
" städt in the Hartz, and in the north-western part of Thüringia.
" At Baden, in the Black Forest, the deepest strata of the
" Rothliegende consist of porphyritic breccia and the middle
" strata of conglomerates."* Even polygenous conglomerates,
such as those above-mentioned, are especially frequent among the
carboniferous and permian strata of Europe.  Naumann thus
briefly characterises the Rothliegende of Germany, which he
considers as equivalent to the English lower New Red Sandstone
and the French *grès rouge*: " The Rothliegende appears in so
" many of the countries of Germany, and in such great thickness,
" that, in its mode of development there, we recognise the normal
" type of this remarkable sandstone formation.  The pigment of
" the sandstone, consisting principally of iron-oxide, the frequent
" occurrence of conglomerates, the often repeated change in the
" size of grain of its rocks, the association with porphyries and
" melaphyres, the very frequent layers of claystones and porphy-
" ritic conglomerates, the great poverty, and often complete
" absence of organic remains,—all these are characters by which
" the Rothliegende is distinguished as quite a peculiar sandstone
" formation."† That not one of the peculiarities here emphasised
by Naumann are absent from the upper group of the Upper
Copper-bearing rocks of Lake Superior, will be evident to any
one who has observed them or carefully gone through the
description above given.  It therefore becomes a matter of much
importance, and deserving of the most careful study, to ascertain
whether this resemblance is a mere coincidence, or whether there
is reason for supposing that any part of these Upper Copper-bearing
rocks are of Permian age.

---

* Zirkel; Petrographie.  Vol. ii., p. 529.
† Naumann; Lehrbuch der Geognosie.  Vol. ii., p. 584.

# BRITISH ASSOCIATION.

## SCIENTIFIC EDUCATION IN SCHOOLS.

MR. GRIFFITH read the report which had been prepared by the Committee on this subject, the members of which were:—The general officers of the Association, the Trustees, the Rev. F. W. Farrar, M.A., F.R.S., the Rev. T. N. Hutchinson, M.A., Professor Huxley, F.R.S., Mr. Payne, Professor Tyndall, F.R.S., and Mr. J. M. Wilson, M.A.

1. A demand for the introduction of science into the modern system of education has increased so steadily during the last few years, and has received the approval of so many men of the highest eminence in every rank and profession, and especially of those who have made the theory and practice of education their study, that it is impossible to doubt the existence of a general, and even national, desire to facilitate the acquisition of some scientific knowledge by boys at our public and other schools.

2. We point out that there is already a *general* recognition of science as an element in liberal education. It is encouraged to a greater or less degree by the English, Scottish, and Irish Universities; it is recognized as an optional study by the College of Preceptors; it forms one of the subjects in the local examinations of Oxford and Cambridge; and it has even been partially introduced into several public schools. We have added an appendix containing information on some of these points. But the means at present used in our schools and universities for making this teaching effective, are, in our opinion, capable of great improvement.

3. That general education in schools ought to include some training in science is an opinion that has been strongly urged on the following grounds:—

1. As providing the best discipline in the observation and collection of facts, in the combination of inductive with deductive reasoning, and in accuracy both of thought and language.

2. Because it is found in practice to remedy some of the defects of ordinary school education. Many boys on whom the ordinary school studies produce very slight effect, are stimulated and improved by instruction in science; and it is found to be a most valuable element in the education of those who show special aptitude for literary culture.

3. Because the methods and results of science have so profoundly affected all the philosophical thought of the age, that an educated man is under a very great disadvantage if he is unacquainted with them.

4. Because very great intellectual pleasure is derived in after life from even a moderate acquaintance with science.

5. On grounds of practical utility as materially affecting the present position and future progress of civilization.

This opinion is fully supported by the popular judgment. All who have much to do with the parents of boys in the upper classes of life are aware that, as a rule, they value education in science on some or all of the grounds above stated.

**4.** There are difficulties in the way of introducing science into schools; and we shall make some remarks on them. They will be found, we believe, to be by no means insuperable. First among these difficulties, is the necessary increase of expense. For if science is to be taught, at least one additional master must be appointed; and it will be necessary in some cases to provide him with additional school-rooms, and a fund for the purchase of apparatus. It is obvious that the money which will be requisite for both the initial and current expenses must in general be obtained by increasing the school fees. This difficulty is a real but not a fatal one. In a wealthy country like England, a slight increase in the cost of education will not be allowed (in cases where it is unavoidable) to stand in the way of what is generally looked upon as an important educational reform; and parents will not be unwilling to pay a small additional fee if they are satisfied that the instruction in science is to be made a reality.

Another ground of hesitation is the fear that the teaching of science will injure the teaching in classics. But we do not think that there need be the slightest apprehension that any one of the valuable results of a classical education will be diminished by the introduction of science. It is a very general opinion, in which school-masters heartily concur, that much more knowledge and intellectual vigour might be obtained by most boys, during the many years they spend at school, than what they do as a matter of fact obtain. It should, we think, be frankly acknowledged, and, indeed, few are found who deny it, that an exclusively classical education, however well it may operate in the case of the very few who distinguish themselves in its curriculum, fails deplorably for the majority of minds. As a general rule, the small proportion of

boys who leave our Schools for the Universities consists undeniably
of those who have advanced furthest in classical studies, and judging the existing system of education by these boys alone, we have
to confess that it frequently ends in astonishing ignorance. This
ignorance, often previously acknowledged and deplored, has been
dwelt on with much emphasis, and brought into great prominence
by the recent Royal Commission for Inquiry into our public schools.
We need not fear that we shall do great damage by endeavouring
to improve a system which has not been found to yield satisfactory results. And we believe, further, that the philological abilities
of the very few who succeed in attaining to a satisfactory knowledge of classics will be rather stimulated than impeded by a more
expansive training.

Lastly, it may be objected that an undue strain will be put
upon the minds of boys by the introduction of the proposed
subjects. We would reply that the same objections were made,
and in some schools are still made, to the introduction of mathematics and modern languages, and are found by general experience
to have been untenable. A change of studies, involving the play
of a new set of faculties, often produces a sense of positive relief;
and at a time when it is thought necessary to devote to games so
large a proportion of a boy's available time, the danger of a general
over pressure to the intellectual powers is very small, while any
such danger in individual cases can always be obviated by special
remissions. We do not wish to advocate any addition to the hours
of work in schools where it is believed that they are already as
numerous as is desirable; but in such schools some hours a week
could still be given up to science by a curtailment of the vastly
preponderent time at present devoted to classical studies, and
especially to Greek and Latin composition.

**5.** To the selection of the subjects that ought to be included in
a programme of scientific instruction in public schools we have
given our best attention, and we would make the following remarks
on the principles by which we have been guided in the selection
that we shall propose.

There is an important distinction between scientific *information*
and scientific *training;* in other words, between general literary
acquaintance with scientific facts, and the knowledge of methods
that may be gained by studying the facts at first hand under the
guidance of a competent teacher. Both of these are valuable; it
is very desirable, for example, that boys should have some general

information about the ordinary phenomena of nature, such as the simple facts of astronomy, of geology, of physical geography, and of elementary physiology. On the other hand, the scientific habit of mind, which is the principal benefit resulting from scientific training, and which is of incalculable value whatever be the pursuits of after life, can better be attained by a thorough knowledge of the facts and principles of one science, than by a general acquaintance with what has been said or written about many. Both of these should co-exist, we think, at any school which professes to offer the highest liberal education; and at *every* school it will be easy to provide at least for giving some scientific information.

1. The subjects that we recommend for scientific *information* as distinguished from training, should comprehend a general description of the solar system; of the form and physical geography of the earth, and of such natural phenomena as tides, currents, winds, and the causes that influence climate; of the broad facts of geology; of elementary natural history, with especial reference to the useful plants and animals; and of the rudiments of physiology. This is a kind of information which requires less preparation on the part of the teacher; and its effectiveness will depend on his knowledge, clearness, method, and sympathy with his pupils. Nothing will be gained by circumscribing these subjects by any general syllabus; they may safely be left to the discretion of the masters who teach them.

2. And for scientific *training* we are decidedly of opinion that the subjects which have paramount claims are experimental physics, elementary chemistry, and botany.

i. The science of experimental physics deals with subjects which come within the range of everybody's experience. It embraces the phenomena and laws of light, heat, sound, electricity and magnetism, the elements of mechanics, and the mechanical properties of liquids and gases. The thorough knowledge of these subjects includes the practical mastery of the apparatus employed in their investigation. The study of experimental physics involves the observation and collation of facts, and the discovery and application of principles. It is both inductive and deductive. It exercises the attention and the memory, but makes both of them subservient to an intellectual discipline higher than either. The teacher can so present his facts as to make them suggest the principles which underlie them, while, once in possession of the

principle, the learner may be stimulated to deduce from it results
which lie beyond the bounds of his experience. The subsequent
verification of his deduction by experiment never fails to excite
his interest and awaken his delight. The effects obtained in the
class-room will be made the key to the explanation of natural
phenomena,—of thunder and lightning, of rain and snow, of dew
and hoar-frost, of winds and waves, of atmospheric refraction and
reflection, of the rainbow and the mirage, of meteorites, of terrestrial magnetism, of the pleasure and buoyancy of water and of
air. Thus the knowledge acquired by the study of experimental
physics is, of itself, of the highest value, while the acquisition of
that knowledge brings into healthful and vigorous play every
faculty of the learner's mind. Not only are natural phenomena
made the objects of intelligent observation, but they furnish
material for them to wrestle with and overcome; the growth
of intellectual strength being the sure concomitant of the enjoyment of intellectual victory. We do not entertain a doubt that
the competent teacher who loves his subject and can sympathise
with his pupils, will find in experimental physics a store of knowledge of the most fascinating kind, and an instrument of mental
training of exceeding power.

ii. Chemistry is remarkable for the comprehensive character of
the training which it affords. Not only does it exercise the memory
and the reasoning powers, but it also teaches the student to gather
by his own experiments and observations the facts upon which to
reason.

It affords a corrective of each of the two extremes against which
real educators of youth are constantly struggling. For on the one
hand, it leads even sluggish or uncultivated minds from simple
and interesting observations to general ideas and conclusions, and
gives them a taste of intellectual enjoyment and a desire for
learning. On the other hand, it checks over-confidence in mere
reasoning, and shows the way in which valid extensions of our
ideas grow out of a series of more and more rational and accurate
observations of external nature.

It must not, however, be supposed that all so-called teaching of
chemistry produces results of this kind. Young men do occasionally come up to public examinations with a literary acquaintance
with special facts and even principles of chemistry, sufficient to
enable them to describe those facts from some one point of view,
and to enunciate the principles in fluent language, and yet who

know nothing of the real meaning of the phrases which they have learnt. Such mere literary acquaintance with scientific facts is in chemistry an incalculable evil to the student if he be allowed to mistake it for science.

Whether the student is to learn much or little of chemistry, his very first lessons must be samples of the science. He must see the chief phenomena which are described to him; so that the words of each description may afterwards call up in his mind an image of the thing. He must make simple experiments, and learn to describe accurately what he has done, and what he has observed. He must learn to use the knowledge which he has acquired before proceeding to the acquisition of more; and he must rise gradually from well examined facts to general laws and theories.

Among the commonest non-metallic elements and their simplest compounds, the teacher in a school will find abundant scope for his chief exertions.

iii. Botany has also strong claims to be regarded as a subject for scientific training. It has been introduced into the regular school course at Rugby, (where it is the first branch of natural science which is studied) ; and the voluntary pursuit of it is encouraged at Harrow, and at some other schools with satisfactory results. It only requires observation, attention, and the acquisition of some new words; but it also evolves the powers of comparison and the colligation of facts in a remarkable degree. Of all sciences it seems to offer the greatest facilities for observation in the fields and gardens ; and to this must be added the fact that boys, from their familiarity with fruits, trees, and flowers, start with a considerable general knowledge of botanical facts. It admits therefore pre-eminently of being taught in the true scientific method. The teaching of science is made really valuable by training the learner's mind to examine into his present knowledge, to arrange and criticise it, and to look for additional information. The science must be begun where it touches his past experience, and this experience must be converted into scientific knowledge. The discretion of the teacher will best determine the range of botany at which it is desirable to aim.

6. The modes of giving instruction in the subjects which we have recommended are reducible to two: —

1. A compulsory system of instruction may be adopted, similar to that which exists at Rugby, where science has now for nearly three years been introduced on precisely the same footing as

mathematics and modern languages, and is necessarily taught to all boys.

2. A voluntary system may be encouraged, as has been done for many years at Harrow, where scientific instruction on such subjects as have been enumerated above, is now given in a systematic series of lectures, on which the attendance of all boys who are interested in them is entirely optional.

Of these systems it is impossible not to feel that the compulsory system is the most complete and satisfactory. The experience of different schools will indicate how it may best be adopted, and what modifications of it may be made to suit the different school arrangements. It will often be very desirable to supplement it by the voluntary system, to enable the boys of higher scientific ability to study those parts of the course of experimental physics which will rarely, if ever, be included in the compulsory school system. Lectures may also be occasionally given by some non-resident lecturer, with a view of stimulating the attention and interest of the boys. We add appendices containing details of these two systems as worked at Rugby and Harrow, and we believe that a combination of the two would leave little or nothing to be desired.

The thorough teaching of the physical sciences at schools, will not, however, be possible, unless there is a general improvement in the knowledge of arithmetic. At present many boys of thirteen and fourteen, are sent to the public schools, almost totally ignorant of the elements of arithmetic, and in such cases they gain only the most limited and meagre knowledge of it; and the great majority enter ill taught. It is a serious and lasting injury to boys so to neglect arithmetic in their early education; it arises partly from the desire of the masters of preparatory schools to send up their boys fitted to take a good place in the classical school, and from the indifference of the public schools themselves to the evil that has resulted.

**7.** With a view to the furtherance of this scheme, we make the following suggestions :—

1. That in all schools natural science be one of the subjects to be taught, and that in every public school at least one natural science master be appointed for the purpose.

2. That at least three hours a week be devoted to such scientific instruction.

3. That natural science should be placed on an equal footing

with mathematics and modern languages in affecting promotions, and in winning honours and prizes.

4. That some knowledge in arithmetic should be required for admission into all public schools.

5. That the universities and colleges be invited to assist in the introduction of scientific education, by making natural science a subject of examination, either at matriculation, or at an early period of a university career.

6. That the importance of appointing lecturers in science, and offering entrance scholarships, exhibitions, and fellowships, for the encouragement of scientific attainments, be represented to the authorities of the colleges.

With reference to the last two recommendations, we would observe that without the co-operation of the universities, science can never be effectively introduced into school education. Although not more than 35 per cent., even of the boys at our great public schools, proceed to the university, and at the majority of schools a still smaller proportion, yet the curriculum of a public school course is almost exclusively prepared with reference to the requirements of the universities and the rewards for proficiency that they offer. No more decisive proof could be furnished of the fact that the universities and colleges have it in their power to alter and improve the whole higher education of England.

## APPENDIX A.

### 1. OXFORD.

The Natural Science School at Oxford was established in the year 1853. By recent changes, the university allows those who have gained a first, second, or third class in this school to graduate without passing the classical school, provided they have obtained honours, or have passed in three books at least at the second classical examination—viz., moderations (which is usually passed in the second year of residence) ; honours in this school are thus placed on an equality with classical honours. The first classical examination, " responsions," is generally passed in the first term of residence. Arithmetic and two books of Euclid, or algebra up to simple equations, are a necessary part of this examination.

The university offers ample opportunities for the study of physics, chemistry, physiology, and other branches of natural science. At present, only a few of the colleges have lecturers on this subject; while for classics and mathematices every college professes to have an adequate staff of teachers. At Christ Church, however, a very complete chemical laboratory has been lately opened.

A junior studentship at Christ Church and a demyship at Magdalen College, tenable for five years, are, by the statutes of those colleges awarded annually for profiency in natural science. A scholarship, tenable for three years, lately founded by Miss Brackenbury, at Balliol College, for the promotion of the study of natural science, will be given away every two years. With the exception of Merton College, where a scholarship is to be shortly given for proficiency in natural science, no college has hitherto assigned any scholarships to natural sciences. The number of scholarships at the colleges is stated to be about 400, varying in annual value from £100 to £60. With these should be reckoned college exhibitions, to the number of at least 220, which range in annual value from £145 to £20, and exhibitions awarded at school, many of which are of considerable value.

The two Burdett-Coutts geological scholarships, tenable for two years, and of the annual value of £75, are open to all members of the university who have passed the examination for the B.A. degree, and have not exceeded the 27th term from their matriculation. Every year a fellowship of £200 a year, tenable for three years (half of which time must be spent on the continent) on Dr. Radcliffe's foundation, is at present competed for by candidates, who, having taken a first class in the school of natural science, propose to enter the medical profession.

At Christ Church, two of the senior studentships (fellowships) are awarded for proficiency in natural science. At the examination for one of these, chemistry is the principal subject, and for the other physiology.

At Magdalen College, it is provided that, for twenty years from the year 1857, every fifth fellowship is assigned to mathematics and physical science alternately. In the statutes of this, and of every college in Oxford (except Corpus, Exeter, and Lincoln) the following clause occurs:—"The system of examinations shall always be such as shall render fellowships accessible, from time to time, to excellence in every branch of knowledge for the time being recognized in the schools of the university." This clause, so far as it relates to the study of natural science, has been acted on

only by Queen's College and at Merton College, where a natural
science fellowship will be filled up during the course of the present
year.

At Pembroke College, one of the two Sheppard fellows must
proceed to the degree of Bachelor and Doctor of Medicine in the
university. At the late election to this fellowship, natural science
was the principal subject in the examination. The number of
college fellowships in Oxford is at present about 400.

## 2. CAMBRIDGE.

It is important to distinguish between the university and the
colleges at Cambridge as at Oxford.

There is a natural science tripos in which the university
examines in the whole range of natural sciences, and grants
honors precisely in the same manner as in classics or mathematics.

The university also recognizes the natural sciences as an
alternative subject for the ordinary degree. As the regulations
on this point are comparatively recent, it will be well to state
them here.

A student who intends to take an ordinary degree without
taking honours has to pass three examinations during his course
of three years,—the first, or previous examination, after a year's
residence, in Paley, Latin, Greek, Euclid, and arithmetic, and one
of the gospels in greek ; the second, or general examination,
towards the end of his second year, in the Acts of the Apostles in
Greek, Latin, Greek, Latin prose composition, algebra, and
elementary mechanics, and the third, or special examination, at
the end of his third year, in one of the following five subjects :—
1. Theology; 2. Moral Science; 3. Law; 4. Natural Science;
5. Mechanism and applied Science.

In the natural science examination, a choice is given of chemistry,
geology, botany, and zoology.

There are only five colleges in Cambridge that take any notice
of natural science—viz., Kings, Caius, Sidney, Sussex, St. John's,
and Downing. At Kings, two exhibitions have been given away
partly for proficiency in this subject; but there are no lectures,
and it is doubtful whether similar exhibitions will be given in
future. At Caius there is a medical lecturer and one scholarship
given away annually for anatomy and physiology. At Sidney
Sussex two scholarships annually are given away for mathematics
and natural science; and a prize of £20 for scientific knowledge.

There is also a laboratory for the use of students. At St. John's there is a chemical lecturer and laboratory ; and though at this college there is no sort of examination in natural science either for scholarships or fellowships, it is believed distinction in the subject may be taken into account in both elections. Downing was founded with "especial reference to the studies of law and medicine;" there is a lecturer here in medicine and natural science, and in the scholarship examinations one paper in these subjects ; no scholarship is appropriated to them, but they are allowed equal weight with other subjects in the choice of candidates. It is believed that the same principle will govern the election to fellowships in this college, though no fellowship has yet been given for honours in natural science. We believe that, owing to the new university regulations (mentioned above), the authorities of Trinity College have determined to appoint a lecturer in natural science; the matter is under deliberation in other colleges, and it is not improbable that the same considerations will induce them to follow this example.

It must always be remembered that the practice is rare in Cambridge of appropriating fellowships and scholarships to special subjects. At present public opinion in the University does not reckon scientific distinction as on a par with mathematical or classical; hence, the progress of the subject seems enclosed in this inevitable circle—the ablest men do not study natural science because no rewards are given for it, and no rewards are given for it because the ablest men do not study it. But it may be hoped that the disinterested zeal of teachers and learners will rapidly break through this circle; in that case the subject may be placed on a satisfactory footing without any express legislative provision.

### 3. The University of London.

At the University of London the claims of science to form a part of every liberal education have long been recognized. At the matriculation examination the student is required to show that he possesses at least a popular knowledge of the following subjects :—

a. In *Mechanics*.—The composition and resolution of forces ; the mechanical powers ; a definition of the centre of gravity ; and the general laws of motion.

b. In *Hydrostatics, Hyrdaulics*, and *Pneumatics*.—The pressure of liquids and gases ; specific gravity ; and the principles of the action of the barometer, the siphon, the common pump and forcing pump, and the air-pump.

*c*. In *Acoustics.*—The nature of sound.

*d*. In *Optics.*—The laws of refraction and reflection. and the formation of images by simple lenses.

*e*. In *Chemistry.*—The phenomena and laws of heat; the chemistry of the non-metallic elements; general nature of acids, gases. &c.; constitution of the atmosphere; composition of water, &c.

At the examination for the degree of B.A. a more extensive knowledge of these subjects is required, and the candidate is further examined in the following branches of science:—

*f. Astronomy.*—Principal phenomena depending on the motion of the earth round the sun, and on its rotation about its own axis; general description of the solar system, and explanation of lunar and solar eclipses.

*g. Animal Physiology.*—The properties of the elementary animal textures; the principles of animal mechanics; the processes of digestion, absorption, assimilation; the general plan of circulation in the great divisions of the animal kingdom; the mechanism of respiration; the structure and actions of the nervous system; and the organs of sense.

Besides the degree examination there is also an examination for *honours* in mathematics and natural philosophy, in which, of course. a much wider range of scientific knowledge is required.

We would venture to remark that, if a similar elementary acquaintance with the general principles of sciences were required for matriculation at Oxford and Cambridge, it is certain that they would at once become a subject of regular teaching in all our great public schools.

There are also two specially scientific degrees, a Bachelor of Science, and a Doctor of Science. For the B. S. there are two examinations of a general but highly scientific character. The degree of D. S. can only be obtained after the expiration of two years subsequent to the taking the degree of B. S. The candidate is allowed to select one *principal subject*, and to prove his thorough practical knowledge thereof, as well as a general acquaintance with other subsidiary subjects.

### 4. The College of Preceptors.

In the diploma examinations at the College of Preceptors, one branch of science—viz., either chemistry, natural history, or physiology—is required as a *necessary* subject for the diploma of *Fellow*. In the examinations for the lower diploma of Associate or Licentiate some branch of science *may* be taken up by candidates at their own option. The council recently decided to offer a prize of three guineas half-yearly for the candidate who showed most proficiency in science, and who at the same time obtained a second class in the other subjects.

In the examinations of pupils of schools, natural philosophy, chemistry, and natural history are optional subjects only, and are not *required* for a certificate for the three classes. Two prizes are given to those candidates who obtain the highest number of marks in these subjects at the half-yearly examinations; and it is an interesting fact that last year, out of a total of 651 candidates, 100 brought up natural history, and 36 brought up chemistry as subjects for examination. Two additional prizes were consequently awarded.

### 5. The French Schools.

In France the "Lycées" correspond most nearly to our public schools, and for many years science has formed a distinct part of their regular curriculum. A strong impulse to the introduction of scientific teaching into French schools was given by Napoleon I., and since that time we believe that no French school has wholly neglected this branch of education. The amount of time given to these subjects appears to average two hours in every week.

The primary education is that which is given to all alike, whatever may be their future destination in life, up to the age of eleven or twelve years. After this period there is "bifurcation" in the studies of boys. Those who are intended for business or for practical professions lay aside Greek and Latin, and enter on a course of "special secondary instruction." In this course, mechanics, cosmography, physics, chemistry, zoology, botany, and geology occupy a large space; and the authorized official programmes of these studies are very full, and are drawn up with the greatest care. The remarks and arguments of the Minister of Public Instruction (Mons. Duruy) and others in the "Programmes officiels, etc., de l'enseignement secondaire spécial," are extremely valuable and suggestive; and we recommend the syllabuses of the various subjects, which have received the sanction of the French government, as likely to afford material assistance to english teachers in determining the range and limits of those scientific studies at which, in any special system of instruction, they may practically aim. The " Enseignement secondaire spécial" might very safely be taken as a model of what it is desirable to teach in the "modern departments" which are now attached to some of our great schools.

The boys who are destined to enter the learned professions continue a classical course, in which, however, much less time is devoted to classical composition than is the case in our public schools. Nor is science by any means neglected in this course, which is intended to cover a period of three years. Besides the " elementary division," there are five great classes in these schools, viz., a grammar division, an upper division, a philosophy class, and classes for elementary and special mathematics.

In the grammar division there is a systematic instruction on the physical geography of the globe.

In the second class of the upper division the boys begin to be taught the elements of zoology, botany, and geology in accordance with the ministerial programmes ; and in the rhetoric class descriptive cosmography (which seems to be nearly co-extensive with the German *Erkunde*) forms the subject of a certain number of weekly lessons.

In the class of philosophy, the young students are initiated into the elementary notions of physics (including weight, heat, electricity, and magnetism, acoustics, and optics) and of chemistry, in which, at this stage, the teaching is confined to " general conceptions on air, water, oxidation, combustion, the conditions and effects of chemical action, and on the forces which result from it."

In the classes of elementary and special mathematics, this course of scientific training is very considerably extended ; and if the authorized programmes constitute any real measure of the teaching, it is clear that no boy could pass through these classes without a far more considerable amount of knowledge in the most important branches of science than is at present attainable in any English Public School.

## 6. THE GERMAN SCHOOLS.

In Germany the schools, which are analogous to public schools in England, are the *Gymnasia*, where boys are prepared for the

universities, and the *Bürgerschulen* or *Realschulen*, which were
established for the most part about thirty years ago, for the
purpose of affording a complete education to those who go into
active life as soon as they leave school.   An account of the Prussian
Gymnasia and Realschulen may be seen in the Public School Commission Report, Appendix G; further information may be obtained
in " Dashöhere Schulwesen in Preussen," by Dr. Wiese, published
under the sanction of the Minister of Public Instruction in Prussia,
and in the programmes issued annually by the school authorities
throughout Germany.

At the Gymnasia natural science is not taught to any great
extent.   According to the Prussian official instructions, in the
highest class two hours, and in the next class one hour a week are
allotted to the study of physics.   In the lower classes, two hours
a week are devoted to natural history, *i. e.*, botany and zoology.

The results of the present training in natural science at the
Gymnasia are considered by many eminent university professors
in Germany to be unsatisfactory, owing to the insufficient time
allotted to it.

In the Realschulen about six hours a week are given to physics
and chemistry in the two highest classes, and two or three hours
a week to natural history in the other classes.   In these schools
all the classes devote five or six hours a week to mathematics, and
no Greek is learnt.   In Prussia there were in 1864 above one
hundred of these schools.


## APPENDIX B.

### On the Natural Science Teaching at Rugby.

Before the summer of 1864, a boy, on entering Rugby, might
signify his wish to learn either modern languages or natural science;
the lessons were given at the same time, and therefore excluded
one another.   If he chose natural science, he paid an entrance
fee of £1 1s., which went to an apparatus fund, and £5 5s.
annually to the lecturer.   Out of the whole school, numbering
from 450 to 500, about one-tenth generally were in the natural
science classes.

The changes proposed by the Commissioners were as follows:—
That natural science should no longer be an alternative with
modern languages, but that all boys should learn some branch of

it; that there should be two principal branches—one consisting of chemistry and physics, the other of physiology and natural history, animal and vegetable—and that the classes in natural science should be entirely independent of the general divisions of the school, so that boys might be arranged for this study exclusively according to their proficiency in it.

Since, owing to circumstances which it would be tedious to detail, it was impossible to adopt literally the proposals of the Commissioners, a system was devised which must be considered as the system of the Commissioners in spirit, adapted to meet the exigencies of the case.

The general arrangement is this—that new boys shall learn botany their first year, mechanics their second, geology their third, and chemistry their fourth.

In carrying out this general plan certain difficulties occur, which are met by special arrangements depending on the peculiarities of the school system. We need not here enter upon these details, because it would be impossible to explain them simply, and because any complications which occur in one school would differ widely from those which are likely to arise in another.

Next, as to the nature of the teaching.

In botany the instruction is given partly by lectures and partly from Oliver's Botany. Flowers are dissected and examined by every boy, and their parts recognized and compared in different plants and then named. No technical terms are given till a familiarity with the organ to be named or described has given rise to their want. The terms which express the cohesion and adhesion of the parts are gradually acquired until the floral schedule, as highly recommended by Henslow and Oliver, can be readily worked. Fruit, seed, inflorescence, the forms of leaf, stem, root, are then treated, the principal facts of vegetable physiology illustrated, and the principle of classification into natural orders explained, for the arrangement of which Bentham's " Handbook of the British Flora" is used. Contrary to all previous expectation, when this subject was first introduced it became at once both popular and effective among the boys.

The lectures are illustrated by Henslow's nine diagrams, and by a large and excellent collection of paintings and diagrams made by the lecturers and their friends, and by botanical collections made for use in lectures. When the year's course is over, such boys as show a special taste are invited to take botanical walks with the principal lecturer, to refer to the school herbarium, and are stimulated by prizes for advanced knowledge and for dried collections, both local and general.

In mechanics, the lecturer is the senior natural science master. The lectures include experimental investigations into the mechanical powers, with numerous examples worked by the boys; into the elements of mechanism, conversion of motion, the steam engine, the equilibrium of roofs, bridges, strength of material, &c. They are illustrated by a large collection of models, and are very effective and popular lectures.

The lectures in geology are undertaken by another master. This subject is only temporarily introduced, on account of the want of another experimental school. When this is built, the third year's course will be some part of experimental physics, for which there already exists at Rugby a fair amount of apparatus. It is very desirable that boys should obtain some knowledge of geology, but it is not so well fitted for school teaching as some of the other subjects, on several grounds. Perhaps a larger proportion of boys are interested in the subject than in any other; but the subject pre-supposes more knowledge and experience than most boys possess, and their work has a tendency to become either superficial, or undigested knowledge derived from books alone. The lectures include the easier part of Lyell's Principles, i. e., the causes of change now in operation on the earth; next, an account of the phenomena observable in the crust of the earth, stratification and its disturbances, and the construction of maps and sections; and lastly, the history of the stratified rocks and of life on the earth. These lectures are illustrated by a fair geological collection, which has been much increased of late, and by a good collection of diagrams and views to illustrate geological phenomena.

For chemistry, the lecturer has a convenient lecture-room, and a small but well-fitted laboratory; and he takes his classes through the non-metallic and metallic elements: the lectures are fully illustrated by experiments. Boys, whose parents wish them to study chemistry more completely, can go through a complete course of practical analysis in the laboratory, by becoming private pupils of the teacher. At present twenty-one boys are studying analysis.

This being the matter of the teaching, it remains to say a few words on the manner. This is nearly the same in all classes, *mutatis mutandis:* the lecture is given, interspersed with questions, illustrations, and experiments, and the boys take rough notes which are re-cast into an intelligible and presentable form in note books. These are sent up about once a fortnight, looked over, corrected and returned; and they form at once the test of how far the matter has been understood, the test of the industry, care and attention of the boy, and an excellent subject for their English composition.

Examination papers are given to the sets every three or four weeks, and to these and to the note books marks are assigned which have weight in the promotion from form to form. The marks assigned to each subject are proportional to the number of school-hours spent on that subject.

There are school prizes given annually for proficiency in each of the branches of natural science above mentioned.

This leads us, lastly, to speak of the results :—

First, as to the value of the teaching itself; secondly, as to its effects on the other branches of study.

The experience gained at Rugby seems to point to these conclusions :—That botany, structural and classificatory, may be taught with great effect, may interest a large number of boys, and is the best subject to start with. That its exactness of terminology, the necessity of care in examining the flowers, and the impossibility of superficial knowledge are its first recommendations; and the successive gradations in the generalizations as to the unity of type of flowers, and the principles of a natural classification, are of great value to the cleverer boys. The teaching must be based on personal examination of flowers, assisted by diagrams, and everything like cram strongly discouraged.

Mechanics are found rarely to be done well by those who are not also the best mathematicians. But it is a subject which in its applications interests many boys, and would be much better done, and would be correspondingly more profitable, if the standard of geometry and arithemetic were higher than it is. The ignorance of arithmetic which is exhibited by most of the new boys of fourteen or fifteen would be very surprising, if it had not long since ceased to surprise the only persons who are acquainted with it ; and it forms the main hinderance to teaching mechanics. Still, under the circumstances, the results are fairly satisfactory.

The geological teaching need not be discussed at length, as it is temporary, at least in the middle school. Its value is more literary than scientific. The boys can bring neither mineralogical, nor chemical, nor anatomical knowledge ; nor have they observed enough of rocks to make geological teaching sound. The most that they can acquire, and this the majority do acquire, is the general outline of the history of the earth, and of the agencies by which that history has been effected, with a conviction that the subject is an extremely interesting one. It supplies them with an object rather than with a method.

Of the value of elementary teaching in chemistry there can be only one opinion. It is felt to be a new era in a boy's mental progress when he has realised the laws that regulate chemical combination and sees traces of order amid the seeming endless variety. But the number of boys who get a real hold of chemistry from lectures alone is small, as might be expected from the nature of the subject.

Of the value of experimental teaching in physics, especially pneumatics, heat, acoustics, optics, and electricity, there can be no doubt. Nothing but impossibilities would prevent the immediate introduction of each of these subjects in turn, into the Rugby curriculum.

Lastly, what are the general results of the introduction of scientific teaching in the opinion of the body of masters? In brief, it is this, that the school as a whole is the better for it, and that the scholarship is not worse. The number of boys whose industry and attention is not caught by any school study is decidedly less ; there is more respect for work and for abilities in the different fields now open to a boy ; and though pursued often with great vigour, and sometimes with great success, by boys distinguished in classics, it is not found to interfere with their proficiency in classics, nor are there any symptoms of over-work in the school.. This is the testimony of the classical masters, by no means specially favourable to science, who are in a position which enables them to judge. To many who have left Rugby with but little knowledge and little love of knowledge, to show as the results of their two or three years in our middle school, the introduction of science into our course has been the greatest possible gain : and others who have left from the upper part of the school, without hope of distinguishing themselves in classics or mathematics, have adopted science as their study at the Universities. It is believed that no master in Rugby School would wish to give up natural science and recur to the old curriculum.

## APPENDIX C.

### ON THE TEACHING OF SCIENCE AT HARROW SCHOOL.

From this time forward, natural science will be made a regular subject for systematic teaching at Harrow, and a natural science master has been appointed.

But for many years before the Royal Commission of Inquiry into the Public Schools had been appointed, a voluntary system for the encouragement of science had been in existence at Harrow. There had been every term a voluntary examination on some scientific subject, which together with the text-books recommended, was announced at the end of the previous term. Boys from all parts of the school offered themselves as candidates for these voluntary examinations, and every boy who acquitted himself to the satisfaction of the examiners (who were always two of the masters) was rewarded with reference to what could be expected from his age and previous attainments. The text-books were selected with great care, and every boy really interested in his subject could and did seek the private assistance of his tutor or of some other master. The deficiencies of the plan, if regarded as a substitute for the more formal teaching of science, were too obvious to need pointing out; yet its results were so far satisfactory that many old Harrovians spoke of it with gratitude, among whom are some who have since devoted themselves to science with distinguished success.

One of the main defects of this plan (its want of all system) was remedied a year ago, when two of the masters drew up a scheme, which was most readily adopted, by which any boy staying at Harrow for three years might at least have the opportunity during that time of being introduced to the elementary conceptions of astronomy, zoology, botany (structural and classificatory), chemistry, and physics. These subjects are entrusted to the responsibility of eight of the masters, who draw up with great care a syllabus on the subject for each term, recommend the best text-books, and give weekly instruction (which is perfectly gratuitous) to all the boys who desire to avail themselves of it; indeed, a boy may receive, in proportion to the interest which he manifests in the subject, almost any amount of assistance which he may care to seek. Proficiency in these examinations is rewarded as before; and to encourage steady perseverance, the boys who do best in the examination during a course of three terms receive more valuable special rewards.

As offering to boys a voluntary and informal method of obtaining much scientific information, this plan (which was originated at Harrow, and has not, so far as we are aware, been ever adopted at any other school) offers many advantages. It is sufficiently elastic to admit of many modifications; it is sufficiently comprehensive to attract a great diversity of tastes and inclinations; it cannot be found oppressive, because it rests with each boy to decide whether he has the requisite leisure or not; it can be adopted with ease at any school where even a small body of the masters are interested in one or other special branch of science; and it may tend to excite in some minds a more spontaneous enthusiasm than could be created by a compulsory plan alone.

We would not, however, for a moment recommend the adoption of any such plan as a substitute for more regular scientific training. Its chief value is purely supplemental, and henceforth it will be regarded at Harrow as entirely subordinate to the formal classes for the teaching of science which will be immediately established. In addition to this, more than a year ago some of the boys formed themselves into a voluntary association for the pursuit of science. This Scientific Society, which numbers upwards of thirty members, meets every ten days at the house and under the presidency of one or other of the masters. Objects of scientific interest are exhibited by the members, and papers are read generally on some subjects connected with natural history. Under the auspices of this Society the nucleus of a future museum has already been formed, and among other advantages the Society has had the honour of numbering among its visitors more than one eminent representative of literature and science. We cannot too highly recommend the encouragement of such associations for intellectual self-culture among the boys of our public schools.—*From a Newspaper Report.*

# MODERN SCIENTIFIC INVESTIGATION : ITS
# METHODS AND TENDENCIES.*

GENTLEMEN OF THE AMERICAN ASSOCIATION FOR THE ADVANCEMENT OF SCIENCE : Every day of our lives we hear that this is an age of progress ; and that it is so we find evidence at every turn. The rapidity with which effects follow causes in human events, the celerity with which the plan is carried into execution, gives to a year in the experience of one of the present generation the practical value of a lifetime in ages past. Much labour has been expended on the exposition of the causes of the mental activity of the present age, and of the grand achievements which have attended it; and yet, the key to the whole enigma is to be found in the universal adoption of the comparatively new system of inductive reasoning. It would be foreign to my purpose to attempt to illustrate or defend this proposition, and I must therefore trust to its acceptance without argument, while we pass to consider that branch of the subject which more immediately demands our attention.

Although the progress of the age to which I have referred has been a matter of wonder and delight to all students of humanity and civilization, many of our best men have been somewhat alarmed and dizzied by it; and while accepting the achievements of modern industry and thought as full of present good and future promise, they are not a little concerned lest our railroad speed of progress should lead to its legitimate consequences, a final crash—not of things material, but of those of infinitely more value—of opinions and of faith. As often as it is boasted that this is pre-eminently an age of progress, that boast is met by the inevitable "but" (which qualifies our praise of all things earthly) "it is equally an age of scepticism." For the truth of this assertion the proof is nearly as palpable as of the other; and in view of the ruthlessness with which the man of the present removes ancient landmarks and profanes shrines hallowed by the faith of centuries, it is not surprising that many of the good and wise among us should deplore a liberty of thought leading, in their view, inevitably to license ; and mourn over this wide-spread scepticism as an

---

* An address delivered before the American Association, at Burlington, August, 1865, by Prof. J. S. Newberry, President of the Association ; from a copy communicated by the Author.

evil and inscrutable disease that has fallen upon the minds and hearts of men.

Now for every consequence there must be an adequate cause; and while confessing the fact of this modern lack of faith, I have thought that a few moments given to an analysis of it, and an attempt to trace it to its source, might not be wholly misspent—might possibly, indeed, result in giving a grain of encouragement to those who look with distrust and dread upon the investigations and discussions which now occupy so large a portion of the time and thought of our men of science.

If the wheels of time could, for our benefit, be rolled back, and we could see in all its details the civilization of Europe three or four hundred years ago, we should find that our so much respected ancestors, who fill so large a space on the page of history, were little better than barbarians. Among the English, the French, the Germans, Spanish and Italians, we should find a phase of civilization which, excepting that it included the elements—as yet but imperfectly developed—of a true religious faith, is scarcely to be preferred to that of the Chinese. Aside from the vast difference perceptible between the civilization of that epoch and ours, as exhibited in the political condition of the people, in their social economy and morals, the general intellectual darkness of the period referred to could not fail to impress us both profoundly and painfully. Out of that darkness and chaos have come, as if by magic, all our modern democracy with its individual liberty and dignity, all our civil and religious freedom, all our philanthropy and benevolence, all our diffused comfort and luxury, most of our good manners and good morals, and all the splendid achievements of our modern scientific investigation.

It is unnecessary for me here to describe in detail the origin and growth of modern science. That has been so well done by Dr. Whewell that all men of education are familiar with the steps by which the grand, beautiful, and symmetrical fabric formed by the grouping of the natural sciences has acquired its present lofty proportions.

Previous to the period when the Baconian philosophy was accepted as a guide in scientific investigation, but one department of science had attained a development which has any considerable claim to our respect. Mathematics, both pure and applied, had been assiduously cultivated from the remotest antiquity, and with a degree of success which has left to modern investigators little more than the elaboration of the thoughts of their predecessors. In Metaphysics—which had claimed even a larger share of the attention of the scholars of antiquity—little progress had been made. Perhaps I am justified in saying little progress was possible, inasmuch as in the light of all the great material discoveries of modern times the metaphysicians of the present day are debating, with as little harmony of opinion, the same questions that divided the rival schools of the Greeks. Each successive generation has had its two parties of idealists and realists, who have discussed the intangible problems which absorbed the great minds of Plato and Aristotle with a degree of enthusiasm and energy—and it may be of acrimony—which seems hardly compensated by any expansion of the human intellect or amelioration of the condition of mankind.

Of the Physical Sciences we may say that, except Astronomy, no one had an existence prior to the time of Bacon. There were men of vast learning, and much that was called science in the mass of reported observation that had been accumulating from century to century, until it had become "*rudis indigestaque moles,*" in which—though it constituted the pride of universities, the intellectual capital with which the savant thought himself rich, and that on which the professional man depended for success—there was far more error than truth, and its study was sure to mislead and likely to injure. In these circumstances the task before the scientific reformer was one far more difficult than that of clearing the Augean stables; no less, in fact, than to seat himself before this great heap of rubbish, this mass of truth and error,—of the sublimest philosophy with the wildest fiction,—to patiently winnow out the grains of truth, and from infiniteismal facts build up a fabric that should have a sure foundation below, and beauty and symmetry above. What more natural, then, than that the process adopted in winnowing this chaff-heap should be that which had given success to the only true science of the period?—that the mathematical touchstone should be the test by which every grain was tried? And such precisely was the course pursued; perhaps we may even say the only one practicable. Provided with this test, the reformer was compelled to rejudge upon its merits every proposition submitted to him, and accepted only as true such as could be demonstrated. The materials which composed the science to be reformed naturally fell into several categories. First,—That which had been demonstrated to be true. Second,

—That which was demonstrable.  Third,—That which was probable.  Fourth,—That which was possible; and fifth,—That which was impossible.  Of these he systematically rejected all but the first and second classes.  And this, in few words, has been the method adopted, not only in the purification of old science, but in the creation of new.

It will be seen at a glance, that in this process all that was contrary to the `order of nature (supernatural or spiritual) was necessarily excluded; and it was taken for granted that the mathematical or logical faculty of the human mind was capable of solving all the problems of the material universe.  Sir William Hamilton and others have demonstrated the inadequacy of mathematical processes as a guide to human reason, and a moment's thought will show us that our boasted intellect is incapable of grasping even all the material truths which are plainly presented to it.  To illustrate : as we scan the heavens of a clear evening, we recognize the fact that we stand as it were on a point in space where our field of vision is limitless ; the heavenly bodies stretching away into the realms of obscurity, and becoming invisible only through the imperfection of our organs of vision.  Bringing to our aid the most powerful telescopes, we are apparently as far as ever from reaching the limits of the universe ; and when we endeavour to conceive of such a limit, the reasoning faculty finds itself incapable of grasping either of the two alternatives offered to it, one or the other of which must be true.  The universe must be either limited or limitless.  But no man can conceive of a universe without a limit ; and if it be regarded as terminated by definite boundaries, the imagination strives in vain to fill the void which reaches beyond.  In fact, we stand here face to face with infinity, and recognize the fact that the infinite exists without the power to comprehend it.

The same is true of time.  We cannot conceive of its beginning or its end.  All things which come within the scope of our senses are limited in duration and circumscribed in space, and though we prate flippantly of the infinite, the pretence that we can grasp it is simply talk.

Conducted on such a plan, it was inevitable that scientific investigations should be narrow and materialistic in their tendency.  No matter how strong the probability in favor of the truth of a certain proposition,—though the whole fabric of society were based upon its acceptance, and it formed the foundation of civil and moral laws, even though it controlled the actions of the philosopher himself,—if not proved consistent with nature's physical and material laws, it must be rejected as unworthy to enter into the construction of the edifice he was erecting. In his great task of undoing the work of blind, unreasoning faith, and wild, illogical speculation, all the fruit of such faith or speculation must be looked upon as matter valueless to his hand. We may even go further and say that were it true that the Supreme Intelligence had created the material universe, and by special providence modified or thwarted the general laws through which that universe was governed, such divine supervision and such miraculous interposition must necessarily have been ignored.

Let it not be inferred, however, that each and all of the great men who have been engaged in this work of scientific reformation were necessarily driven to be impious iconoclasts, or that in their efforts to emancipate themselves from time-honored errors, they necessarily prostituted the liberty they gained to selfish or sensual purposes. On the contrary, the most important advances which the human intellect has made within these latter centuries have been due to the efforts of men of the purest and most conscientious character; men whose lives were devoted with the utmost singleness of purpose to determine "what is truth;" men who, knowing that all truth must be consistent with all other truth, were willing to go whithersoever it should lead. If it shall prove that they have been occupied with "mint, anise, and cumin," omitting the "weightier matters of the law," it is also true that in no other way could the material laws of the universe be thoroughly investigated than by making them the subjects of an absorbed and undivided attention. It would be as just to impugn the motives and decry the merits of the maker of our almanacs because his mathematical calculations were not interlarded with moral maxims, as to reproach the student of natural phenomena because he did his work so well, and left to others the co-ordination of the results of his efforts with the accepted dogmas of religious faith. And it is not true, in any sense, that these devotees of science have lived in vain; for to them we mainly owe the fact that man is not only wiser now than formerly, but that he is better and happier.

In justice to the man of science we must claim for him the position of co-laborer with, and indispensable ally to, the philanthropists and moralists: for from no source have they drawn richer lessons, stronger arguments, or more eloquent illustrations than from his discoveries.

And yet, while conceding conscientiousness and purity of motive to the vast majority of our men of science, and acknowledging the contributions they have made and are making to human happiness, compelled by my sense of justice to defend their spirit, approve their methods, admire their devotion, and assert their usefulness, I cannot deny that the tendency of modern investigation is decidedly materialistic. All natural phenomena being ascribed to material and tangible causes, the search for and analysis of these causes have begotten a restless activity and an indomitable energy which will leave no stone unturned for the attainment of their object. But while this is apparent, and, indeed, inevitable, as has been seen from the sketch of the growth of modern science, I am far from sharing the alarm which it excites in the minds of many good men. Nor would I encourage or excuse that spirit of conservatism—to call it by no harsher term—which, for the safety of a popular creed, would by any and all means repress, and, if possible, arrest investigations that, it is feared, may become revolutionary and dangerous.

Such opposition, in the first place, must be fruitless. All history has proved that persecution by physical coercion or obloquy is powerless to arrest the progress of ideas, or quench the enthusiasm of the devotees of a cause approved by their moral sense. The problems before our men of science must be solved in the manner proposed, if human wisdom will suffice for the task. In every department of science are men actuated simply by a thirst for truth, whom neither heat nor cold, privation nor opposition, will hold back from their self-appointed tasks. We may, therefore, accept it as a finality, that this problem will be carried to its logical conclusion.

In the second place, if possible, the arrest of scientific investigation would be not only undesirable, but an infinite calamity to our race. As has been so often said, truth is consistent with itself. If, therefore, our faith in this or that is based on truth, we have no cause for fear that this truth will be proved untrue by other truths. And more than this: it seems to me that, in the reach and thoroughness of this material investigation, we may hope for such demonstration of the reality of things immaterial as shall produce a deeper and more universal faith than has ever yet prevailed.

Through this very spirit of scepticism which pervades the

modern sciences we are compelled to exhaust all material means before we can have recourse to the supernatural. When, however, that has been done, and men have tried patiently and laboriously, but in vain, to refer all natural phenomena to material causes, then, having proved a negative, they will be compelled to accept the existence of truth not reached by their touchstone, and faith be recognized as the highest and best knowledge.

That such will be the result is the confident expectation of many of the wisest of the scientific men whose influence is looked upon with such alarm by those who, in their anxiety for their faith, demonstrate its weakness.

Already, as it seems to me, scientists have reached the wall of adamant—the inscrutable—that surrounds them on every side, and, ere long, we may expect to see them return to that heap of chaff from which the germs of modern science were winnowed, with the conviction that there were there left buried other germs of other and higher truths than those they gleaned; truths without which human knowledge must be a dwarfed and deformed thing.

A few illustrations from the many that might be cited will suffice to show the materialistic tendency of modern science. In "Pure Philosophy,"—as the students of Psychology are fond of styling their science,—the names alone of Compte, Buckle, Herbert Spencer, Mill, and Draper, will suggest the more prominent characters of the school they may be said to represent. The most conspicuous feature in the "Positive Philosophy" of Compte is the effort it exhibits to co-ordinate the laws of mind with those of matter. Spencer is a thorough-going mental Darwinist, who considers the highest attributes of the human mind, the loftiest aspirations of the soul, as only developed instincts, as these were but developed sensations. Mill, more guarded, more fully inspired with the spirit of the age,—which believes nothing, and is a foe to speculation,—leaves the history of our faculties to be written, if at all, by others; takes them as they are, but reasons of conscience and free-will with an independence of popular belief that savors more of the material than the spiritual school. Buckle wore himself out in a vain chase after an *ignis fatuus*, an inherent, inflexible law of human progress, and hence of human history. Draper is a developmentist, but not a Darwinian. With him civilization is a definite stage in the growth of mind; a degree of development to which it is impelled by a *vis a tergo*, not unlike, in kind, to that which evolves from the germ, the bud, the leaf, the flower, and the fruit in plant-life,—a development which, when unchecked and free, will be regular and inevitable, but which is so modified by the accidents of race, climate, soil, geographical position, etc., as to render it difficult to say whether the rule or the exception has, in his judgment, greatest potency.  If he were a consistent Darwinist, the accidents of development would be its law.

Among the students of "Social Science,"—a new and important member of the sisterhood of sciences,—as in most of the other departments of modern investigation, two groups of devotees are found; one patiently and conscientiously studying the problems of social organization, inspired with the true spirit of the Baconian Philosophy, ready to follow whithersoever the facts shall lead, and having for their object that noblest of all objects, the increase of human happiness.  The other class of investigators, in whom the bump of destructiveness is largely developed, would be delighted to tear down the whole fabric of society, and abrogate all laws, both human and divine.  Looking upon man as literally the creature of circumstances, as an inert atom driven about by material forces, conscience and responsibility are by them repudiated, and laws and penalties regarded simply as relics of barbaric despotism.  The dreary soul-killing creed of these fatalists is fortunately so repugnant to the reason and feelings of the majority of men, that there is little danger that their efforts will reach their legitimate conclusion in throwing society into a state of anarchy and chaos.

In Theology or Biblical Science the tendency of modern investigation is so distinctly felt, that I need only refer to it.  The spirit of independent criticism, so noticeable elsewhere, is still more conspicuous here; assuming sometimes the form of derisive scepticism, but oftener of cold, passionless judgment on the reported facts of sacred history, or the psychological phenomena of religious faith, studied simply as scientific problems.

The names of Strauss, Renan, and Colenso, will suggest the results to which men, possibly honest, are led by this so-styled "enlightened and emancipated spirit of enquiry;" while "Ecce Homo" and cognate productions may be considered as the fruit of this spirit, tempered by a very liberal but apparently sincere faith.

Aside from these more marked examples of the decided "set" in the tide of modern religious opinions, we everywhere see evidences that no part of the religious world is unmoved by it. In every sect and section an impulse is felt to substitute for abstract faith, the " faith without works "—rather a characteristic of the religion of our fathers, and not unknown at présent—that other faith which is evidenced by works.  In other words ; in our day more and more value is being attached to this life, as a sphere for religious effort and experience.  With what propriety, I leave to the individual judgment of my auditors ; the faith of every sect and man is coming to be respected and valued precisely in the ratio of the purity, unselfishness, and active sympathy in the life produced by it.

While, therefore, we have less now than formerly of the self-centred and fruitless piety of the old deacon whom I chanced to know, who excused his avarice by proclaiming that ' business was one thing and religion another, and he never allowed them to interfere;' in place of that we have many an Abou Ben Adhem, and all the splendid exhibitions of modern philanthropy.

Though the golden mean displayed in the life and words of Christ is far better than either extreme, I cannot but think the present religious condition of the world is better than any which has preceded it.

In Ethnology—the pre-historic history of the human race—the researches of the large number of investigators who are devoted to its study have made interesting and important additions to our knowledge ; but it cannot be denied that the result of such investigation has been to create general distrust of our previously accepted chronology, and give an antiquity to man such as the scholars of a previous generation would have looked upon as not only unwarranted but impious.  It should be said, however, that our preconceived opinions of the antiquity of the human race—like those of the age of the earth itself—were based upon no solid foundation in nature, history, or revelation ; and that our system of chronology was a matter of convention, about which there has been a wide latitude of opinion among the scholars of all ages.

In regard to the origin of man—whether by special creation or development—we may confidently assert, that modern investigation has given us no new light.  Among those who have accepted the theory of a special creation, and have differed only in regard to the number of species and their places of origin or centres of creation, there has been such a diversity of opinion that all confidence in the reality and value of the bases of their reasoning has been lost. Among the advocates of a multiplicity of species and diversity of origin we have from Blumenbach to Agassiz almost every number between fifteen and three as that of distinct species of the human race, scarcely any two writers advocating the same number. We may, therefore, very fairly infer that the facts upon which their conclusions are founded are not of a very clear and unmistakeable character.

The subject of the origin of the human race brings us into the domain of zoölogy, and opens the wide question of the origin of species, which, of late years, has been shaking the moral and intellectual world as by an earthquake. While the various writers upon the origin of the human race were gathering with so much industry, and reporting with so much eloquence, the proofs of a diversity of origin, the Darwinian hypothesis comes in and refers, not only all the human family, but all classes of animals and plants, to an initial point in a nucleated cell.

It would be impossible for any one to discuss, in a fair and intelligent manner, the great question of the origin of species, in anything less than a bulky volume. The merest mention is, therefore, all we can give to it at the present time. Although the appearance of Darwin's book on the Origin of Species communicated a distinct shock to the prevalent creeds, both religious and scientific, the hypothesis which it suggests, though now for the first time distinctly formularized, was by no means new; as it enters largely into the less clearly stated development theories of Oken, Lamarck, De Maillet, and the author of the Vestiges of Creation. There was this difference, however, that in the developmental theories of the older writers the element of evolution had a place; the process of development had its main spring in an inherent growth, or tendency, such as produces the evolution of the successive parts in plant-life, while, according to Darwin, the beautiful symmetry and adaptation which we see in nature is simply the form assumed by plastic matter in the mold of external circumstances.

Although this Darwinian hypothesis is looked upon by many as striking at the root of all vital faith, and is the *bête noire* of all those who deplore and condemn the materialistic tendency of modern science, still the purity of life of the author of the Origin of Species, his enthusiastic devotion to the study of truth, the industry and acumen which have marked his researches, the candor and caution with which his suggestions have been made, all combine to render the obloquy and scorn with which they have been received in many quarters peculiarly unjust and in bad-taste. It should also be said of Mr. Darwin that his views on the origin of species are not inconsistent with his own acceptance of the doctrine of Revelation ; and that many of our best men of science look upon his theory as not incompatible with the religious faith which is the guide of their lives, and their hope for the future. To these men it seems presumption that any mere man should restrict the Deity in His manner of vitalizing and beautifying the earth.  To them it is a proof of higher wisdom and greater power in the Creator that He should endow the vital principle with such potency that, pervaded by it, all the economy of nature, in both the animal and vegetable worlds, should be so nicely self-adjusting that, like a perfect machine from the hands of a master-maker, it requires no constant tinkering to preserve the constancy and regularity of its movements.

This much I have said in view of the possible acceptance of the Darwinian theory by the scientific world.  I should have stated *in limine*, however, that the Darwinian hypothesis is not accepted and can never be fully accepted by the student of science who is inspired with the spirit of the age.  From the nature of things it can be proved only to a certain point, and while we accept that which is proven,—and for it sincerely thank Mr. Darwin,—that which is hypothesis, or based only upon probabilities, we reject, as belonging in the category of mere theories, to disprove or purify which the modern scientific reform was inaugurated.  Much, too, may be said against the sufficiency of ' natural selection in the struggle of life,' from observations made upon the phenomena of the economy of nature.  Necessarily, the action of the Darwinian principle must be limited to the individual, be literally and purely selfish ; and if it can be proved that a broader influence pervades the created world, that something akin to benevolence enters into the organization of the individual, something which benefits others and not himself, one single fact establishing this truth would hurl the entire Darwinian fabric to the ground ; or rather restrict it to its proper bearing upon the limits of variation, and the mooted question of ' what is a species ? '

One of the most potent influences in the perpetuation of species is fecundity in the individual, whereas we see in social insects the economy of the community is best served by a total loss of this power in the great majority of the individuals which compose it.

This objection will perhaps be met by the Darwinians with the assertion that the community, in fact, constitutes an individual; but I must confess that I find it difficult to comprehend how the sterility of the workers in ants and bees was ever introduced through the medium of modified descent, the Darwinian method, or how it is kept up from generation to generation among those individuals which have no posterity to inherit their peculiarities of structure.

The Honey Ants of Mexico offer additional difficulties. Among them a portion of the community secrete honey in the abdominal cavity until they resemble small grapes, and these individuals, during the winter, are despatched in succession to furnish food for the other members of the colony. How, by modified descent, is this honey-making faculty transmitted, when those who possess it are systematically destroyed?

A still harder nut for the Darwinians to crack is furnished in a fact stated by Dr. Stimpson, that among the crustacea, which do not live in communities, a very large proportion of the individuals of a numerically powerful species pass their lives as neuters, or undeveloped females.

Another element in nature's economy, which at first sight suggests an objection to the Darwinian theory, is that of beauty, which affects others far more than the possessor. This is considered by the Darwinians simply as an attraction to the opposite sex, but as a fact we find that in the larval condition of some insects—a condition in which no propagation is effected—varieties of form and combinations of color exist which appeal to our sense of beauty scarcely less forcibly than in the perfect insects.

Again, the origin of life is left completely untouched by the Darwinian hypothesis, and so long as the vital principle resists, as it has done, all efforts of theorists and experimenters to bring it within the category of material forces, so long we must regard the world of life as including elements not amenable to the laws which control simple inert matter.

Upon this question of the origin of life so much is being done and said that you will expect a word of reference to it at my hands, yet little more can be reported as the result of modern research than that the origin of life is as great a mystery as ever. You will all remember how, a few years since, we were startled by the announcement of the discovery of the generation of the *Acarus Crossii;* and, while our original distrust of the accuracy of the observations of Mr. Crosse was strengthened by the failure of subsequent experimenters to reproduce his results, our belief is further confirmed by the unanimity of all the more modern and intelligent devotees of spontaneous generation in the assertion that life can only originate in its simplest form, that of a unicellular organism. There is no Darwinist who will concede the possibility of an animal as highly organized as an Acarus, with body, head, limbs, digestion, and senses, all more or less complete, being the product of spontaneous generation and not the result of slow and gradual development.

Still farther; it is known that the animal kingdom rests upon the vegetable as a base. Animals being incapable of assimilating inorganic matter could not exist without plants. Plants must therefore have preceded animals, and the fruit of spontaneous generation must be a prototype and not a protozoan.

Strange as it may seem, there are, however, men, respectable by their numbers and attainments, who are believers in spontaneous generation; but it is with this proviso—which leaves the mystery as great as ever—that only from organic matter can organisms be produced. So that to the original and primary appearance of life upon the earth modern science has given us not the slightest clue.

As I have said, the materialists have so far utterly failed to co-ordinate the vital force with those which we designate as material. The beautiful and important discoveries which have followed researches into the correlation and conservation of forces, by pointing to a unity of all the forces in the material world, have naturally prompted efforts to centralize, with electricity, magnetism, and chemical affinity, that which we know as vital force. But a moment's reflection will show us how far removed is this vital force from all others with which it has been compared.

The nicest manipulations of chemical science will probably fail to detect a difference in composition between the microscopic germs of two cryptogamous plants. Each consists of the same elements, carbon, nitrogen, hydrogen, and oxygen, in nearly or quite the same proportions. Both may be planted in a soil which laborious mixture has rendered homogenous, and subsequently supplied with the same pabulum, and yet, in virtue of some inscrutable, inherent principle, one develops a humble moss, and the other rises into the beauty, symmetry, and even grandeur of a tree fern. The same may be said of the spermatozoa of the mouse and the elephant. Indeed all the phenomena which attend the reproduction of species are totally at variance and incompatible with those which mark the action of material laws. Why, in physical circumstances differing *toto cœlo*, does the germ produce a plant or animal so closely copying the parent ? and whence this tenacity of purpose in the germ which reproduces, through a long line of posterity, the trivial characteristics of a remote ancestor ? Even within our limited observation we have been struck by the reappearance in the grandchild of the voice, the gesture, the stature, the features, or some other marked peculiarity of his grandsire. Whence comes the force of the axiom that ' blood will tell ' ?—and how incomprehensible that, by the action of only material laws, mental force, or, it may be, moral infirmity, is transmitted from generation to generation, in spite of the system of infinitesimal dilution through which it passes !

And now, even with this hurried and sadly imperfect exposition of the tendency of modern science, the time at our command has been consumed. Before leaving the subject, however, I crave your indulgence for a word to those who, wholly absorbed in the study of the laws which regulate the material universe, are so deeply impressed with their universality and potency, that they forget that law is but another name for an order of sequence, and has in itself no force. These are they who, in their pride in the achievements of the human intellect, fail to realize that the universe furnishes conclusive proof that all our philosophy, all our logic, all our observation are utterly inadequate to solve the problems that are presented to us; inadequate not simply from the limited nature of our powers of observation, but because the human mind, though forced to confess the existence of the infinite, is utterly unable to grasp it; and that while the logic of reason and the logic of numbers suffice for a qualified understanding of the manner in which material forces work, of the origin and nature of these forces we are and must ever remain ignorant, unless gifted with higher powers than we now possess. As has been stated, seen from the stand-point of our modern materialists, and judged by the criteria which they have adopted, spiritual existence and supernatural phenomena, even if as all-pervading as the most devout religionist believes, must, from *a priori* considerations, be utterly ignored. Of those who are thus led by their regard for the dignity of material laws to reject the idea of a creative and overruling Deity, I would ask, Is not man himself a disturbing element in your universe ? Whatever may be said in regard to man's free-agency, and however confidently it may be asserted that his will is but the resultant of the various motives that operate as distinct forces upon it, consciousness lies at the base of all reasoning; and the conduct of every man proves that he accepts this axiom. As he issues from his door he is conscious, beyond all argument, that it is in his power to turn to the right or to the left; and while he holds himself responsible for his volition, he cannot blame us if we ascribe to him free-agency. Man is therefore an independent power in the universe. He wills and creates. The locomotive is as truly his creation as himself, fashioned from the dust of the earth and vitalized by the breath of the Almighty, is the work of His hands. If, therefore, all the realm of nature is controlled through material laws, by forces that, like attraction, electricity, chemical affinity, etc., act in an invariable and inflexible way, in this universe man is a stupendous anomaly; and unless he can be degraded from his position of pre-eminence in this material world, the boldest and most irreverant of modern philosophers will strive in vain to dethrone the great Creator from the rule of the universe, or from His place in the hearts and minds of men.

---

# AMERICAN ASSOCIATION

## FOR THE ADVANCEMENT OF SCIENCE.

The sixteenth annual meeting of the American Association for the Advancement of Science, was held in Burlington, Vermont; under the presidency of Professor J. S. Newberry, of Columbia College, New York; commencing on Wednesday, Aug. 21, and closing on Monday, Aug. 26, 1867.

The attendance was larger than at the Buffalo meeting in 1866; but still below that at the meetings held before the suspension of the Association's active work, rendered necessary by the late American war.

The papers presented were not very numerous; but nearly all were of great scientific value; bearing on controverted questions, applying the results of investigation to the determination of natural laws, or suggesting new fields and methods of research. A few comprised nothing beyond local details, which, if published, might have been of some service; but which should not have come before the Association. We give above the annual address of the President, and the following abstract of most of the papers read in the Natural History section, acknowledging our indebtedness to Prof. Newberry, and other members; and to the editors of the *American Naturalist*, for notes of papers, and proofs, kindly furnished us.

CONSIDERATIONS DRAWN FROM THE STUDY OF THE OR-THOPTERA OF NORTH AMERICA; by S. H. SCUDDER, of Boston. – This paper was a lengthened, comparative view of the North American and European orthopterous faunas. The groups characteristic of each continent were detailed, and the conclusion inferred, that, under similar climatic and other conditions, this family of insects is much richer in species and individuals in North America than in Europe.

TRACES OF ANCIENT GLACIERS IN THE WHITE MOUNTAINS OF NEW HAMPSHIRE; with a few remarks upon the geological structure of that part of the group.—The author recounted the observations made by him on the geological structure and grouping of the rocks in the region of the Androscoggin, Peabody, and other valleys in the White Mountains, and the traces they bear of glacial action. His observations tended to confirm the opinion that these valleys have been occupied by local glaciers, as well as by a general one.

ON THE ORIGIN OF THE LIGNILITES, OR EPSOMITES; by Prof. O. C. MARSH, Yale College, New Haven.—These names have been applied to the columnar markings, and more or less detached columns, occurring in the seams between strata, among limestone rocks of all ages. Prof. Marsh, after stating the conflicting opinions hitherto held and published by geologists, on the cause of the structure, exhibited a fine series of specimens, showing it to be due to pressure. He has found that a shell, or other foreign substance, often forms the nucleus of one of these columns.

THE FOSSIL INSECTS OF NORTH AMERICA; by S. H. SCUDDER.—This paper was a summary of all that is yet known on the subject. Eighty species have been determined and described; while a few fragments are so badly damaged that it is impossible to identify them. The Orthoptera have the greatest number of representatives in the North American rocks; and no species of coleoptera has yet been found. The oldest of our known fossil insects are from the Devonian strata of New Brunswick; while in Tertiary rocks their remains have been found in only one locality, near Green River, Colorado. No fossil spiders have been discovered in North America.

ON THE WINOOSKI MARBLE OF COLCHESTER, VT.; by Prof. C. H. HITCHCOCK.—This beautiful marble, which is found in Potsdam rocks, near Burlington, consists of a silicious dolomite, containing imbedded nodules of silica, enclosed in calcite. The prevailing color is red, mottled, and veined with white, brown, chocolate, yellowish, and whitish tints. So highly is it valued abroad, that considerable quantities are exported to Italy for the use of the sculptors of that country. The presence of the quartz, however, renders it somewhat difficult to work.

ON THE ZOOLOGICAL AFFINITIES OF THE TABULATE CORALS; by Prof. A. E. VERRILL.—The questions discussed in this paper were the position of the tabulate corals among Polyps, and the true value of the tabulate structure in classification. Coral-like forms are produced by Protozoa (Eozoon, Polytrema, Sponges, etc.), Molluscan corals (Bryozoa), Hydroid corals (Sertularia, etc.), Polyp corals (Gorgonia, Tubipora, Madrepora, etc.), and by vegetable corals (Nullipora, Corallina). Most of these have been carefully studied. Two important groups, however, are still involved in considerable doubt,—the Cyathophylloid corals (Rugosa, Edw.), and the Tabulate corals. The former are entirely extinct, and their structure may long remain somewhat uncertain. The latter are represented in tropical seas by several genera and numerous species. Usually they have been considered true Polyps; but some zoologists urge their affinity with the Bryozoan mollusks, while Agassiz, after examining the genus Millepora, places the whole group among Hydroids. Prof. Verrill considered the point as only settled so far as Millepora and its allies were concerned, and requested Mr. F. H. Bradley, while collecting, at Panama, for the Yale College Museum, to study the structure and habits of a species of Pocilopora found at the place. The descriptions and figures of the animal show it to be a true Polyp, scarcely differing from Porites, except in the position of the tentacles. The animals are exsert when expanded, and have twelve equal cylindrical tentacles surrounding the margin, in a single circle, six of them being held horizontally, and the alternate ones erect. Prof. Verrill, therefore, concludes that the tabulate structure is of secondary importance as a character, in fixing their affinities, and that the Tabulata must be dismembered,—Halisites, Millepora, and their allies, being classed as Hydroids; and Pocillopora and Favosites with other extinct tabulated genera, as true Polyps.

ON THE COAL MEASURES OF ILLINOIS; by Prof. A. H. WORTHEN, State Geologist.—In prosecuting the geological survey of Illinois, it seemed desirable to identify the coal seams of that State with those of Kentucky, which occupy the same basin. To effect this, a section was constructed along the valley of the Illinois River, which traverses the coal-field from S.W. to N.E. for about 100 miles. Six beds of workable coal, and four or five thin seams, were met with in the section. After correcting an error, which he thinks had been made in constructing the Kentucky section, by considering the outcrop of the same sandstone at Mahoning and Anvil Rock as different beds, Prof. W. found a very close resemblance between the Illinois and Kentucky strata. From his observations, he infers the existence of coal seams over wide geographical areas. The fact was also stated that many of the fossils of the carboniferous limestone, in this region, are identical with those described by Hayden and Meek, from the so-called sub-carboniferous rocks of Eastern Kansas.

ON RECENT GEOLOGICAL DISCOVERIES IN THE ACADIAN PROVINCES OF BRITISH AMERICA; by J. W. DAWSON, LL.D., F.R.S., Principal of McGill University. The object of the paper was to notice some recent discoveries, which, though of interest, might have escaped the notice of members of the Association.

In New Brunswick, the older rocks in the vicinity of the city of St. John have been reduced to order, and their probable ages ascertained, principally through the labors of Mr. Matthew, Mr. Hartt, and Professor Bailey. The first step toward the knowledge of their precise date was the discovery of a rich land flora in some of the upper beds, next below the Lower Carboniferous rocks which overlie them unconformably. These fossil plants he was enabled to recognize as of the Devonian Period, and the zealous researches, more especially of Mr. Hartt, have brought to light no less than forty to fifty species, or half of the whole number known in the Devonian of Eastern America, as well as six species of insects, four of which have been described by Mr. Scudder.[*]

---

[*] Canadian Naturalist. 1867.

These insects are the first ever found in rocks older than the Carboniferous.

These rocks, consisting chiefly of hard shales and sandstones, having been ascertained to be Devonian, there still remained an immense thickness of underlying rocks of uncertain age. In the upper member of these rocks, the same active observers already mentioned have observed a rich primordial fauna, embracing species of Conocephalites, Paradoxides, Microdiscus, and Agnostus, as well as an Orthis, and a new type of Cystidians. These fossils are regarded by Mr. Hartt and Mr. Billings as of the age of Barrande's " Etage C," and as marking a new and older period of the " Silurian Primordial" than any other as yet recognized in America, with the exception of the slates holding Paradoxides in Massachusetts, and the similar slates of the "Older Slate Formation" of Jukes, in Newfoundland. Descriptions of these fossils, by Mr. Hartt, will be published in the edition of "Acadian Geology" now in press. It is proposed to call this series, represented in New Brunswick by the St. John slates, the *Acadian Series*.

Below these primordial beds are highly metamorphosed rocks, at least 9,000 feet in thickness, which have afforded no fossils. A portion of these, consisting principally of conglomerate and trappean beds, is regarded by Messrs. Matthew and Bailey as of the age of the Huronian. The remainder, containing much gneiss and a bed of crystalline limestone, they regard as Laurentian. If this view is correct, and it certainly seems to be probable, these rocks, thus rising through the oldest members of the Lower Silurian, and forming a stepping-stone between the Laurentian of Newfoundland and that of New Jersey, show that the foundations of the north-east and south-west line of the east side of North America were already laid in the Laurentian period. Still, it is not here, but farther west, that we are to look for the dividing line between the great inland Silurian basin of America, and that of the Atlantic coast; the latter has been pointed out by Professor Hall and Sir. W. E. Logan, as remarkably distinguished by the predominance of mechanical sediments, and by a development of the lower rather than the upper members of the Lower Silurian.

To ascend from these rocks to the Carboniferous,—recent labors of Mr. Davidson, Mr. Hartt, and the author, had led to the division of the Lower Carboniferous into successive subordinate stages, and to the determination of most of the marine fossils, and also to the explanation of the curious and apparently anomalous fact that some forms allied to Permian species actually exist in the Lower Carboniferous, under the productive coal-measures. These researches had also shown that no distinction between Sub-carboniferous and Carboniferous proper, can fairly be made in Nova Scotia, notwithstanding the grand development of the Carboniferous in thickness.

After noticing the large advances made in the fossil botany of Nova Scotia and New Brunswick, the paper referred to the discovery by Mr. Barnes of two new species of insects, and to the discovery by the author of a new pulmonate mollusk, described by Dr. P. P. Carpenter as *Conulus priscus*. There are thus in the coal formation of Nova Scotia a Pupa and a Conulus or Zonites, generically allied to living pulmonates, and representing already in that early period two of the principal types of these creatures.[*]

Specimens of these fossils were exhibited, and also specimens and a photograph of the Laurentian fossil *Eozoön Canadense* sent by Sir. W. E. Logan. Special attention was drawn to the specimen recently found by the Canadian Survey at Tudor, which shows this organism in a state of preservation comparable with that of ordinary Silurian fossils.

ON SOME REMARKABLE FOSSIL FISHES, FROM THE "BLACK SHALE"(DEVONIAN) AT DELAWARE, OHIO; by J. S. NEWBERRY. —Dr. Newberry exhibited to the Section different portions of the head of a gigantic fish, to which he had given the name of *Dinichthys Herzeri*; and which, he said, from its size and structure, deserved the same distinction among fishes that *Dinotherium* and *Dinornis* enjoy among mammals and birds. Most of the bones obtained as yet belonged to the head, which was over three feet long, by one and a-half broad, and wonderfully strong and massive. All parts of the head had been procured, and many different individuals were represented in the collections made by Mr. Herzer. The cranium was composed of a number of plates firmly anchylosed together, and strengthened near the occiput by internal ribs or ridges, nearly as large as one's arm. The external surface was covered with a very fine vermicular ornamentation. The anatomical structure was more wonderful than the size, and was such as to separate this quite widely from any fishes known, living or

---

* Acadian Geology.  Second Edition.

fossil. The most marked peculiarity was in the structure of the jaws and teeth, both as regards the form and texture. The form of the jaws will be best understood by the following figures.

Fig. 1.

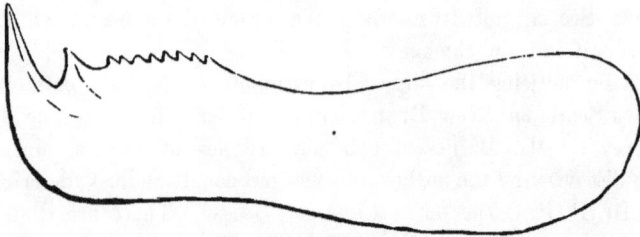

Lower Jaw—one-eighth natural size.

Fig. 2.

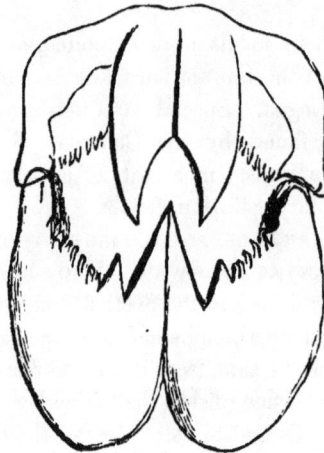

Front view of Head—one-eighth natural size.

The head terminated anteriorly and above in two great incisors, representing the premaxillary, behind which on either side were the maxillaries—broad, flattened bones of very dense tissue—along the lower edge of which was set one row of small robust teeth, which were neither implanted in sockets nor cemented to the jaw, but were formed by the consolidation and prolongation of the jaw tissue. The mandibles are over two feet long by six inches deep, laterally flattened and very massive, being without any medullary cavity. The anterior extremity was turned up in a huge triangular tooth, composed of dense ivory-like tissue, which alternated with (passing between) the divergent incisors of the upper jaw. Back of this terminal tooth, on some specimens, was another triangular summit, behind which was a row of small teeth corresponding to those of the maxillaries. Such was the power of this tremendous dental apparatus, that the bodies of our largest living fishes would be instantly pierced and crushed by it, if exposed to its action. Behind the head were large and thick plates, one of which corresponded to the "*os medium dorsi*" of *Heterostius* (of Pander) being at least of equal size.

These interesting fossils were found in the calcareous concretions, which occur so abundantly near the base of the "Black Shale" (Hamilton) at Delaware, in Central Ohio, by Mr. Herzer, a clergyman, who, while performing his pastoral duties, and living on a very small salary, had been a most zealous and remarkably successful student of the local geology.

ON SOME FOSSIL REPTILES AND FISHES FROM THE CARBONIFEROUS STRATA OF OHIO, KENTUCKY, AND ILLINOIS; by J. S. NEWBERRY.—The specimens exhibited and described in this communication consisted of reptiles and fishes from the cannel stratum underlying the main coal seam at Linton, Ohio; of fishes from the coal measures of Illinois, collected by the State Geologist; and of a group of fishes collected by Dr. Patterson from a stratum of bituminous shale lying in the Waverly group, 125 feet above its base at Vanceburg, Kentucky. Of these, the first series included *Raniceps Lyellii* (Wyman) with several as yet undescribed reptiles, some of which apparently belong to Prof. Huxley's new genera, *Ophiderpeton* and *Urocordylus*. Associated with these were some twenty species of fossil fishes, most of which have been described by Dr. Newberry, but were now represented by new and more perfect specimens. Among these were eight species of *Eurylepis*, a genus created by Dr. Newberry to receive a group of small lepidoids, allied to *Palæoniscus*, but distinguished by the scales of the sides, which are much higher than long. The scales on several of these species are very highly ornamented. The specimens exhibited were preserved in cannel coal, and covered with a film of sulphide of iron, by which they were brilliantly gilded. With these were two species of *Coelacanthus*, some of the specimens of which showed that the fishes of this genus were furnished with a supplemental caudal fin, as in *Undina*. This, Dr. Newberry stated, was an interesting fact, confirmatory of Prof. Huxley's view of the relations of *Undina*,

*Macropoma* and *Coelacanthus.* The numerous and very complete specimens of *Coelacanthus* exhibited, supply much that was wanting to a perfect knowledge of the anatomy of this genus. The bones of the head are similar in form to those of *Macropoma*, are highly ornamented with tubercles above and thread lines below; the jugular plates are double and long-elliptical as in *Undina* and *Macropoma*, but the teeth are conical and curved. The position and form of the fins is as in *Undina*, but the anterior dorsal is stronger. The fins are supported on palmated interspinous bones, similar in a general way to those of the other genera of the family. The paired fins are slightly lobed. The supplemental caudal has been referred to; the scales are ornamented with curved and converging raised lines. In many specimens the earbones (otolites) are distinctly visible. Besides the fishes found at Linton already enumerated, there were scales and teeth of *Rhizodus*, two species, at least one of which (*R. angustus*) has teeth of two forms,—one large, flattened, and double-edged; the others smaller, more numerous, slender, striated, and conical, with a circular section throughout; two species of *Diplodus*, consisting of bony base and enamelled crown,—the latter distinctly and beautifully serrated; so that there can scarcely be a question that they were teeth, and not, as claimed by Mr. Atthey, of Newcastle, England, dermal tubercles.

In the Linton fauna is one species of *Palæoniscus* (*P. Scutigerus, N.*); one of *Pygopterus*; one of *Megalichthys*, represented by scales; and numerous spines of placoid fishes of the genera *Compsacanthus* and *Pleuracanthus.*

The fish remains from Illinois consisted of a splendid specimen of *Edestus vorax* (Leidy) from the coal at Bellville, opposite St. Louis, and of several individuals of a new species of *Platysomus* from the concretions of iron ore at Mazon Creek. The *Edestus* was said by Dr. Newberry to have been described as a jaw, but the specimen exhibited was much more complete than any before found, and there could scarcely be a doubt that it was the spine of a Selachian. *Platysomus*, he said, though common in the coal measures of England, had not been before found in America.

The fishes from the Waverly were from a new locality, and from a horizon that had furnished very few fossils of any kind, and no fishes except a *Palæoniscus* (*P. Brainerdi*) found in northern Ohio. The specimens collected at Vanceburg, by Dr. Patterson, consisted of teeth of *Cladodus* and *Orodus*, with spines of *Ctenacanthus*, with the tail of one of these Selachians distinctly preserved.   This Dr. Newberry said was a great rarity, as the soft, and even the cartilaginous parts of plagiostomous fishes had usually disappeared, the teeth, spines, and dermal tubercles—the only bony parts—alone remaining.   The only similar case of which he had any knowledge was the discovery of the tail and fins of *Chondostreus*, in the Lias of Lyme Regis, England, and the preservation of *Thyalina* in the Solenhofen slate.   The specimen shown was greatly older than these, being from the base of the Carboniferous, and was the only figure that nature has yet given us of the external form of these ancient sharks.   This tail was very heterocercal, had the form of the caudal fins of some living sharks, and indicated a fish of seven or eight feet in length. In the specimen exhibited, the vertebral column had entirely disappeared, but the impressions of the spinous bones were distinctly visible, those of the lower lobe of the tail being ossified throughout.   Dr. Newberry said that he hoped to gather data from this collection for uniting teeth and spines, which, though described under different names, were parts of one fish.

ON SOME NEW FOSSIL SPONGES FROM THE LOWER SILURIAN ; by Prof. O. C. MARSH.—The author exhibited and described some specimens of the new genus Brachiospongia, from the Lower Silurian rocks of Kentucky.   These sponges, of which a full account will shortly be published by Prof. Marsh, differ widely from all the species hitherto known, and are of great interest to science.

ON THE OCCURRENCE OF FOSSIL SPONGES IN THE SUCCESSIVE GROUPS OF THE PALÆOZOIC SERIES ; by Prof. JAS. HALL.— This paper was an epitome of all that is known of the sponges of the Silurian, Devonian, Carboniferous, and Permian formations. Sponges with calcareous skeletons, and coral-like forms, were among the earliest inhabitants of the earth, being found in the Lower Silurian strata.   In the Devonian age they were still more abundant ; but from this period diminished in numbers, and became more like the horny sponges of the present day.

ON THE AMERICAN BEAVER; by LEWIS H. MORGAN, Rochester, N.Y.—The Beaver appears to be rapidly becoming extinct wherever civilization advances.   It is still found, however, in certain localities, from Virginia to the parallel of 60 N. lat., though most abundant in the Hudson Bay Territory.   Mr. Morgan had examined the dams constructed by them around the southern shore of Lake Superior. Some of these show an astonishing amount of instinct in the way of engineering. Trees many feet in diameter have been cut down by them; canals are often constructed from their ponds to the localities of the trees on the bark of which they feed—in one instance the canal measuring over 60 yards in length. Natural obstacles are overcome by means of bridges, tunnels, etc., built with great ingenuity.

ON THE METAMORPHOSIS AND DISTORTION OF PEBBLES IN CONGLOMERATE ; by C. H. HITCHCOCK, State Geologist, Vermont.—Geologists have noticed that in certain highly disturbed localities, when a band of conglomerate can be traced from its normal position to that in which it is contorted and folded, the undisturbed stratum is simply a loosely cemented gravel with rounded pebbles, while in the plicated rocks the pebbles are distorted and flattened. Examples of this occur at Middleton, R. I., Plymouth, Vt., Nagelflue, in Switzerland, and the Permian conglomerate in England. The pebbles are not only distorted, but often changed in their chemical composition, impure lime-stones or schists being displaced by quartz, and probably the original sandstone and conglomerate changed into schists, gneiss, and granite. Prof. Hitchcock thinks that both the metamorphism and warping are due to the agency of infiltrated water under enormous pressure.

ON THE LOWER SILURIAN BROWN HAEMATITE BEDS OF AMERICA; by B. S. LYMAN.—Thirty exposures of the four beds of this ore have been studied in Western Virginia. Of these, three or four show the solid bed; the others only have weathered boulders of the ore, mixed with other detrital matter. In comparing these with other Brown Haematite deposits in the United States, the author infers that the lumps of ore, sometimes found mixed with the *debris* of other rocks, mark the proximity of beds of the Haematite, from which the blocks have been separated by denudation. From the frequent occurrence of carbonate of iron, he regards this as the original composition of the ore,—the carbonic acid having been driven off by heat, or other causes, and the protoxide changed to a sesquioxide.

EXPLANATIONS OF THE GEOLOGICAL MAP OF MAINE ; by Prof. C. H. HITCHCOCK. —The author showed the large geological map, which embodied the results of work done by the State Survey during 1861-62, and called attention to several points of interest settled during that period.

ON THE GEOGRAPHICAL DISTRIBUTION OF THE RADIATES ON THE WEST COAST OF AMERICA; by Prof. A. E. VERRILL.— Eleven distinct marine zoological provinces have been recognized along the coast, each characterized by the existence or prevalence of peculiar genera and species. These provinces were discussed by Prof. Verrill, in detail, the characteristic species, and the conditions under which they exist stated, the number of species of each class of Radiates known to exist in the several provinces, and number peculiar to the respective provinces given, and each Pacific shore region compared with parallel regions on the Atlantic coasts of America and Europe. Distribution is effected mainly by temperature, less by the nature of the bottom and shore. Depth of water exerts principally an indirect influence by diminishing the temperature as we descend. A few Holothurians are the only Radiates recorded as common to the Atlantic and Pacific. The Polyps and corals of the two seas differ widely. The mollusca, crustacea, fishes, and echinoderms are usually specifically distinct, but the genera and families of these groups are often identical. No direct evidence exists of a water communication across the Isthmus later than the cretaceous period. Prof. Verrill concludes that all the phenomena observed in the distribution of identical species may be accounted for by supposing a former depression of about 300 feet, which would cause a connection across the Isthmus by means of a shallow, brackish estuary, capable of sustaining the life of many mollusca, crustacea, and fishes, but not the genera of corals and other Radiates. In the case of distinct, but similar species, we must suppose different centres of creation, or a descent from common ancestors, the distribution having taken place at a very early period, when an extensive connection existed between the two oceans. The animals on the latter supposition have subsequently become distinct, by natural selection, or otherwise.

CONSIDERATIONS RELATING TO THE CLIMATE OF THE GLACIAL EPOCH IN NORTH AMERICA; by Prof. E. HUNGERFORD. —The object of this paper was to discuss the growth, and climatic influence of such an accumulation of ice and snow as the glacial hypothesis supposes to have once existed. The result of an elevation of the northern part of the continent would be to lower the snow line by depressing the mean summer temperature. If the surface were raised by the accumulation of frozen snow, instead of by an upheaval of the land, the frigorific effect would

be similar, but greatly intensified. Then every addition to this
icy accumulation would depress still farther the temperature of
the continent, and extend the area of perennial snow. The great
northern ice plateau would thus increase in height and superficial
extent until prevented by some reactionary cause. Meteorological
considerations all show us that the interior of such a plateau
would be intensely cold,—so cold as to prevent the simultaneous
moving of the continental glaciers in one determined direction.
Hence, the erosive effects which we witness are due to glacial
motion along the southern and seaward edge of the glacier, where
the snow is softened by the sun, or sea-breezes, and a slope
supplied by the glacial front itself.

DEPRESSION OF THE SEA DURING THE GLACIAL PERIOD; by
Colonel CHARLES WHITTLESEY.—The level of the ocean is
maintained by the evaporated water being returned through
rivers, etc. If part of this vapor, instead of returning, accumulate on the land as permanent snow and ice, the result will be a
depression of the sea level, proportionate to the extent of the ice-
fields. A decrease of one degree annually in the earth's
temperature would lower the snow line 300 feet, extend the area
of ice and snow, and diminish evaporation; while additions
would be constantly made to the thickness of the ice beds. Now,
as one-fifth of the earth must have been covered by ice-fields
during the glacial period, and the extent of the ocean at the time
is known with considerable certainty, by knowing the thickness,
and, consequently, the mass of the beds of ice, we can easily
determine the decrease in the water of the sea. Ice etchings are
observed on rocks in British America and New England, at heights
varying from 1,500 to 5,300 feet above the present sea level. Admitting an average of 2,000 feet, and an expansion of one-tenth in
freezing, we have a sufficient amount of congealed water to cover
the above area to a depth of 1,800 feet. As nearly the entire
remaining surface of the earth was covered with water, the surface
would sink about one-fifth of the above, or 360 feet. The weight
of such a mass of ice would probably be sufficient to cause a
sinking of the land on which it rested, while that adjacent to it
would be elevated; just as we see Greenland settling down, and
Newfoundland rising, at the present day. These facts should be
kept in mind in studying fresh water and marine terraces, and
drift-beds. From the absence of these elevations on the Rocky
Mountains above a height of 2,000 feet, that part of the continent

seems to have been sinking, during the past glacial period, while the eastern sea-coast was rising,—the line of rest being near the middle of Lake Ontario.

On the Ripton Sea-beaches; by Prof. E. Hungerford.— This paper described a series of terraces, situated at a height of 2,196 feet above the sea, on the west flank of the Green Mountains, on the pass from Ripton to Hancock. They consist of a modified drift, overlying the true boulder drift of the region, and arranged in this present form by the action of waves and currents. As the configuration of the country would not allow the accumulation of a large body of fresh water at this point, these deposits are regarded as strongly confirming other evidence that this area has suffered a depression of at least 2,000 feet since the glacial epoch. The author regards the following as the successive geological events by which the drift phenomena have been produced:

1. The formation of a continental glacier, to whose partial movements, always limited to a comparatively narrow belt upon the southern or seaward margin, are due the erosive phenomena, and the transportation of the drift over limited areas.

2. A depression of the continent, bringing the ocean into contact with the long glacial border, which on its retreat sends off icebergs and ice-rafts into the ocean. To these are attributed the further transportation of detritus and boulders.

3. Emergence of the continent, the higher beaches marking the earlier, and the Champlain terraces the later stages of this process.

On certain effects produced upon Fossils by Weathering; by Prof. O. C. Marsh.—Prof. Marsh has discovered that certain peculiarities observed in *Ceratites nodosus*, and other fossil shells, especially cephalopods, and which have long perplexed German geologists, are due to the action of the elements, the layers of the shell differing in composition, hardness, and markings. In some cases the markings characteristic of two distinct genera may be observed on the same specimen.

On the Geology of Vermont; by Prof. C. H. Hitchcock. —Prof. Hitchcock exhibited a large geological map of the State, showing the great progress made in determining the structure of its rocks, since the publication of his final Report upon the Geology of Vermont, in 1861. This is largely due to the extension southward of the recent discoveries of the Canadian survey.

On the Ichthyological Fauna of Lake Champlain;

by F. W. PUTNAM, Superintendant of the Essex Institute.—A list numbering 45 species of true lake fishes, obtained by the author from Lake Champlain, was given; of these, 41 were found by him in Lake Erie. As Lake Champlain was a salt-water bay at a period subsequent to the glacial epoch, while the lakes above Niagara Falls contained fresh-water, the weight of evidence goes to support the conclusion that the fishes of Lake Champlain have been chiefly derived from those higher lakes.

Among other business transacted before the close of the meeting, the following resolution was moved by Prof. O. C. Marsh, and adopted:—

"*Resolved*, That the chair appoint a commission of nine members to examine the Linnean rules of Zoological Nomenclature by the light of the suggestions and examples of recent writers, and to prepare a code of laws and recommendations in conformity with past modern usage, to be submitted to the Association at the next annual meeting; the committee to have authority to fill vacancies and increase the number to twelve, if deemed advisable."

The committee appointed consists of:—Prof. J. D. Dana, of Yale College; Prof. Jeffries Wyman, of Howard University; Prof. S. F. Baird, of the Smithsonian Institution; Prof. Joseph Leidy, of the Philadelphia Academy of Natural Sciences; Prof. J. F. Newberry, of Columbia College; Principal Dawson, of McGill University, Montreal; Dr. Wm. Stimpson, of the Chicago Academy of Science; S. H. Scudder, of the Boston Natural History Society; and F. W. Putnam, of the Essex Institute.

The next meeting will be held at Chicago, commencing on the first Wednesday of August, 1868.          II.

---

## ON NEW SPECIMENS OF EOZOON.

By Sir W. E. LOGAN, F.R.S., F.G.S.[*]

SINCE the subject of Laurentian fossils was placed before this Society in the papers of Dr. Dawson, Dr. Carpenter, Dr. T. Sterry Hunt, and myself, in 1865, additional specimens of *Eozoon* have been obtained during the explorations of the Geological Survey of Canada. These, as in the case of the specimens first discovered, have been submitted to the examination of Dr. Dawson; and it will be observed, from his remarks contained in the paper which is

* From the Quar. Jour. Geol. Soc. for August, 1867.  Read before the Geological Society, May 8, 1867.

to follow, that one of them has afforded farther, and what appears to him conclusive, evidence of their organic character. The specimens and remarks have been submitted to Dr. Carpenter, who coincides with Dr. Dawson; and the object of what I have to say in connexion with these new specimens is merely to point out the localities in which they have been procured.

The most important of these specimens was met with last summer by Mr. G. H. Vennor, one of the assistants on the Canadian Geological Survey, in the township of Tudor and county of Hastings, Ontario, about forty-five miles inland from the north shore of Lake Ontario, west of Kingston. It occurred on the surface of a layer, three inches thick, of dark grey micaceous limestone or calc-schist, near the middle of a great zone of similar rock, which is interstratified with beds of yellowish-brown sandstone, grey close grained siliceous limestone, white coarsely granular limestone, and bands of dark bluish compact limestone and black pyritiferous slates, to the whole of which Mr. Vennor gives a thickness of 1,000 feet. Beneath this zone are grey and pink dolomites, bluish and greyish mica slates, with conglomerates, diorites, and beds of magnetite, a red orthoclase gneiss lying at the base. The whole series, according to Mr. Vennor's section, which is appended, has a thickness of more than 21,000 feet; but the possible occurrence of more numerous folds than have hitherto been detected, may hereafter render necessary a considerable reduction.

These measures appear to be arranged in the form of a trough, to the eastward of which, and probably beneath them, there are rocks resembling those of Grenville, from which the former differ considerably in lithological character; it is therefore supposed that the Hastings series may be somewhat higher in horizon than that of Grenville. From the village of Madoc, the zone of grey micaceous limestone, which has been particularly alluded to, runs to the eastward on one side of the trough, in a nearly vertical position into Elzivir, and on the other side to the northward, through the township of Madoc into that of Tudor, partially and unconformably overlaid in several places by horizontal beds of Lower Silurian limestone, but gradually spreading, from a diminution of the dip, from a breadth of half a mile to one of four miles. Where it thus spreads out in Tudor it becomes suddenly interrupted for a considerable part of its breadth by an isolated mass of anorthosite rock, rising about 150 feet above the general plain, and

supposed to belong to the unconformable Upper Laurentian, thus
showing that the specimens of *Eozoon* of this neighbourhood, like
those previously discovered and described, belong to the Lower
Laurentian series.

The Tudor limestone is comparatively unaltered; and, in the
specimen obtained from it, the general form or skeleton of the
fossil (consisting of white carbonate of lime) is imbedded in the
limestone, without the presence of serpentine or other silicate, the
colour of the skeleton contrasting strongly with that of the rock.
It does not sink deep into the rock, the form having probably
been loose and much abraded on what is now the under part,
before being entombed. On what was the surface of the bed, the
form presents a well-defined outline on one side; in this and in
the arrangement of the septal layers it has a marked resemblance
to the specimen first brought from the Calumet, eighty miles to
the north-east, and figured in the 'Geology of Canada,' p. 49;
while all the forms from the Calumet, like that from Tudor, are
isolated, imbedded specimens, unconnected apparently with any
continuous reef, such as exists at Grenville and the Petite Nation.
It will be seen, from Dr. Dawson's paper, that the minute
structure is present in the Tudor specimen, though somewhat
obscure; but in respect to this, strong subsidiary evidence
is derived from fragments of *Eozoon* detected by Dr. Dawson in a
specimen collected by myself from the same zone of limestone
near the village of Madoc, in which the canal-system, much more
distinctly displayed, is filled with carbonate of lime, as quoted
from Dr. Dawson by Dr. Carpenter in the Journal of this Society
for August, 1866.

In Dr. Dawson's paper mention is made of specimens from
Wentworth, and others from Long Lake. In both of these localities the rock yielding them belongs to the Grenville band, which
is the uppermost of the three great bands of limestone hitherto
described as interstratified in the Lower Laurentian series. That
at Long Lake, situated about twenty-five miles north of Côte St.
Pierre in the Petite Nation Seigniory, where the best of the
previous specimens were obtained, is in the direct run of the
limestone there; and like it the Long Lake rock is of a serpentinous
character. The locality in Wentworth occurs on Lake Louisa,
about sixteen miles north of east from that of the first Grenville
specimens, from which Côte St. Pierre is about the same
distance north of west, the lines measuring these distances running

across several important undulations in the Grenville band
in both directions. The Wentworth specimens are imbedded in
a portion of the Grenville band which appears to have escaped
any great alteration, and is free from serpentine, though a mixture of serpentine with white crystalline limestone occurs in the
band within a mile of the spot. From this grey limestone, which
has somewhat the aspect of a conglomerate, specimens have been
obtained resembling some of the figures given by Gümbel in his
· Illustrations' of the forms met with by him in the Laurentian
rocks of Bavaria.

In decalcifying by means of a dilute acid some of the specimens
from Côte St. Pierre, placed in his hands in 1864–65, Dr. Carpenter found that the action of the acid was arrested at certain
portions of the skeleton, presenting a yellowish-brown surface; and
he showed me, two or three weeks ago, that in a specimen recently
given him, from the same locality, considerable portions of the
general form remained undissolved by such an acid. On partially
reducing some of these portions to a powder, however, we immediately observed effervescence by the dilute acid; and strong acid
produced it without bruising. There is little doubt that these portions of the skeleton are partially replaced by dolomite, as more
recent fossils are often known to be, of which there is a noted instance in the Trenton limestone of Ottawa. But the circumstance
is alluded to for the purpose of comparing these dolomitized portions of the skeleton with the specimens from Burgess, in which
the replacement of the septal layers by dolomite appears to be the
general condition. In such of these specimens as have been examined the minute structure seems to be wholly, or almost wholly,
destroyed; but it is probable that upon a further investigation of
the locality some spots will be found to yield specimens in which
the calcareous skeleton still exists unreplaced by dolomite; and I
may safely venture to predict that in such specimens the minute
structure, in respect both to canals and tubuli, will be found as
well preserved as in any of the specimens from Côte St. Pierre.

It was the general form on weathered surfaces, and its strong
resemblance to *Stromatopora*, which first attracted my attention to
*Eozoon* ; and the persistence of it in two distinct minerals, pyroxene
and loganite, emboldened me, in 1857, to place before the Meeting
of the American Association for the Advancement of Science specimens of it as probably a Laurentian fossil. After that, the form
was found preserved in a third mineral, serpentine; and in one of

the previous specimens it was then observed to pass continuously
through two of the minerals, pyroxene and serpentine. Now we
have it imbedded in limestone, just as most fossils are. In every
case. with the exception of the Burgess specimens, the general form
is composed of carbonate of lime ; and we have good grounds for
supposing it was originally so in the Burgess specimens also. If.
therefore, with such evidence, and without the minute structure, I
was, upon a calculation of chances, disposed, in 1857, to look upon
the form as organic, much more must I so regard it when the
chances have been so much augmented by the subsequent accumulation of evidence of the same kind, and the addition of the minute
structure. as described by Dr. Dawson, whose observations have
been confirmed and added to by the highest British authority
upon the class of animals to which the form has been referred,
leaving in my mind no room whatever for doubt of its organic
character. Objections to it as an organism have been made by
Professors King and Rowney ; but these appear to me to be based
upon the supposition that because some parts simulating organic
structure are undoubtedly mere mineral arrangement, therefore all
parts are mineral. Dr. Dawson has not proceeded upon the
opposite supposition, that because some parts are, in his opinion,
undoubtedly organic, therefore all parts stimulating organic
structure are organic; but he has carefully distinguished between
the mineral and organic arrangements. I am aware, from having
supplied him with a vast number of specimens prepared for the
microscsope by the lapidary of the Canadian Survey, from a series
of rocks of Silurian and Huronian, as well as Laurentian age, and
from having followed the course of his investigation as it proceeded,
that nearly all the points of objection of Messrs. King and Rowney
passed in review before him prior to his coming to the conclusions
which he has published ; and his reply to these objections forms a
part of the succeeding paper.

---

*Ascending Section of the Laurentian Rocks in the County of
Hastings, Ontario.* By Mr. H. G. VENNOR.

   1. Reddish and flesh-coloured granitic gneiss, the thickness of   Feet.
which is unknown ; estimated at not less than ................. 2,000
   2. Greyish and flesh-coloured gneiss, sometimes hornblendic,
passing towards the summit into a dark mica-schist, and including
portions of greenish-white diorite ; mean of several pretty closely
agreeing measurements. ..................................... 10,400

3. Crystalline limestone, sometimes magnesian, including lenticular patches of quartz, and broken and contorted layers of quartzo-felspathic rock, rarely above a few inches in thickness. This limestone, which includes in Elzivir a one-foot bed of graphite, is sometimes very thin, but in other places attains a thickness of 750 feet; estimated as averaging.................. 400

4. Hornblendic and dioritic rocks, massive or schistose, occasionally associated near the base with dark micaceous schists, and also with chloritic and epidotic rocks, including beds of magnetite; average thickness................................ .. ........ 4,200

5. Crystalline and somewhat granular magnesian limestone, occasionally interstratified with diorites, and near the base with silicious slates and small beds of impure steatite.......... ..... 330

This limestone, which is often siliceous and ferruginous, is metalliferous, holding disseminated copper pyrites, blende, mispickel, and iron pyrites, the latter also sometimes in beds of two or three feet. Gold occur in the limestone at the village of Madoc, associated with an argentiferous grey copper ore, and in irregular veins with bitter-spar, quartz, and a carbonaceous matter at the Richardson mine in Madoc.

6. Grey silicious or fine-grained mica-slates, with an interstratified mass of about sixty feet of yellowish-white dolomite divided into beds by thin layers of the mica-slate, which, as well as the dolomite, often becomes conglomerate, including rounded masses of gneiss and quartzite from one to twelve inches in diameter.... 400

7. Bluish and greyish micaceous slate, interstratified with layers of gneiss, and occasionally holding crystals of magnetite. The whole division weathers to a rusty brown................ 500

8. Gneissoid micaceous quartzites, banded grey and white, with a few instratified beds of silicious limestone, and, like the last division, weathering rusty brown...... .................... 1,900

9. Grey micaceous limestone, sometimes plumbaginous, becoming on its upper portion a calc-schist, but more massive towards the base, where it is interstratified with occasional layers of diorite, and layers of a rusty-weathering gneiss like 8.......... 1,090

This division in Tudor is traversed by numerous N.W. and S.E. veins, holding galena in a gangue of calcite and barytine. The Eozoon from Tudor here described was obtained from about the middle of this calcareous division, which appears to form the summit of the Hastings series.

* Total thickness................ ............... 21,130

*In explanation of the apparent discrepancies between the above section and the one given in the *Quarterly Journal of the Geological Society*, it is to be said that 8 and 9 of the latter section are repetitions of 1 and 2 on the other side of a synclinal, and that 2 in that section represents but a small exposed portion of the great mass of 8, whose measured thickness, as there stated, is 15,000 feet, and includes divisions 2, 3, and 4 of the present section.—EDS.

# ON EOZOON CANADENSE.*

By J. W. DAWSON, LL.D., F.R.S., F.G.S. With Notes by W. B. CARPENTER, M.D., F.R.S.

## I. SPECIMEN OF EOZOON FROM TUDOR, ONTARIO.

This very interesting specimen, submitted to me for examination by Sir. W. E. Logan, is, in my opinion, of great importance, as furnishing a conclusive answer to all those objections to the organic nature of *Eozoon* which have been founded on comparisons of its structures with the forms of fibrous, dendritic, or concretionary minerals,—objections which, however plausible in the case of highly crystalline rocks, in which organic remains may be simulated by merely mineral appearances readily confounded with them, are wholly inapplicable to the present specimen.

1. GENERAL APPEARANCE.—The fossil is of a clavate form, six and a half inches in length, and about four inches broad. It is contained in a slab of dark-colored, coarse, laminated limestone, holding sand, scales of mica, and minute grains and fibres of carbonaceous matter. The surface of the slab shows a weathered section of the fossil (Pl. II.); and the thickness remaining in the matrix is scarcely two lines, at least in the part exposed. The septa, or plates of the fossil, are in the state of white carbonate of lime, which shows their form and arrangement very distinctly, in contrast to the dark stone filling the chambers. The specimen lies flat in the plane of stratification, and has probably suffered some compression. Its septa are convex towards the broad end, and somewhat undulating. In some places they are continuous half-way across the specimen; in other places they divide and re-unite at short distances. A few transverse plates, or connecting columns, are visible; and there are also a number of small veins or cracks passing nearly at right angles to the septa, and filled with carbonate of lime, similar in general appearance to the septa themselves.

On one side, the outline of the fossil is well preserved. The narrow end, which I regard as the basal portion, is rounded. The outline of the side first bends inward, and then outward, forming a graceful double curve, which extends along the greater part of the length. Above this is an abrupt projection, and then a sudden

---

* From the Quar. Jour. Geol. Soc., Aug. 1867. Read before the Geological Society, May 8, 1867.

narrowing; and in the middle of the narrow portion, a part has the chambers obliterated by a white patch of carbonate of lime, below which some of the septa are bent downward in the middle. This is probably an effect of mechanical injury, or of the interference of a calc-spar vein.

With the exception of the upper part above referred to, the septa are seen to curve downward rapidly toward the margin, and to coalesce into a lateral wall, which forms the defined edge or limit of the fossil, and in which there are some indications of lateral orifices opening into the chambers. It is worthy of remark that, in this respect, the present specimen corresponds exactly with that which was originally figured by Sir W. E. Logan in the ' Geology of Canada,' p. 49, and which is the only other specimen that exhibited the lateral limit of the form.

On the side next the matrix, the septa terminate in blunt edges, and do not coalesce; as if the organism had been attached by that surface, or had been broken before being imbedded.

2. MICROSCOPIC CHARACTERS.—Under the microscope, with a low power, the margins of the septa appear uneven, as if eroded or tending to an acervuline mode of growth; but occasionally the septa show a distinct and regular margin. For the most part merely traces of structure are presented, consisting of small parts of canals, filled with the dark colouring-matter of the limestone. In a few places (Pl. III. fig. 1), however, these appear as distinct bundles, similar to those in the Grenville specimens, but of fine texture.

[In fig. 2 is represented a portion of the canal-system in a Grenville specimen, in which the canals, which are transparent in one side (being infiltrated with carbonate of lime only) are seen on the other to be partially filled with black matter, probably a carbonaceous residuum of the sarcode which they originally contained.—W. B. C.]

In a few rare instances only can I detect, with a higher power, in the margin of some of the septa, traces of the fine tubulation characteristic of the proper chamber wall of *Eozoon*. For the most part this seems to have been obliterated by the infiltration of the tubuli with colourless carbonate of lime, similar to that of the skeleton.

In comparing the structure of this specimen with that of those found elsewhere, it would appear that the chambers are more continuous, and wider in proportion to the thickness of the septa, and that the canal-system is more delicate and indistinct than usual.

In the two former respects the specimens from the Calumet and from Burgess approach that now under consideration more nearly than do those from Grenville and Petite Nation; but it would be easy, even in the latter, to find occasional instances of a proportion of parts similar to that in the present example. General form is of little value as a character in such organisms; and so far as can be ascertained, this may have been the same in the present specimen and in that originally obtained from the Calumet, while in the specimens from Grenville a massive and aggregative mode of growth seems to have obliterated all distinctness of individual shape. Without additional specimens, and in the case of creatures so variable as the Foraminifera, it would be rash to decide whether the differences above noticed are of specific value, or depend on age, variability, or state of preservation. For this reason I refer the specimen for the present to *Eozoon Canadense*, merely distinguishing it as the Tudor variety.

From the state of preservation of the fossil, there are no crystalline structures present which can mislead any ordinarily skilful microscopist, except the minute veins of calcareous spar traversing the septa, and the cleavage-planes which have been developed in some portions of the latter.

I would remark that, as it seemed desirable not to injure any more than was absolutely necessary a unique and very valuable specimen, my observations of the microscopic structure have been made on a few slices of small size,—and that, as the microscopic structures are nearly the same in kind with those of specimens figured in former papers, I have not thought it necessary to prepare numerous drawings of them; while the admirable photograph executed for Sir W. E. Logan by Mr. Notman illustrates sufficiently the general form and arrangement of parts (see Pl. II.).

3. CONCLUDING REMARKS.—In a letter to Dr. Carpenter, quoted by him in the 'Quarterly Journal of the Geological Society' for August 1866, p. 228, I referred to the occurrence of *Eozoon* preserved simply in carbonate of lime. The specimens which enabled me to make that statement were obtained at Madoc, near Tudor, this region being one in which the Laurentian rocks of Canada appear to be less highly metamorphosed than is usual. The specimens from Madoc, however, were mere fragments, imbedded in the limestone, and incapable of showing the general form. I may explain, in reference to this, that long practice in the examination of these limestones has enabled me to detect the smallest

fragments of *Eozoon* when present, and that in this way I had ascertained the existence of this fossil in one of the limestones of Madoc before the discovery of the fine specimen now under consideration.

I am disposed to regard the present specimen as a young individual, broken from its attachment and imbedded in a sandy calcareous mud. Its discovery affords the hope that the comparatively unaltered sediments in which it has been preserved, and which also contain the worm-burrows described by me in the ' Quarterly Journal of the Geological Society ' for November,* will hereafter still more largely illustrate the Laurentian fauna.

## II. SPECIMENS FROM LONG LAKE AND WENTWORTH.

Specimens from Long Lake, in the collection of the Geological Survey of Canada, exhibit white crystalline limestone with light-green compact or septariiform† serpentine, and much resemble some of the serpentine-limestones of Grenville. Under the microscope the calcareous matter presents a delicate areolated appearance, without lamination ; but it is not an example of acervuline *Eozoon*, but rather of fragments of such a structure, confusedly aggregated together, and having the interstices and cell-cavities filled with serpentine. I have not found in any of these fragments a canal-system similar to that of *Eozoon Canadense*, though there are casts of large stolons, and, under a high power, the calcareous matter shows in many places the peculiar granular or cellular appearance which is one of the characters of the supplemental skeleton of that species. In a few places a tubulated cell-wall is preserved, with structure similar to that of *Eozoon Canadense*.

Specimens of Laurentian limestone from Wentworth, in the collection of the Geological Survey, exhibit many rounded siliceous bodies, some of which are apparently grains of sand, or small pebbles ; but others, especially when freed from the calcareous matter by a dilute acid, appear as rounded bodies, with rough surfaces, either separate or aggregated in lines or groups, and having minute vermicular processes projecting from their surfaces (Pl. III. fig. 3). At first sight these suggest the idea of spicules ;

* Vol. xxii. p. 608.

† I use the term ' septariiform' to denote the *curdled* appearance so often presented by the Laurentian serpentine.

but I think it on the whole more likely that they are casts of cavities and tubes belonging to some calcareous Foraminiferal organism which has disappeared. Similar bodies, found in the limestone of Bavaria, have been described by Gümbel, who interprets them in the same way.* They may also be compared with the silicious bodies mentioned in a former paper as occurring in the Loganite filling the chambers of specimens of *Eozoon* from Burgess.

### III. SPECIMENS FROM MADOC.

I have already referred to fragments of *Eozoon* occurring in the limestone at Madoc, one of which, found several years ago, I did not then venture to describe as a fossil. It projected from the surface of the limestone, being composed of a yellowish dolomite, and looking like a fragment of a thick shell. When sliced, it presents interiorly a crystalline dolomite, limited and separated from the enclosing rock by a thin wall having a granular or porous structure and excavated into rounded recesses in the manner of *Eozoon*. It lies obliquely to the bedding, and evidently represents a hollow flattened calcareous wall filled by infiltration. The limestone which afforded this form was near the beds holding the worm-burrows described in the Society's Journal for November, 1866.

[A thin section of this body, carefully examined microscopically, presents numerous and very characteristic examples of the canal system of *Eozoon*, exhibiting both the large widely branching systems of canals and the smaller and more penicillate tufts (Pl. III. figs. 4, 5) shown in the most perfect of the serpentinous specimens—but with this difference, that the canals, being filled with a material either identical with or very similar to that of the substance in which they are excavated, are so transparent as only to be brought into view by careful management of the light. —W. B. C.]

### IV. OBJECTIONS TO THE ORGANIC NATURE OF EOZOON.

The discovery of the specimen from Tudor, above described, may appear to render unnecessary any reference to the elaborate attempt made by Profs. King and Rowney to explain the structures of *Eozoon* by a comparison with the forms of fibrous and

---

* Proceedings of Royal Academy of Munich, 1866; Q. J. G. S. vol. xxii. pt. i. p. 185 *et seq.*; also, Can. Naturalist, vol. iii. p. 81.

† Quart. Journ. Geol. Soc. vol. xxii. pt. ii. p. 23.

dendritic minerals,* more especially as Dr. Carpenter has already shown their inaccuracy in many important points. I think, however, that it may serve a useful purpose shortly to point out the more essential respects in which this comparison fails with regard to the Canadian specimens—with the view of relieving the discussion from matters irrelevant to it, and of fixing more exactly the limits of crystalline and organic forms in the serpentine-limestones and similar rocks.

The fundamental error of Messrs. King and Rowney arises from defective observation—in failing to distinguish, in the Canadian limestones themselves, between organic and crystalline forms. This is naturally followed by the identification of all these forms, whether mineral or organic, with a variety of purely crystalline arrangements occurring in other rocks, leading to their attaching the term 'Eozoonal' to any rock which shows any of the characters, whether mineral or organic, thus arbitrarily attached to the Canadian *Eozoon*. This is obviously a process by which the structure of any fossil might be proved to be a mere *lusus naturæ*.

A notable illustration of this is afforded by their regarding the veins of fibrous serpentine, or chrysotile, which occur in the Canadian specimens, as identical with the tubulated cell-wall of *Eozoon* —although they admit that these veins traverse all the structures indifferently and do not conform to the walls of the chambers. But any microscopist who possesses specimens of *Eozoon* containing these chrysotile veins may readily satisfy himself that, under a high power, they resolve themselves into *prismatic crystals in immediate contact with each other;* whereas, under a similar power, the true cell-wall is seen to consist of *slender, undulating, rounded threads of serpentine, penetrating a matrix of carbonate of lime.* Under polarized light, more especially, the difference is conspicuously apparent. It is true that, in many specimens and parts of specimens, the cell-wall of *Eozoon* is badly preserved and fails to show its structure; but in no instance does it present the appearance of chrysotile, or of any other fibrous mineral, when examined with care under sufficiently high powers. In my original examination of Sir William Logan's specimens from Grenville and the Calumet, I did not detect the finely tubulated cell-wall, which is very imperfectly preserved in those specimens; but the veins of

---

* I do not include here the 'septariiform' structure referred to above, which is common in the Canadian serpentine and has no connexion with the forms of the chambers.

fibrous serpentine were well known to me; and when Dr. Carpenter discovered the tubulation of the cell-wall in the specimens from Petite Nation, I compared this structure with that of these veins, and satisfied myself of its distinctness before acceding to his conclusions on this point.

It would also appear that the radiating and sheaf-like bundles of crystals of tremolite, or similar prismatic minerals, which occur in the Canadian serpentines, and also abound in those of Connemara, have been confounded with the tubulation of *Eozoon ;* but these crystals have no definite relation to the forms of that fossil, and often occur where these are entirely absent; and in any case they are distinguished by their straight prismatic shape and their angular divergence from each other. Much use has also been made of the amorphous masses of opaque serpentinous matter, which appear in some parts of the structure of *Eozoon*. These I regard as, in most cases, simply results of alteration or defective preservation, though they might also arise from the presence of foreign matters in the chambers, or from an incrustation of mineral matter before the final filling up of the cells. Generally their forms are purely inorganic; but in some cases they retain indications of the structures of *Eozoon*.

With reference to the canal-system of *Eozoon*, no value can be attached to loose comparisons of a structure so definite with the forms of dendritic silver and the filaments of moss-agates; still less can any resemblance be established between the canal-system and vermicular crystals of mica. These occur abundantly in some serpentines from the Calumet, and might readily be mistaken for organic forms; but their rhombic or hexagonal outline when seen in cross section, their transverse cleavage planes, and their want of any definite arrangement or relation to any general organic form, are sufficient to undeceive any practised observer. I have not seen specimens of the metaxite from Reichenstein referred to by Messrs. King and Rowney; but it is evident, from the description and figure given of it, that, whether organic or otherwise, it is not similar to the canals of *Eozoon Canadense*. But all these and similar comparisons are evidently worthless when it is considered that they have to account for definite, ramifying, cylindrical forms, penetrating a skeleton or matrix of limestone, which has itself a definite arrangement and structure, and, further, when we find that these forms are represented by substances so diverse as serpentine, pyroxene, limestone, and carbonaceous matter.

This is intelligible on the supposition of tubes filled with foreign matters, but not on that of dendritic crystallization.

If all specimens of *Eozoon* were of the acervuline character, the comparisons of the chamber-casts with concretionary granules might have some plausibility. But it is to be observed that the laminated arrangement is the typical one; and the study of the larger specimens, cut under the direction of Sir W. E. Logan, shows that these laminated forms must have grown on certain strata-planes before the deposition of the overlying beds, and that the beds are, in part, composed of the broken fragments of similar laminated structures. Further, much of the apparently acervuline *Eozoon* rock is composed of such broken fragments, the interstices between which should not be confounded with the chambers; while the fact that the serpentine fills such interstices as well as the chambers shows that its arrangement is not concretionary.[*] Again, these chambers are filled in different specimens with serpentine, pyroxene, loganite, calcareous spar, chondrodite, or even with arenaceous limestone. It is also to be observed that the examination of a number of limestones, other than Canadian, by Messrs. King and Rowney, has obliged them to admit that the laminated forms in combination with the canal-system are 'essentially Canadian,' and that the only instances of structures clearly resembling the Canadian specimens are afforded by limestones Laurentian in age, and in some of which (as, for instance, in those of Bavaria and Scandinavia) Carpenter and Gümbel have actually found the structure of *Eozoon*. The other serpentine-limestones examined (for example, that of Skye) are admitted to fail in essential points of structure; and the only serpentine believed to be of eruptive origin examined by them is confessedly destitute of all semblance of *Eozoon*. Similar results have been attained by the more careful researches of Prof. Gümbel, whose paper is well deserving of study by all who have any doubts on this subject.

In the above remarks I have not referred to the disputed case of the Connemara limestones; but I may state that I have not been able to satisfy myself of the occurrence of the structures of *Eozoon* in such specimens as I have had the opportunity to examine.[*] It is perhaps necessary to add that there exists in

---

[*] Such Irish specimens of serpentine limestone as I have seen, appear much more highly crystalline than the beds in Canada which contain *Eozoon*.

Canada abundance of Laurentian limestone which shows no indication of the structures of *Eozoon*. In some cases it is evident that such structures have not been present. In other cases they may have been obliterated by processes of crystallization. As in the case of other fossils, it is only in certain beds, and in certain parts of those beds, that well-characterized specimens can be found. I may also repeat here that in the original examination of *Eozoon*, in the spring of 1864, I was furnished by Sir W. E. Logan with specimens of all these limestones, and also with serpentine-limestones of Silurian age, and that, while all possible care was taken to compare these with the specimens of *Eozoon*, it was not thought necessary to publish notices of the crystalline and concretionary forms observed, many of which were very curious and might afford materials for other papers of the nature of that criticised in the above remarks.

[The examination of a large number of sections of a specimen of *Eozoon*, recently placed in my hands by Sir William Logan, in which the canal-system is extraordinarily well preserved, enables me to supply a most unexpected confirmation of Dr. Dawson's statements in regard to the occurrence of dendritic and other forms of this system, which cannot be accounted for by the intrusion of any foreign mineral ; for many parts of the calcareous lamellæ in these sections, which, when viewed by ordinary transmitted light, appear quite homogeneous and structureless, are found, when the light is reduced by Collin's ' graduating diaphragm,' to exhibit a most beautiful development of various forms of canal-system (often resembling those of Dr. Dawson's Madoc specimen represented in Pl. III. figs. 4, 5), which cross the cleavage-planes of the shell-substance in every direction. Now these parts, when subjected to decalcification, show no trace of canal-system ; so that it is obvious, both from their optical and from their chemical reactions, that the substance filling the canals must have been *carbonate of lime*, which has thus completely solidified the shell layer, having been deposited in the canals previously excavated in its interior, just as crystalline carbonate of lime fills up the reticular spaces of the skeleton of Echinodermata fossilized in a calcareous matrix. This fact affords conclusive evidence of *organic structure*, since no conceivable process of crystallization could give origin to dendritic extensions of carbonate of lime disposed on exactly the same crystalline system with the calcite which includes it, the two substances being

mineralogically homogeneous, and only structurally distinguishable,
by the effect of their junction-surfaces on the course of faint rays
of light transmitted through them.— W. B. C.]

# MISCELLANEOUS.

## NOTE ON SUPPOSED BURROWS OF WORMS IN THE LAURENTIAN ROCKS OF CANADA.

### By J. W. DAWSON, LL.D., F.R.S., &c.

Among other indications of fossils in the Laurentian rocks, mentioned in my paper on the structure of Eozoon, are certain perforations resembling burrows of worms, found in a calcareous
quartzite or impure limestone from Madoc, in Upper Canada.
They occur in specimens in the Museum of the Geological Survey,
and also in specimens subsequently collected by myself at the same
place.

The beds at Madoc, containing these impressions, underlie, unconformably, the Lower Silurian limestones, and are regarded by
Sir W. E. Logan as belonging to a somewhat higher horizon in the
Laurentian, than the Eozoon Serpentines of Grenville. They are
also less highly metamorphosed than the Laurentian rocks gener-

rally. They are described in Sir W. E. Logan's Report on the Geology of Canada, 1863, at p. 32.

The impressions referred to consist of perforations approaching to a cylindrical form, and filled with rounded siliceous sand, more or less stained with carbonaceous and ferruginous matter, more especially near the circumference of the cylinders. These superficial portions being harder than the containing rock, and of darker colour, and also harder than the interior of the cylinders, project as black rings from the weathered surfaces; but in their continuation into the interior of the mass, they appear only as spots or lines of a slightly darker colour, or stained with iron-rust.

When sliced transversely and examined under the microscope, they appear as round, oval, or semicircular holes drilled through the rock, and lined around their circumference with dense and dark-coloured siliceous matter, while the axis, which is often of a bilobate form, is comparatively transparent and of softer texture. The perforations are often at right angles to the bedding, but in some cases nearly parallel with it.

In regard to the origin of these perforations, I suppose that they may have been either (1) burrows of worms filled with sand subsequently hardened and stained at the surface, or (2) tubes composed of sand, like those of Sabella, or (3) cavities left by the decay of Algæ and filled with sand. The first I think the most probable view.

I may add that the beds at Madoc, containing these supposed fossils, hold also, on their weathered surfaces, impressions with rude casts of concentric laminæ like those of Stromapotora or Eozoon, but too obscure for determination. The limestones interstratified with these beds also contain fragments of Eozoon not fossilized by serpentine but simply by carbonate of lime, carbonaceous fibres, spicules like those of sponges, and lenticular bodies of unknown nature. - *Journal of the Geological Society of London.*

---

## OBITUARY.

### THE RIGHT HONOURABLE SIR EDMUND HEAD, BART., K.C.B., LL.D., F.R.S., &c.

By the sudden death of the able and patriotic man whose name stands at the head of this article, Canada loses one of the few statesmen in the mother country interested in her welfare, and having influence to make their good wishes effectual. Sir Edmund, after a brilliant career at Oriel College, Oxford, where he took a first-class, and afterwards a fellowship, entered into educational and literary pursuits as a tutor of Mereton College and a writer of articles in the Reviews. Having attracted attention by the ability displayed in the latter, he was appointed an Assistant Poor Law Commissioner, and subsequently Chief Commissioner. On the reconstruction of the Poor Law Board in 1847, he received the government of New Brunswick, and in 1854 was promoted to be the Governor-General of Canada, from which office he retired in 1861.

Both in New Brunswick and Canada Sir Edmund was, as might have been expected, conspicuous as a patron of education, literature, and science; and was remarkable, not only for his readiness to give his countenance to every worthy undertaking, but for the judicious advice which he gave, and his willingness to devote time and thought to the consideration of the best means for advancing the interests in view.

In New Brunswick he more especially took a warm interest in the Provincial University, then in a languishing condition ; and procured the appointment of a Commission to inquire into its deficiencies and difficulties, and the means for their remedy. The labours of this Commission (which consisted of the Honble. J. H. Gray of New Brunswick, Rev. Dr. Ryerson, Principal Dawson--then Superintendent of Education in Nova Scotia,—and the Honbles. J. H. Saunders and James Brown of New Brunswick) resulted in the preparation of a scheme which, if fully carried out, would have placed New Brunswick far in advance of the other colonies in this respect. Sir Edmund was, however, soon after removed to Canada, and the plan devised was only partially acted on ; but it has already given a new stimulus to higher education in New Brunswick, and has resulted in placing the University in a very satisfactory condition.

In Canada, though checked by the unsettled condition of political affairs and by the want of sympathy with his large views on the part of most of our public men, Sir Edmund did much for the promotion of his own favourite pursuits and for laying the foundation of a high educational culture. The educational measures adopted during his administration all more or less bore the impress of his mind, and the various Scientific and Literary Societies, and the Geological Survey, owe much to his personal influence. In this community, the McGill University, the

Natural History Society, and the Normal Schools, specially owe him a debt of gratitude.

While in Canada he met with the most severe calamity of his life, the death, by drowning, of his only son, a young man of excellent parts, who had already made much progress in scientific attainments, and who bade fair to follow in the footsteps of his father.

Sir Edmund's largest literary work was his "Handbook of Spanish Painting." He also published a clever little book on "Shall and Will," and an important memoir on the celebrated "Temple of Serapis at Pozzuoli," in which he brings his classical and antiquarian lore to aid the geologist in explaining the wonderful alternations of elevation and subsidence to which this building and the neighbouring coast have been subjected.

Sir Edmund died suddenly at his town residence, Eaton Square, London, on the 25th of January, 1868.

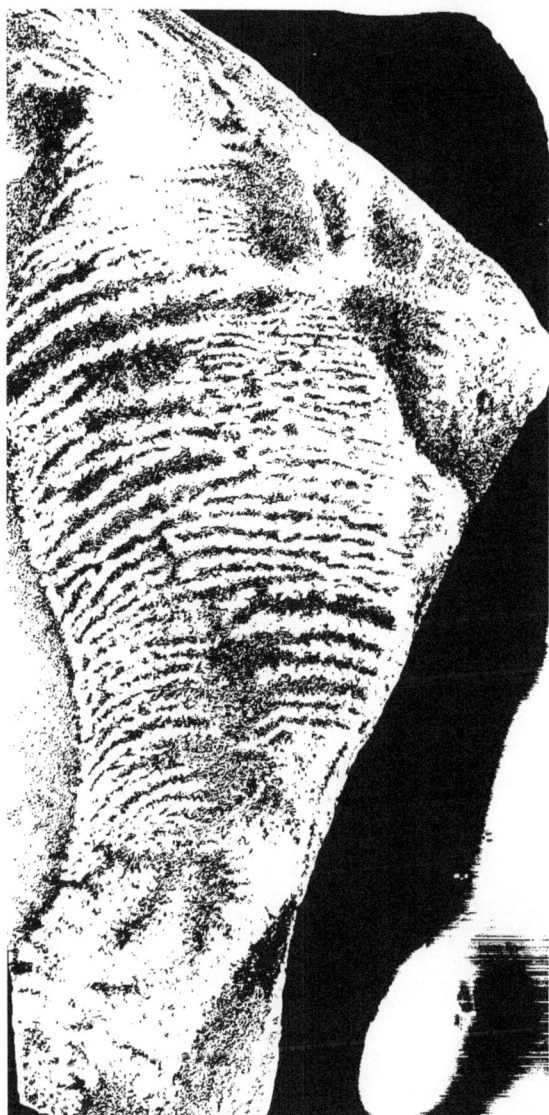

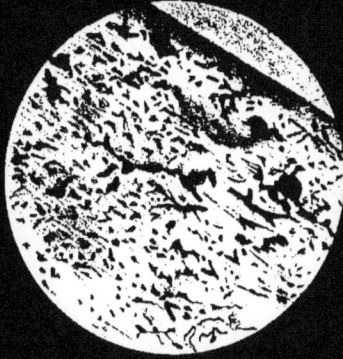

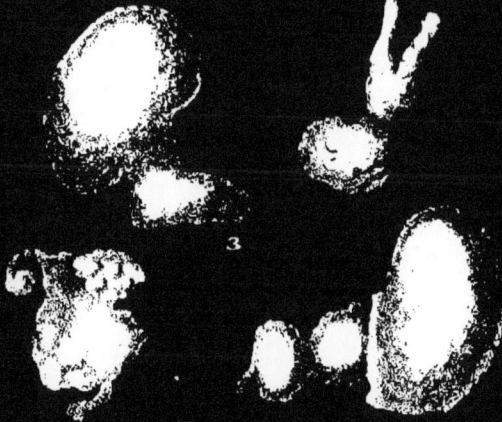

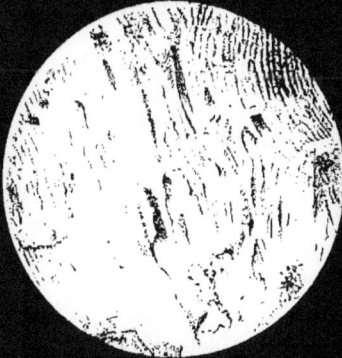

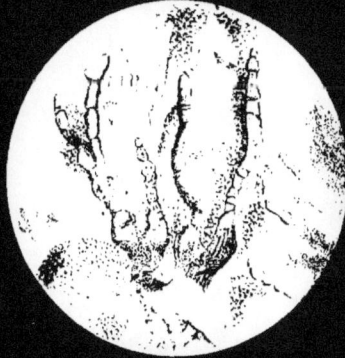

THE

# CANADIAN NATURALIST.

## SECOND SERIES.

## OUTLINES OF THE DISTRIBUTION OF ARCTIC PLANTS.

### By Jos. D. Hooker, M.D., F.R.S., &c.[*]

I shall endeavour in the following pages to comply, as far a I can, with a desire expressed by several distinguished Arctic voyagers, that I should draw up an account of the affinities and distribution of the flowering plants of the North Polar Regions. The method I have followed has been, first to ascertain the names and localities of all plants which appear on good evidence to have been found north of the arctic circle in each continent; then to divide the polar zone longitudinally into areas characterized by differences in their vegetation; then to trace the distribution of the arctic plants, and of their varieties and very closely allied forms, into the temperate and alpine regions of both hemispheres. Having tabulated these data, I have endeavoured to show how far their present distribution may be accounted for by slow changes of climate during and since the glacial period.

The arctic flora forms a circumpolar belt of 10° to 14° latitude, north of the arctic circle. There is no abrupt break or change in the vegetation anywhere along this belt, except in the meridian of Baffin's Bay, whose opposite shores present a sudden change from an almost purely European flora on its east coast, to one with a large admixture of American plants on its west.

The number of flowering plants which have been collected within

---

[*] Read before the Linnean Society, London, June 21st, 1860, and reprinted (by permission of the President) from its Transactions, Vol xxiii., pp. 251-291: with some corrections by the Author.
Vol. III.                              V                              No. 5

the arctic circle is 762 (Monocot. 214; Dicot. 548). In the present state of cryptogamic botany it is impossible to estimate accurately the number of flowerless plants found within the same area, or to define their geographical limits; but the following figures give the best approximate idea I have obtained:—

| | | | | | |
|---|---|---|---|---|---|
| Filices | 28 | Characeæ | 2 | Fungi | 200? |
| Lycopodiaceæ | 7 | Musci | 250 | Algæ | 100 |
| Equisetaceæ | 8 | Hepaticæ | 80 | Lichenes | 250 |
| | | Total Cryptogams | | 925 | |
| | | " Phænogams | | 762 | |
| | | | | 1687 | |

Regarded as a whole, the arctic flora is decidedly Scandinavian; for Arctic Scandinavia, or Lapland, though a very small tract of land, contains by far the richest arctic flora, amounting to three-fourths of the whole; moreover, upwards of three-fifths of the species, and almost all the genera, of Arctic Asia and America are likewise Lapponian, leaving far too small a percentage of other forms to admit of the Arctic Asiatic and American floras being ranked as anything more than subdivisions, which I shall here call districts, of one general arctic flora.

Proceeding eastwards from Baffin's Bay, there is, first, the Greenland district, whose flora is almost exclusively Lapponian, having an extremely slight admixture of American or Asiatic types: this forms the western boundary of the purely European flora. Secondly, the Arctic European district, extending eastward to the Obi river, beyond the Ural range, including Nova Zembla and Spitzbergen; Greenland would also be included in it, were it not for its large area and geographical position. Thirdly, the transition from the comparatively rich European district to the extremely poor Asiatic one is very gradual; as is that from the Asiatic to the richer fourth or West American district, which extends from Behring's Straits to the Mackenzie River. Fifthly, the transition from the West to the East American district is even less marked; for the lapse of European and West American species is trifling, and the appearance of East American ones is equally so: the transition in vegetation from this district, again, to that of Greenland is, as I have stated above, comparatively very abrupt.

The general uniformity of the arctic flora, and the special differences between its subdivisions, may be thus estimated: the arctic Phænogamic flora consists of 762 species; of these, 616 are Arctic European, many of which prevail throughout the polar area, being distributed in the following proportions through its different longitudes :—

| | | | | |
|---|---|---|---|---|
| Arctic Europe..... 616 : | Scandinavian forms 586 ; | Asiatic and American 30 = 1 : 19·57 |
| "   Asia........ 233 | "      " 189 | "      44 = 1 :  4·2 |
| "   W. America. 364 | "      " 254 | "      110 = 1 :  2·3 |
| "   E. America . 379 | "      " 269 | "      110 = 1 :  2·4 |
| "   Greenland .. 207 | "      " 195 | "      12 = 1 : 16·2 |

This table places in a most striking point of view the anomalous condition of Greenland, which, though so favourably situated for harbouring an Arctic American vegetation, and so unfavourably for an Arctic European one, presents little trace of the botanical features of the great continent to which it geographically belongs, and an almost absolute identity with those of Europe. Moreover, the peculiarities of the Greenland flora are not confined to these ; for a detailed examination shows that it differs from all other parts of the arctic regions in wanting many extremely common Scandinavian plants which advance far north in all the other polar districts, and that the general poverty of its flora in species is more due to an abstraction of arctic types than to a deficiency of temperature. This is proved by an examination of the temperate portion of the Greenland peninsula, which adds very few plants to the entire flora, as compared with a similar area south of any other arctic region ; and these few are chiefly arctic plants and almost without exception Arctic Scandinavian species.

There is nothing in the physical features of the arctic regions, their oceanic or aerial currents, their geographical relations, nor their temperature, which, in my opinion, at all accounts for the exceptional character of the Greenland flora ; nor do I see how it can be explained, except by assuming that extensive changes of climate, and of land and sea, have exerted great influence, first, in directing the migration of the Scandinavian species over the whole polar zone, and afterwards in introducing the Asiatic and American species with which the Scandinavian are so largely associated in all the arctic districts except those of Europe and Greenland. It is inconceivable to me that, under existing conditions of sea, land, and temperature, so many Scandinavian plants should have found their way westward to Greenland, by migration across the Atlantic, and stopped short on its west coast, net crossing to America;—or that so many American types should terminate as abruptly on the west coast of Baffin's Bay, and not cross to Greenland and Europe;—or that Greenland should contain actually much fewer species of European plants than have found their way eastwards from Lapland by Asia into Western and Eastern Arctic America;—or that the Scandinavian vegetation should in every longitude have migrated southward across the tropics of Asia and America, whilst the typical genera of Asia and America which have found their way into the arctic regions have remained restricted to these continents.

It appears to me difficult to account for these facts, unless we admit Mr. Darwin's * hypothesis, first, that the existing Scandinavian flora is of great antiquity, and that previous to the glacial epoch it was more uniformly distributed over the polar zone than it is now; secondly, that during the advent of the glacial period this Scandinavian vegetation was driven southward in every longitude, and even across the tropics into the south temperate zone; and that on the succeeding warmth of the present epoch, those species that survived both ascended the mountains of the warmer zones, and also returned northward, accompanied by aborigines of the countries they had invaded during their southern migration. Mr. Darwin shows how aptly such an explanation meets the difficulty of accounting for the restriction of so many American and Asiatic arctic types to their own peculiar longitudinal zones; and that far greater difficulty, the representation of the same arctic genera by most closely allied species in different longitudes. To this representation, and the complexity of its character, I shall have to allude when indicating the sources of difficulties I have encountered, whether in limiting the polar species, or in determining to what southern forms many are most directly referable. Mr. Darwin's hypothesis accounts for many varieties of one plant being found in various alpine and arctic regions of the globe, by the competition into which their common ancestor was brought with the aborigines of the countries it invaded: different races survived the struggle for life in different longitudes; and these races again, afterwards converging on the zone from which their ancestor started, present there a plexus of

---

* This theory of a southern migration of northern types being due to the cold epochs preceding and during the glacial, originated, I believe, with the late Edward Forbes; the extended one, of their transtropical migration, is Mr. Darwin's, and is discussed by him in his ' Origin of Species,' chap. xi.

closely allied but more or less distinct varieties or even species, whose geographical limits overlap, and whose members very probably occasionally breed together.

Nor is the application of this hypothesis limited to this inquiry ; for it offers a possible explanation of a general conclusion at which I had previously arrived * and shall have again to discuss here— viz. : that the Scandinavian flora is present in every latitude of the globe, and is the only one that is so; and it also helps to explain another class of most interesting and anomalous facts in arctic distribution, at which I have now arrived from an examination of the vegetation of the several polar districts, and especially that of Greenland.

A glance at a circumpolar chart will show how this theory bears upon the Greenland flora, explaining the identity of its existing vegetation with that of Lapland, and accounting for its paucity of species, for the rarity of American species, of peculiar species, and of marked varieties of European species.  If it be granted that the polar area was once occupied by the Scandinavian flora, and that the cold of the glacial epoch did drive this vegetation south- wards, it is evident that the Greenland individuals, from being confined to a peninsula, would be exposed to very different con- ditions to those of the great continents.  In Greenland many species would, as it were, be driven into the sea, that is, exterminated ; and the survivors would be confined to the southern portion of the peninsula, and not being there brought into competition with other types, there could be no struggle for life amongst their progeny, and consequently no selection of better adapted varieties.  On the return of heat, these survivors would simply travel northwards, unaccompanied by the plants of any other country.

In Arctic America and Asia, on the other hand, where there was a free southern extension and dilatation of land for the same Scandinavian plants to occupy, these would multiply enormously in individuals, branching off into varieties and subspecies, and occupy a larger area the further south they were driven ; and none need be altogether lost in the southern migration over plains, though many would in the struggle that ensued when they reached the mountains of those continents and were brought into competi- tion with the alpine plants, which the same cold had caused to descend to the plains.  Hence, on the return of warmth, many

* Introductory Essay to the ' Flora of Tasmania,' p. ciii.

more Scandinavian species would return to Arctic America and
Asia than survived in Greenland; some would be changed in form,
because only the favoured varieties could have survived the struggle;
some of the species of Alpine Siberia and of the Rocky Mountains
would accompany the Scandinavian in their return to the arctic
zone; while many arctic species would ascend those mountains,
accompanying the alpine species in their reascent.

Again, as the same species may have been destroyed in many
longitudes, or at most elevations, but not at all, we should expect
to find some of those Arctic Scandinavian plants of Greenland
which have not returned to Arctic America still lurking in remote
corners of that great continent; and we may account for *Draba
aurea* being confined to Greenland and the Rocky Mountains,
*Potentilla tridentata* to Greenland and some scattered localities from
the Alleghanies northward, and *Arenaria Granlandica* to
Greenland, Labrador and the Mountains of New England, by supposing that these were originally Scandinavian plants, which were
driven south by the cold of the glacial epoch, but which on the return
of warmth, being exterminated on the plains of the American continent, found a refuge among its mountains, where they now exist.

It appears, therefore, to be no slight confirmation of the general
truth of Mr. Darwin's hypothesis, that, besides harmonizing with
the distribution of arctic plants within and beyond the polar zone,
it can also be made, without straining, to account for that distribution and for many anomalies of the Greenland flora, viz., i.—its
identity with the Lapponian; ii.—its paucity of species; iii.—the
fewness of temperate plants in temperate Greenland, and the still
fewer plants that area adds to the entire flora of Greenland;
iv.—the rarity of both Asiatic and American species or types in
Greenland; and v.—the presence of a few of the rarest Greenland
and Scandinavian species in remote and often alpine localities of
West America and the United States.

### I.—ON THE LOCAL DISTRIBUTION OF PLANTS WITHIN THE ARCTIC CIRCLE.

The greatest number of plants occurring in any given arctic
district is found in the European, where 616 flowering plants
have been collected from the verge of the circle to Spitzbergen.
From this region vegetation rapidly diminishes in proceeding eastwards and westwards, especially the latter. Thus, in Arctic Asia
only 233 flowering plants have been collected; in Arctic Green-

land, 207 species ; in the American continent east of the Mackenzie
River, 379 species ; and in the area westward from that river to
Behring's Straits, 364 species.

A glance at the annual and monthly isothermal lines will show
that there is little relation between the temperature and vegetation
of the areas they intersect, beyond the general feature of the scantiness of the Siberian flora being accompanied by a great southern
bend of the annual isotherm of 32° in Asia, and the greatest
northern bend of the same isotherm occurring in the longitude of
west Lapland, which contains the richest flora. On the other
hand, the same isotherm bends northwards in passing from Eastern
America to Greenland, the vegetation of which is the scantier of
the two ; and passes to the northward of Iceland, which is much
poorer in species than those parts of Lapland to the southward of
which it passes.

The June isothermals, as indicating the most effective temperatures in the arctic regions (where all vegetation is torpid for nine
months, and excessively stimulated during the three others), might
have been expected to indicate better the positions of the most
luxuriant vegetation : but neither is this the case ; for the June
isothermal of 41°, which lies within the arctic zone in Asia, where
the vegetation is scanty in the extreme, descends to 54° N. lat. in
the meridian of Behring's Straits, where the flora is comparatively
luxuriant ; and the June isothermal of 32°, which traverses Greenland north of Disco, passes to the north, both of Spitzbergen and
the Parry Islands. In fact, it is neither the mean annual, nor the
summer (flowering), nor the autumn (fruiting) temperature that
determines the abundance or scarcity of the vegetation in each
district, but these combined with the ocean temperature and consequent prevalence of humidity, its geographical position, and its
former conditions both climatal and geographical. The relations
between the isothermals and floras in each longitude being therefore special, and not general, I shall consider them further when
defining the different arctic floras.

The northern limits to which vegetation extends varies in every
longitude ; and its extreme limits are still unknown ; it may, indeed,
reach to the pole itself. Phænogamic plants, however, are probably
nowhere found far north of lat. 81°. 70 flowering plants are found
in Spitzbergen ; and Sabine and Ross collected 9 on Walden Island,
towards its northern extreme, but none on Ross's Islet, fifteen miles
further to the north. Sutherland, a very careful and intelligent

collector, found 23 at Melville Bay and Wolstenholme and Whale
Sounds, in the extreme north of Baffin's Bay (lat. 76°, 77° N.).
Parry, James Ross, Sabine, Beechey, and others, together, found
60 species on Melville Island, and Lyall 50 on the islands north
of Barrow Straits and Lancaster Sound. About 80 have been
detected on the west shores of Baffin's Bay and Davis' Straits,
between Pond Bay and Home Bay. To the north of Eastern
Asia, again, Seemann collected only 4 species on Herald Island,
lat. 71½° N., the northernmost point attained in that longitude.
On the east coast of Greenland, Scoresby and Sabine found only
50 between the parallels of 70° and 75° N.; whilst 150 inhabit
the west coast, between the same parallels.

The differences between the vegetations of the various polar
areas seem to be to a considerable extent constant up to the
extreme limits of vegetation in each. Thus *Ranunculus glacialis*
and *Saxifraga flagellaris*, which are all but absent in West
Greenland*, advance to the extreme north in East Greenland and
Spitzbergen. *Caltha palustris, Astragalus alpinus, Oxytropis
Uralensis, O. nigrescens, Parrya arctica, Sieversia Rossii,
Nardosmia corymbosa, Senecio palustris, Deschampsia cæspitosa,
Saxifraga hieraciifolia* and *S. Hirculus*, all of which are absent in
West Greenland, advance to Lancaster Sound and the polar
American islands, a very few degrees to the westward of Greenland.

On the other hand, *Lychnis alpina, Arabis alpina, Stellaria
cerastioides, Potentilla tridentata, Cassiopeia hypnoides, Phyllodoce taxifolia, Veronica alpina, Thymus Serphyllum, Luzula
spicata,* and *Phleum alpinum,* all advance north of 70° in West
Greenland, but are wholly unknown in any part of Arctic Eastern
America or the polar islands.

The most arctic plants of general distribution that are found far
north in all the arctic areas are the following; all inhabit the
Parry Islands, or Spitzbergen, or both :—

| | | |
|---|---|---|
| anunculus nivalis. | Draba hirta. | Stellaria longipes. |
| —— auricomus. | —— muricella. | Cerastium alpinum. |
| —— pygmæus. | —— incana. | Potentilla nivea. |
| Papaver nudicaule. | —— rupestris. | —— frigida. |
| Cochlearia officinalis. | Cochlearia anglica. | Dryas octopetala. |
| Braya alpina. | —— officinalis. | Epilobium latifolium. |
| Cardamine bellidifolia. | Silene acaulis. | Sedum Rhodiola. |
| —— pratensis. | Lychnis apetala. | Chrysos. alternifolium. |
| Draba alpina. | Arenaria verna. | Saxifraga oppositifolia. |
| —— androsacea. | —— arctica. | —— cæspitosa. |

Both were found by Kane's Expedition, but by no previous one.

Saxifraga cernua.
—— rivularis.
—— nivalis.
—— stellaris.
—— flagellaris.
—— Hirculus (E. Green-
land only.)
Antennaria alpina.
Erigeron alpinus.
Taraxacum Dens-leonis.
Cassiopeia tetragona.
Pedicularis hirsuta.

Pedicularis sudetica.
Oxyria reniformis.
Polygonum viviparum.
Empetrum nigrum.
Salix herbacea.
—— reticulata.
Luzula arcuata.
Juncus bigiumis.
Carex fuliginosa (not yet found
in Arctic Asia, but no doubt
there.)
—— aquatilis (do.)

Eriophorum capitatum.
—— polystachyon.
Alopecurus alpinus.
Deyeuxia Lapponica.
Deschampsia cæspitosa (E.
Greenland only).
Phippsia algida.
Colpodium latifolium.
Poa flexuosa.
—— pratensis.
—— nemoralis.
Festuca ovina.

Of the above, *Saxifraga oppositifolia* is probably the most ubiquitous, and may be considered the commonest and most arctic flowering plant.

The following are also inhabitants of all the five arctic areas, but do not usually attain such high latitudes as the foregoing :—

Ranunculus Lapponicus.
Draba rupestris.
Viola palustris.
Honkenya peploides.
Epilobium angustifolium.
—— alpinum.
Hippuris vulgaris.
Artemisia borealis.

Vaccinium uliginosum.
—— Vitis-idæa.
Ledum palustre.
Pyrola rotundifolia.
Polemonium cæruleum, and
vars. (E. Greenland only.)
Pedicularis Lapponica.
Armeria vulgaris.

Betula nana.
Salix lanata.
—— glauca.
—— alpestris.
Luzula campestris.
Carex vesicaria.
Eriophorum vaginatum.
Atropis maritima.

The absence of *Gentiana* and *Primula* in these lists is very unaccountable, seeing how abundant and very alpine they are on the Alps and Himalaya, and *Gentiana* on the South American Cordilleras also.

The few remaining plants, which are all very northern and almost or wholly confined to the arctic zone, are the following. † indicates those species absolutely peculiar; ‡ the only peculiar genus.

Ranunculus Palasii.
—— hyperboreus.
Trollius Asiaticus.
Corydalis glauca,
Cardamine purpurea.
Turritis mollis.
Cochlearia sisymbrioides.
Hesperis Pallasii.
†Braya pilosa
Eutrema Edwardsii.
Parrya arctica.
† —— arenicola.
Odontarrhena Fischeriana.
Sagina nivalis.
Stellaria dicranoides.
Oxytropis nigrescens.
Sieversia Rossii.
—— glacialis.
Rubus arcticus.
Parnassia Kotzebuei.

Saxifraga Eschscholtzii.
—— serpyllifolia.
† —— Richardsoni.
Cœnolophium Fischeri.
†Nardosmia glacialis.
Artemisia Richardsoniana.
—— glomerata.
† —— androsacea.
Erigeron compositus.
Chrysanthemum arcticum.
Pyrethrum bipinnatum.
†Saussurea subsinuata.
Campanula uniflora.
Gentiana arctophila.
—— aurea.
Eutoca Franklinii.
Pedicularis flammea.
†Douglasia arctica.
†Monolepis Asiatica.

Betula fruticosa.
Salix speciosa.
† —— glacialis.
—— phlebophylla.
—— arctica.
Orchis cruenta.
Platanthera hyperborea.
Carex nardina.
—— glareosa.
—— rariflora.
Hierochloe pauciflora.
Deschampsia atropurpurea.
Phippsia algida.
Dupontia Fisheri.
Colpodium pendulinum.
—— fulvum.
—— latifolium.
‡Pleuropogon Sabini.
†Festuca Richardsoni.

II.—ON THE DISTRIBUTION OF ARCTIC FLOWERING PLANTS IN
VARIOUS REGIONS OF THE GLOBE.

There is but one distinct genus confined to the arctic regions,
the monotypic and local*Pleuropogon Sabini;* and there are but seven
other peculiarly arctic species, together with one with which I am
wholly unacquainted, viz., *Monolepis Asiatica.* The remaining
762 species are all of them found south of the circle ; and of these all
but 150 advance south of the parallel of 40° N. lat., either in the
Mediterranean basin, Northern India, the United States, Oregon,
or California ; about 50 are natives of the mountainous regions of
the tropics ; and just 105 inhabit the south temperate zone.

The proportion of species which have migrated southwards in
the Old and New World also bear a fair relation to the facilities
for migration presented by the different continents.   Thus,

Of 616 Arctic European species,
   496 inhabit the Alps, and
   450 cross them ;
   126 cross the Mediterranean ;
   26 inhabit South Africa.
Of 379 Arctic East American,
   203 inhabit the United States.
   34 inhabit tropical American mountains.
   50 inhabit temperate South America.

Of 233 Arctic Asiatic species,
   210 reach the Altai, Soongaria, etc. ;
   106 reach the Himalaya ;
   0 are found on the tropical mts. of Asia;
   5 inhabit Australia and New Zealand.
Of 346 Arctic West American species,
   274 are north temperate ;
   24 on tropical mountains ;
   37 in south temperate zone.

These tables present in a very striking point of view the fact of
the Scandinavian flora being the most widely distributed over the
globe.   The Mediterranean, South African, Malayan, Australian,
and all the floras of the New World have narrow ranges compared
with the Scandinavian, and none of them form a prominent feature
in any other continent than their own ; but the Scandinavian not
only girdles the globe in the arctic circle, and dominates over all
others in the north temperate zone of the Old World, but intrudes
conspicuously into every other temperate flora, whether in the
northern or southern hemisphere, or on the alps of tropical
countries.

The severest test to which this observation could be put is that
supplied by the Arctic Scandinavian forms ; for these belong to
the remotest corner of the Scandinavian area, and should of all
plants be the most impatient of temperate, warm, and tropical
climates.   The following will, approximately, express the result :—

Total Arctic Scandinavian forms........ 586   Cross Alps. etc..................... 480
In North United States, Canada, etc.... 360   Reach South Africa................ 20
In Tropical America .................. 40   Himalaya, etc................... 300
In Temperate South America.......... 70   Tropical Asia ..................... 20
In Alps of Middle Europe, Pyrenees, etc. 490   Australia, etc....... .............. 60

In one respect this migration is most direct in the American meridian, where more arctic species reach the highest southern latitudes. This I have accounted for (Flora Antarctica, p. 230) by the continuous chain of the Andes having favoured their southern dispersion.

But the greatest number of arctic plants are located in Central Europe, no fewer than 530 out of 762 inhabiting the Alps and Central and Southern Europe, of which 480 cross the Alps to the Mediterranean basin. Here, however, their further spread is apparently suddenly arrested; for though many, doubtless, are to be found in the Alps of Abyssinia and the western Atlas; these are few compared with what are found further east in Asia; and fewer still have found their way to South Africa.

The most continuous extension of Scandinavian forms is in the direction of the greatest continental extension; namely that from the North Cape in Lapland to Tasmania*; for no less than 350 Scandinavian plants have been found in the Himalaya, and 53 in Australia and New Zealand; whereas there are scarcely any Himalayan and no Australian or Antarctic forms in Arctic Europe. Now that Mr. Darwin's hypotheses are so far accepted by many botanists, in that these concede many species of each genus to have had in most cases a common origin, it may be well to tabulate the generic distribution of arctic plants as I have done the specific; and this places the prevalence of the Scandinavian types of vegetation in a much stronger light :—

| Scandinavian Arctic Genera in Europe.. | 280 | Cross Alps (approximately) | 260 |
| Found in N. U. S. (approximately)... | 270 | Found in South Africa (approximately) | 110 |
| " Tropical American Mts. " ... | 100 | " Himalaya, etc. | 270 |
| " Temperate South America " ... | 120 | " Tropical Asia | 80 |
| " Alps " ... | 280 | " Australia, etc. | 100 |

The most remarkable anomaly is the absence of *Primula* in Tropical America, that genus being found in Extra-tropical South

---

* The line which joins these points passes through Siberia, Eastern China, the Celebes Islands, and Australia, but the glacial migration has no doubt been due south from the arctic and north temperate regions in various longitudes to the Pyrenees, Alps, Carpathians, Caucasus, Asia Minor, Persian and North Indian mountains, etc. The further migration south to the distant and scattered alpine heights of the tropics, and thence to South Australia, Tasmania, and New Zealand, is, in the present state of our knowledge, to me quite unaccounted for. Mr. Darwin assumes for this purpose a cooled condition of the globe that must have been fatal to all such purely tropical vegetation as we are now familiar with.

America ; and its absence in the whole southern temperate zone of the Old World, except the Alps of Java.

*Thalictrum, Delphinium, Impatiens, Prunus, Circæa, Chrysos-plenium, Parnassia, Bupleurum, Heracleum, Viburnum, Valeri-ana, Artemisia, Vaccinium, Rhododendron, Pedicularis,* and *Salix,* are all arctic genera found on the tropical mountains of Asia (Nilghiri, Ceylon, Java, etc.), but not yet in the south temperate zones of Asia, and very few of them in Temperate South Africa.

There are, however, a considerable number of Scandinavian plants which are not found in the Alps of Middle Europe, though found in the Caucasus, Himalaya, etc. ; and conversely there are several Arctic Asiatic and American plants found in the Alps of Central Europe, but nowhere in Arctic Europe. In other words, certain species extend from Arctic America through Central Asia and North India to Central Europe, which do not extend from Arctic America westward to Arctic Europe; and there are certain other species which extend from Arctic Europe to the Caucasus and Central Asia, which do neither exist on the Alps of Central Europe nor extend eastward to Arctic America : thus,

*Common to Arctic Europe and Temperate Asia, etc., but not to Alps of Europe.*

| | | |
|---|---|---|
| Ranunculus nivalis. | Cœnolophium Fischeri. | Eritrichium villosum. |
| —— hyperboreus. | Conioselinum Fischeri. | Gymnandra borealis. |
| Trollius Asiaticus. | Ligusticum Scoticum. | Castilleja pallida. |
| Cardamine bellidifolia ? | Chærophyllum bulbosum. | Veronica macrostemon. |
| Parrya macrocarpa. | Cornus suecica. | Pedicularis Lapponica. |
| —— arctica. | Galium triflorum. | —— hirsuta. |
| Draba alpina. | Valeriana capitata. | —— Sudetica. |
| —— muricella. | Nardosmia frigida. | Pinguicula villosa. |
| —— hirta. | —— palmata. | Primula stricta. |
| —— rupestris. | Chrysanthemum arcticum. | —— Sibirica. |
| Eutrema Edwardsii. | Pyrethrum bipinnatum. | Koenigia Islandica. |
| Silene turgida. | Artemisia borealis. | Betula alpestris. |
| Lychnis apetala. | Antennaria alpina. | Salix lanata. |
| Sagina nivalis. | Senecio frigidus. | —— polaris. |
| Arenaria lateriflora. | Ligularia Sibirica. | Picea orientalis. |
| —— arctica. | Aster Sibiricus. | Larix Ledebourii. |
| Stellaria borealis. | —— Tataricus. | Platanthera obtusata. |
| —— humifusa. | Mulgedium Sibiricum. | Calypso borealis. |
| —— longipes | Campanula uniflora. | Sparganium natans. |
| —— crassifolia. | Cassiopeia hypnoides. | Luzula arcuata. |
| Rubus arcticus. | Cassandra calyculata. | Juncus biglumis. |
| —— chamæmorus. | Diapensia Lapponica. | Carex glareosa. |
| Rosa blanda. | Rhododendron Lapponicum. | —— Norvegica. |
| Saxifraga rivularis. | Ledum palustre. | —— festiva. |
| —— nivalis. | Gentiana detonsa. | —— loliacea. |
| —— flagellaris. | Pleurogyne rotata. | —— rariflora. |
| —— bronchialis. | Myosotis sparsiflora. | —— livida. |

| Carex laxa. | Alopecurus alpinus. | Hierochloe alpina. |
|---|---|---|
| —— salina. | Deyeuxia deschampsioides. | Colpodium latifolium. |
| —— aquatilis. | —— Lapponica. | —— pendulinum. |
| —— globularis. | —— strigosa. | —— fulvum. |
| Blysmus rufus. | —— Langsdorffii. | Dupontia Fisheri.* |

It is curious to remark how many of these boreal European plants, which are absentees in the Alps, have a very wide range, not only extending to the Himalaya and North China, but many of them all over Temperate North America; only one is found in the south temperate zone. In the present state of our knowledge we cannot account for the absence of these in the Alps; either they were not natives of Arctic Europe immediately previous to the glacial period, or if so, and they were then driven south to the Alps, they were afterwards there exterminated; or, lastly, they still inhabit the Alps under disguised forms, which pass for different species. Probably some belong to each of these categories. I need hardly remark that none inhabit Europe south of the Alps, or any part of the African continent.

The list of Arctic American and Asiatic species which do inhabit the Alps of Europe, but not Arctic Europe, is much smaller. Those marked † are Scandinavian, but do not there enter the arctic circle.

| Anemone patens. | †Astragalus hypoglottis. | Alnus viridis. |
|---|---|---|
| —— alpina. | †Spiræa salicifolia. | Pinus cembra. |
| —— narcissiflora. | †Potentilla fruticosa. | †Sparganium simplex. |
| †Ranunculus sceleratus. | —— sericea. | †Typha latifolia. |
| †Aconitum Napellus. | †Ceratophyllum demersum. | Carex ferruginea. |
| †Arabis petræa. | Bupleurum ranunculoides. | —— supina. |
| †Cardamine hirsuta. | †Viburnum Opulus. | —— stricta. |
| Draba stellata. | Galium rubioides. | †—— pilulifera. |
| †Thlaspi montanum. | †—— saxatile. | †Scirpus triqueter. |
| †Lepidium ruderale. | Ptarmica alpina. | Deyeuxia varia. |
| †Sagina nodosa. | Aster alpinus. | Spartina cynosuroides. |
| †Linum perenne. | Gentiana prostrata. | †Glyceria fluitans. |
| Phaca alpina. | Polygonum polymorphum. | Hordeum jubatum. |
| | Corispermum hyssopifolium. | |

### III.—BOTANICAL DISTRICTS WITHIN THE ARCTIC CIRCLE.

The following are the prominent features, botanical, geographical, and climatal, of the five districts of the arctic zone :—

1. ARCTIC EUROPE.—The majority of its plants are included

---

* The following species were included in this list as first published, but have since been found in Switzerland :—

| Naumbergia thyrsiflora. | Calla palustris. | Carex vulgaris. |
|---|---|---|
| Salix myrtilloides. | Carex fuliginosa. | —— cæspitosa. |
| | —— capillaris. | |

[Cardamine bellidifolia has been found on the Pyrenees by Lange !—ED.]

in the Lapland and Finland floras; and, owing to the temperature of the Gulf Stream, which washes its coasts, Lapland is by far the richest province in the arctic regions. The mean annual temperature at the polar circle, where it cuts the coast-line, is about 37°, and the June and September temperatures throughout Lapland are 40° and 37° respectively; thus rendering the climate favourable both to flowering and fruiting. Spitzbergen belongs to this flora, as do Nova Zembla and the arctic countries west of the river Obi, which forms its eastern boundary; for the Ural Mountains do not limit the vegetation, any more than do the Rocky Mountains in America. Gmelin observed more than a century ago that the river Obi in lower latitudes indicates the transition longitude from the European to the Asiatic flora.

Even in this small area, however, there are two floras, corresponding to the Arctic Norwegian and Arctic Russian. The latter, commencing at the White Sea, though comparatively excessively poor in species, contains nearly twenty that are not Lapponian, including *Braya rosea, Dianthus alpinus, D. Seguieri, Spiraea chamaedrifolia, Saxifraga hieraciifolia, Heracleum Sibiricum, Ligularia Sibirica, Ptarmica alpina, Gentiana verna, Pleurogyne rotata,* and *Larix Sibirica.*

There are further several Scandinavian plants which cross the arctic circle on the east shores of the White Sea, but do not do so in Lapland, as *Athamanta Libanotis, Chrysanthemum Leucanthemum, Bidens tripartita,* and others.

Iceland and Greenland also botanically belong to the Arctic Lapland province, but I have here excluded both: the former because it lies to the south of the arctic circle; the latter because both its magnitude, position, and other circumstances, require that it should be treated of separately.

As far as I can ascertain, 616 species (Monocotyledons, 183; Dicotyledons, 433 = 1 : 2·3) enter the arctic circle in this region, of which 70 advance into Spitzbergen; but no phænogamic plant is found in Ross's Islet, which lies to the north of Spitsbergen. The proportion of genera to species 266 : 616=1 : 2·3. Of these Arctic European plants, 453 cross the Alps or Pyrenees to the Mediterranean basin, a few occur on the mountains of Tropical Africa, (including *Luzula campestris* and *Deschampsia cæspitosa*), and 23 are found in South Africa.

No fewer than 264 species do not enter the arctic circle in any other longitude, and 184 are almost exclusively natives of the

Old World, or of this and of Greenland; not being found in any part of North America; 24 are confined to Arctic Europe and Greenland.

The following Arctic European plants are of sporadic occurrence in North America:—

Ranunculus acris. (Rocky Mountains).
Arabis alpina, (Greenland and Labrador).
Lychnis alpina, (Greenland and Labrador).
Arenaria arctica, (Greenland and Rocky Mountains).
—— verna, (Greenland, Arctic Islands, and Rocky Mountains).
Alchemilla vulgaris, (Greenland and Labrador).
Gnaphalium sylvaticum, (Greenland and Labrador).
—— supinum, (Greenland, Labrador, and White Mountains).
Vaccinium myrtillus, (Rocky Mountains and shores of great lakes).
Cassiopeia hypnoides, (Greenland, U. States Mountains, Canada, and Labrador).

Phyllodoce taxifolia, (Greenland, New England Mountains. and Labrador).
Gentiana nivalis. (Greenland and Labrador).
Veronica alpina, (Greenland and White Mountains).
Bartsia alpina, (Greenland and Labrador).
Pedicularis palustris, (Lab'r & Newfoundl'd).
Primula farinosa, (Labrador, Canada, Maine and shores of the Great Lakes).
Salix phylicifolia, (U. States Mountains).
—— herbacea, (Greenland, Labrador, and White Mountains, etc.).
Juncus trifidus,     (do.     do.).
Carex capitata, (Greenland and White Mountains).
Phleum alpinum, (Greenland, White Mountains, Canada, and Labrador).
Calamagrostis lanceolata, (Labrador).

There are besides a considerable number of Arctic European plants, which, in the New World, are confined to Greenland, being nowhere found in East America: these will be enumerated when treating of the Greenland flora.

The plants which are widely distributed in Temperate America or Asia, but almost exclusively Arctic in Europe, are the following:—

Ranunculus Pallasii, (Asia and America).
Trollius Asiaticus, (Asia).
Parrya macrocarpa, (Asia and America).
—— arctica. (Asia and America).
Stellaria longipes, (Asia and America).
Potentilla emarginata, (America).
Epilobium latifolium, (Asia and America).
Sedum quadrifidum. (Asia).
Saxifraga bronchialis, (Asia and America).
Senecio resedæfolius, (Asia and America).
Ligularia Sibirica, (Asia).
Mulgedium Sibiricum. (Asia)
Cassiopeia tetragona. (Asia and America).
Gentiana detonsa, (Asia and America).
Pleurogyne rotata, (Asia and America).

Eritrichium aretioides. (Asia and America).
Gymnandra Pallasii, (Asia).
Castilleja pallida, (Asia and America).
Veronica macrostemon, (Asia).
Pedicularis flammea, (America).
Pinguicula villosa, (Asia and America).
Koenigia Islandica, (Asia and America).
Salix polaris, (Asia and America).
Picea orientalis, (Asia).
Larix Ledebourii, (Asia).
Platanthera hyperborea, (America).
—— obtusata, (America).
Deyeuxia deschampsioides, (Asia and N. W. America).
Dupontia Fisheri, (America).

The works upon which I have mainly depended for the habitats of the Arctic European plants are Wahlenberg's 'Flora Lapponica.' Ledebour's 'Flora Rossica,' Fries's 'Summa Vegetabilium Scandinaviæ,' and 'Mantissæ,' and various admirable treatises by

Andersson, Nylander, Hartmann, Lindblöm, Wahlberg, Blytt, C. Martins, Ruprecht, and Schrenk.

For Spitzbergen plants I have depended on Hooker's enumeration of the Spitzbergen collections made during Parry's attempt to reach the north pole, Capt. Sabine's collection made in the same island, and on Lindblöm and Beilschmied's ' Flora von Spitzbergen' (Regensburg, Flora, 1842).

For the southern distribution of the Arctic European plants, I have further consulted Nyman's excellent ' Sylloge,' Ledebour's 'Flora Rossica,' Grisebach's ' Flora Rumelica,' Grenier and Godron's ' Flore de France,' Parlatore's 'Flora Italiana,' Koch's 'Synopsis Floræ Germanicæ,' Munby's ' Catalogue of Algerian Plants,' A. Richard's of those of Abyssinia, Visiani's ' Flora Dalmatica,' Delile's ' Flora Ægyptiaca,' Boissier's noble ' Voyage Botanique dans l'Espagne,' and Tchihatcheff's 'Asia Minor,' besides numerous local floras of the Mediterranean region, Madeira, the Azores, and Canaries.

2. ARCTIC ASIA.—This, which for its extent, contains by far the poorest flora of any on the globe, reaches from the Gulf of Obi eastwards to Behring's Straits, where it merges into the West American. The climate is marked by excessive mean cold ; at the Obi the isotherm of 18° cuts the arctic circle in its S.E. course, and at the eastern extremity of the province the isotherm of 20° cuts the same circle, while the centre part of the district is all north of the isotherm of 9°. The whole of the district is hence far north of the isotherm of 32°, which descends to 52° N. lat. in its middle longitude. The extremes of temperature are also very great ; the June isotherm of 41° ascending eastward through its western half to the Polar Sea, whilst the September isotherm of 41° descends nearly to 6° N. lat. ; whence the low autumn temperature must present an almost insuperable obstacle to the ripening of seeds within this segment of the polar circle.

The warming influence of the Atlantic currents being felt no further east than the Obi, and the summer desiccation of the vast Asiatic continent, combine to render the climate of this region one of excessive drought as well as cold; whence it is in all ways most unfavourable to every kind of vegetation.

The total number of species hitherto recorded from this area is 233 (Monocotyledons, 42 ; Dicotyledons, 191 = 1 : 4·5.) The proportion of genera to species is 1 : 2. Of the 233 species, 217 inhabit Siberia as far south as the Altai, or Japan, etc. ; 104

extend southwards to the Himalaya or mountains of Persia ; none are found on the mountains of the two Indian peninsulas, but 5 are found on those of Australia and New Zealand.  All but the following 37 are European.  Those marked with a † are almost exclusively arctic :—

Delphinium Menziesii (West America).
†Cochlearia sisymbrioides (Boreal ditto).
Hesperis Pallasii (East and West America).
Odontharrena Fischeriana.
Cardamine macrophylla.
†Arenaria macrocarpa (West America).
—— laricina.
†—— Rossii (Rocky Mountains).
Cerastium maximum (West America).
†Oxytropis nigrescens (Boreal E. and W. America).
Hedysarum Sibiricum.
†Sieversia glacialis (Boreal W. America).
Potentilla stipularis.
—— fragiformis.
Claytonia lanceolata.
†Sedum euphorbioides (Arctic Asia only).
Saxifraga Escholtzii (Boreal W. America).
—— punctata (West America).

Saxifraga serpyllifolia (W. America).
†Nardosmia glacialis (Arctic Asia only).
—— Gmelini.
†Artemisia Steveniana (Arctic Asia only).
—— glomerata (West America).
—— biennis (E. and W. America).
Osmothamnus fragrans.
Pedicularis capitata (E. and W. America).
—— euphrasioides (E. and W. America).
†Monolepis Asiatica (Arctic Asia only).
Rumex salicifolius (E. and W. America).
—— graminifolius.
Salix ovalifolia (West America).
Abies alba (E. and W. America).
Larix Americana (E. and W. America).
Tofieldea coccinea (E. and W. America).
Fritillaria Kamtchatkensis (West America).
Carex concinna (West America).
Elymus mollis (E. and W. America).

Thus out of 37 non-European species, only 12 are confined to Asia, the remaining 25 being American.  On the other hand, there are only 22 European species in Arctic Asia which are not also American ; which scarcely establishes a nearer relationship between Arctic Asia with Europe than with America.  These are :—

Dianthus Seguieri.
—— superbus.
Silene inflata.
Arenaria uliginosa.
Phaca alpina.
Hedysarum obscurum.
Rubus Idæus.

Sedum quadrifidum.
Gaya simplex.
Leontodon autumnalis.
Hieracium alpinum.
Veronica longifolia.
Pedicularis Sceptrum.
Pinguicula alpina.
Polygonum Sibiricum.

Salix Lapponum.
—— nigricans.
—— hastata.
Picea orientalis.
Larix Ledebourii.
Cypripedium Calceolus.
Carex ferruginea.

In other words, of the 233 Asiatic species, 196 are common to Asia and Europe, 22 are confined to Asia and Europe, 25 are confined to Asia and America, and 12 are confined to Asia (three of which are peculiar to the arctic circle).

The rarity of Gramineæ and especially of Cyperaceæ in this region is its most exceptional feature; only 21 of the 138 arctic species of these orders having hitherto been detected in it.  Cryptogamic plants seem to be even more rare; *Woodsia Ilvensis* and *Lastrea fragrans* being the only Filices hitherto enumerated.

Further researches along the edge of the arctic circle would, doubtless, add more Siberian species to this flora, as the examination of the north-east extreme would add American species, and possibly lead to the flora of the country of the Tchutchis being ranked with that of West America.

The works which have yielded me most information regarding this flora, are Ledebour's 'Flora Rossica,' and the valuable memoirs of Bunge, C. A. Meyer, and Trautvetter, on the vegetation of the Taimyr and Boganida rivers; and on the plants of Jenissei River in Von Middendorff's Siberian 'Travels'. For their southern extension Trautvetter and Meyer's 'Flora Ochotensis,' also in Middendorff's 'Travels;' Bunge's enumeration of North China and Mongolian plants; Maximovicz's 'Flora Amurensis;' Asa Gray's paper on the botany of Japan (Mem. Amer. Acad. N.S. vi.); Karelin and Kiriloff's enumeration of Soongarian plants: Regel, Bach, and Herder on the East Siberian and Jakutsk collections of Paullowsky and Von Stubendorff. For the Persian and Indian distribution, I have almost entirely depended on the herbarium at Kew, and on Boissier's and Bunge's numerous works.

3. ARCTIC WEST AMERICA.—The district thus designated is analogous in position, and to a considerable extent in climate, to the Arctic European, but is much colder; as is indicated both by the mean temperature, and by the position of the June isotherm of 41°, which makes an extraordinary bend to the south, nearly to 52° N. lat., in the longitude of Behring's Straits.

It extends from Cape Prince of Wales, on the east shore of Behring's Straits, to the estuary of the Mackenzie river, and as a whole it differs from the flora of the province to the eastward of it by its far greater number both of European and Asiatic species, by containing various Altai and Siberian plants which do not reach so high a latitude in more western meridians, and by some temperate plants peculiar to West America. This eastern boundary is, however, quite an artificial one; for a good many eastern plants cross the Mackenzie and advance westwards to Point Barrow, but which do not extend to Kotzebue Sound; and a small colony of Rocky Mountain plants also spread eastwards and westwards along the shores of the Arctic Sea, which further tend to connect the floras; such are *Aquilegia brevistylis, Sisymbrium humile, Hutchinsia calycina, Heuchera Richardsonnii, Crepis nana, Gentiana arctophila, Salix speciosa;* none of which are

generally diffused arctic plants, or natives of any other parts
of Temperate America but the Rocky Mountains.

The arctic circle at Kotzbue Sound is crossed by the isotherm
of 23°, and at the longitude of the Mackenzie by that of 12° 5';
whilst the June isotherm of 41° ascends obliquely from S.W. to
N.E., from the Aleutian Island to the mouth of the Mackenzie,
and passes south of this province; the June and the September
isotherms of 41° and 32° both traverse it obliquely, ascending to
the N.E.

The vast extent of the Pacific Ocean and its warm northerly
currents greatly modify the climate of West Arctic America,
causing dense fogs to prevail, especially throughout the summer
months, whilst the currents keep the ice to the north of Behring's
Straits. The shallowness of the ocean between America and Asia,
north of lat. 60°, together with the identity of the vegetation in
the higher latitudes of these continents, suggests the probability of
the land having been continuous at no remote epoch.

The number of phænogamic plants hitherto found in Arctic
West America is 364 (Monocotyledons, 76; Dicotyledons, 288
= 1 : 3·7.) The proportion of genera to species is 1 : 1·7. Of
these 364 species, almost all but the littoral and purely arctic
species are found in West Temperate North America, or in the
Rocky Mountains, 26 in the Andes of Tropical or Subtropical
America, and 37 in Temperate or Antarctic South America. Comparing this flora with that of Temperate and Arctic Asia, I find
that no less than 320 species are found on the north-western shores
and Islands of that continent, or in Siberia, many extending to
the Altai and the Himalaya. A comparison with Eastern Arctic
America shows that 281 are common to it, and the following 38
are found in Temperate, but not Arctic East America :—

| | |
|---|---|
| Anemone alpina. | Saxifraga bronchialis (Eu., As., & R. M.) |
| —— Pennsylvanica. | Archangelica officinalis ( Eu., As., & A.). |
| Hutchinsia calycina (Rocky Mountains only and Asia.) | Ligusticum Scoticum (Eu., Asia, Am.). |
| Sisymbrium humile (Rky. Mts. and As.). | Cornus Suecica (Europe, Asia, Am.). |
| Draba oligosperma ( Rocky Mounts. only). | Galium rubioides (Europe, Asia, Am.). |
| Lathyrus palustris (Europe, Asia, East and West America). | Senecio resedæfolius (Europe, Asia, Am.). |
| Spiræa salicifolia (Eu., As., E. & W. Am.). | —— Pseudo-Arnica ( Asia and America.) |
| Potentilla fruticosa (Eu., As., E. & W. Am.). | Cassandra calyculata (Europe, Asia, Am.). |
| —— Pennsylvanica (E., A., E. & W.Am.). | Gentiana arctophila (Rocky Mounts. only). |
| Comarum palustre (Eu., A., E. & W. Am.). | —— prostrata (Europe, Asia. Am.). |
| Montia fontana (Eur., As., and W. Am.). | —— tenella (Europe, Asia, Am.). |
| Saxifraga Sibirica (Asia and Labrador only). | Veronica scutellata (Europe, Asia, Am.). |
| ——Dahurica (Asia and Rky. Mts. only). | Pedicularis palustris (Europe. Asia, Am.). |
| | Atriplex patula (Europe, Asia, Am.). |
| | Corispermum hyssopifolium (E., A., Am.). |

Corallorhiza innata (Europe, Asia, Am.).
Luzula spadicea (Europe, Asia, Am.).
——— spicata (Europe, Asia, Am.).
——— pilosa (Europe, Asia, Am.).
Juncus balticus (Europe, Asia, Am.).

Carex lagopina (Europe, Asia, Am.).
——— Gmelina (America only).
———— cryptocarpa (Europe, Asia, Am.).
——— stricta (Europe, America).
Hierochloe borealis (Europe. Asia, Am.).

These, it will be seen, are for the most part north temperate plants, common in many parts of the globe, and which are only excluded from Eastern Arctic America by the greater rigour of its climate.

The best marked European and Asiatic species that are not found further east in Temperate or Arctic America are the following :—

Anemone narcissiflora.
Ranunculus Pallasii.
Aconitum Napellus.
Parrya macrocarpa.
Dianthus alpinus.
Cerastium vulgatum.

Spiræa chamædrifolia.
Pyrethrum bipinnatum.
Gentiana prostrata.
Eritrichium aretioides.
Pedicularis verticillata.
Primula nivalis.

Atriplex littoralis.
Pinus cembra.
Carex Norvegica.
Deyeuxia strigosa.
——— Langsdorffii.
Colpodium fulvum.

Hence it appears that of the 364 species found in Arctic West America, 319 inhabit East America (arctic or temperate, or both), and 320 are natives of the Old World—a difference hardly sufficient to establish a closer affinity of this flora with one continent rather than with the other.

The species peculiar to this tract of land are :—

Braya pilosa.
Saxifraga Richardsoni.

Artemisia androsacea.
Saussurea subsinuata.

Salix glacialis.

The rarity of monocotyledons, and especially of the glumaceous orders, is almost as marked a feature of this as of the Asiatic flora : of the 138 arctic species of Glumaceæ only 54 are natives of West Arctic America.

The materials for this flora are principally the plants of Chamisso, collected during Kotzebue's voyage, and described by himself and Schlechtendahl ; Lay and Collie's collections, described in Beechey's voyage; the ' Flora Boreali-Americana ;' and Seemann's plants, described in the ' Botany of the Herald.' Most of the above collections are from Behring's Straits. For the arctic coast flora I am mainly indebted to Richardson's researches, and to Pullen's and other collections enumerated by Seemann in his account of the flora of Western Eskimo Land. For the southern extension of the flora I have had recourse to the ' Flora Boreali-Americana;' Ledebour's ' Flora Rossica,' which includes the Sitcha plants ; the American floras of Nuttall, Pursh, Torrey, Gray, etc. ; and to the collections of Drs. Lyall and Wood formed in Vancouver Island and British Columbia; for the Californian, Mexican, and Cordillera floras generally, to the herbarium at Kew, the works above mentioned, and the various memoirs of Torrey and of Gray on the plants of the American Surveying Expeditions.

4. ARCTIC EAST AMERICA (EXCLUSIVE OF GREENLAND).— This tract of land is analogous to the Arctic Asiatic in many respects of position and climate, but is very much richer in species. It extends from the estuary of the Mackenzie River to Baffin's Bay, and its flora differs from that of the western part of the continent, both in the characters mentioned in the notice of that province, and in possessing more East American species. The western boundary of this province is an artificial one; the eastern is very natural, both botanically and geographically; for Baffin's Bay and Davis Straits (unlike Behring's Strait) have very deep water and different floras on their opposite shores.

The arctic circle is crossed in the longitude of the Mackenzie River by the isotherm of 12°, which thence trends south-eastward to the middle of Hudson Bay ; and in the longitude of Davis Straits it is crossed by the isotherm of 18½°. The June isotherm of 41° descends obliquely from the shores of the Arctic Sea, near the mouths of the Mackenzie, to the northern parts of Hudson Bay, south of the arctic circle; and the September isotherm of 41° is everywhere south of the circle. Hence, the western parts of this province are very much warmer than the eastern; so much so, that the whole west coast and islands of Baffin's Bay lie north of a southern inflection of the June isotherm of 32° which passes north of all the other polar islands; the Parry Islands have an analogous temperature of 40°. The warmth of the western portion of this tract is no doubt mainly due to the influence of the Pacific Ocean being felt across the continent of West America ; though possibly also to the presence of a comparatively warm polar ocean, or to Atlantic currents crossing the pole between Nova Zembla and Spitzbergen, of which nothing certain is known[*]. Be this as it may, the comparative luxuriance of the flora of Melville Island is a well-known fact, and one inexplicable by considerations of temperature, if unaccompanied by a humid atmosphere. The

---

[*] It is a well-known fact that the temperature always rises rapidly with the north (as well as other) winds over all this Arctic American area.

whole region is of course far north of the isotherm of 32°, which, in the longitude of its middle district, descends to Lake Winnipeg, in lat. 52°.

That portion of this province which is richest in plants is the tract which intervenes between the Coppermine and Mackenzie Rivers ; east of this, vegetation rapidly diminishes, as also to the northward. The flora of the Boothian Peninsula, surrounded as it is with glacial straits, and placed centrically among the arctic islands, is perhaps the poorest of any part of the area ; those of Banks Land and Melville Island to the N.W. being considerably richer, as are those of the shores of Lancaster Sound and Barrow's Strait, and the shores of Baffin's Bay to the north and east. *

The phænogamic flora of Arctic East America contains 379 species (Monocotyledons, 92 ; Dicotyledons, 287 = 1 : 3·1). The proportion of genera to species is 1 : 2·0. Of these 379 species, 323 inhabit Temperate North America, east of the Rocky Mountains ; 35 the Cordillera ; and 49 Temperate or Antarctic South America. Comparing this flora with that of Europe, I find that 239 (or two-thirds) species are common to the arctic regions of both continents, whilst but little more than one-third of the Arctic European species are Arctic East American. Of 105 non-European species in Arctic East America, 32 are Asiatic ; leaving 73 species confined to America, of which the following are further-more confined to the eastward of the Rocky Mountains and Mackenzie River :—

| | | |
|---|---|---|
| Corydalis glauca. | Prunus Virginiana. | Urtica dioïca. |
| Sarracenia purpurea. | Heuchera Richardsoni. | Salix cordata. |
| Viola cucullata. | Cornus stolonifera. | Populus tremuloides. |
| Silene Pennsylvanica. | Grindelia squarrosa. | Picea nigra. |
| Arenaria Michauxii. | Vaccinium Canadense. | Spiranthes gracilis. |
| Polygala Senega. | Dracocephalum parviflorum. | Cypripedium acaule. |
| Lathyrus ochroleucus. | Douglasia arctica. | Carex oligosperma. |
| Rubus triflorus. | Elæagnus argentea. | Pleuropogon Sabini. |

Of these *Douglasia* and *Pleuropogon* are the only ones abso-lutely peculiar to Arctic East America. It is a noticeable fact that not one of them is found in any part of Greenland. Com-pared with Greenland, the Arctic East American flora is rich ; containing, besides those just enumerated, no less than 165 other

---

* Details of these florulas will be found in the volume of the ' Linnean Journal,' under the notice of Dr. Walker's Collections, made during the voyage of the *Fox*.

species not found in Greenland. The following are found on the arctic islands, and many of them on the west coast of Baffin's Bay, but not in West Greenland :—

| | | |
|---|---|---|
| Caltha palustris. | Nardosmia corymbosa. | Castilleja pallida. |
| Parrya arctica. | Ptarmica vulgaris. | Pedicularis capitata. |
| Merkia physodes. | Chrysanthemum arcticum. | —— versicolor. |
| Stellaria crassifolia. | Artemisia vulgaris. | Androsace septentrionalis. |
| Astragalus alpinus. | Senecio frigidus. | —— Chamæjasme. |
| Oxytropis campestris. | —— palustris. | Salix phlebophylla. |
| —— Uralensis. | —— pulchellus. | Lloydia serotina. |
| —— nigrescens. | Solidago Virga-aurea | Hierochloe pauciflora. |
| Sieuersia Rossii. | Aster salsuginosus. | Deschampsia cæspitosa (East |
| Saxifraga hieracifolia. | Crepis nana. | Greenland only). |
| —— Virginiensis. | Saussurea alpina. | Glyceria fluitans. |
| —— Hirculus (East | Arctostaphylos alpina. | Pleuropogon Sabini. |
| Greenland only). | Kalmia glauca. | Bromus purgans. |
| Valeriana capitata. | Phlox Sibirica. | Elymus mollis.* |

There are thus no fewer than 184 of the 379 Arctic East American species (fully half) which are absent in West Greenland, whilst only 105 (much less than one-third) are absent in Europe. This alone would make the limitation of species in the meridian of Baffin's Bay more decided than in any other arctic longitude; and I shall show that it is rendered still more decisive by the number of Arctic Greenland plants that do not cross to Arctic East America.

Of the 379 Arctic East American species only 56 are not found in Temperate East America, of which two are absolutely confined to this area; two others (*Parrya arenicola* and *Festuca Richardsoni*) to Arctic East and West America; 25 are found in Temperate West America, and about 20 are Rocky Mountain species, and not found elsewhere in Temperate America.

For our knowledge of this flora I am principally indebted to the 'Flora Boreali-Americana,' and to Richardson's† botanical appendix to Franklin's first voyage—and his 'Boat Journey through Rupert's Land.' I have also examined the materials upon which the above works were founded, and the collections of almost every subsequent journey and voyage, up to those of Dr.

---

* *Andromeda polifolia* has been found in Greenland by Mr. Taylor, an intelligent surgeon of whale ships, who has spent many seasons in Baffin's Bay. He has given me a list of all the plants he knows.

† I am indebted to Sir John Richardson for some corrections to this list, which account for a few discrepancies between his lists of Arctic American plants and my own ; these refer chiefly to genera and species introduced into his lists, but here excluded.

Walker in the *Fox.* To enumerate the numerous botanical appendices to voyages, and separate opuscules to which these have given rise, from Ross's first voyage to the present time, would be out of place here. I have endeavoured to embody in the essay the information gleaned from all of them. For the southern distribution of these plants in the United States, etc., I have had recourse primarily to Asa Gray's excellent 'Manual of the Botany of the Northern United States,' to Chapman's ' Flora of the Southern States,' and to the reports on the Botany of various Exploring Expeditions.

5. ARCTIC GREENLAND.—In area Arctic Greenland exceeds any other arctic district except the Asiatic, but ranks lowest of all in number of contained species. In many respects it is the most remarkable of all the provinces, containing no peculiar species whatever, scarcely any peculiarly American ones, and but a scanty selection of European. A further peculiarity is that the flora of its temperate regions is extremely poor, and adds very few species to the whole flora, and, with few exceptions, only such as are arctic in Europe also. Being the only arctic land that contracts to the southward, forming a peninsula, which terminates in the ocean in a high northern latitude, Greenland offers the key to the explanation of most of the phenomena of arctic vegetation; and as I have already made use of it for this purpose, I shall be more full in my description of its flora than of any other.

The east and west coasts of Greenland differ in many important features ; the eastern is the largest in extent, the least indented by deep bays, is perennially encumbered throughout its entire length by icefields and bergs, which are carried south by a branch of the arctic current that sets between Iceland and Greenland ; and is hence excessively cold, barren, and almost inaccessible. The west coast, again, is generally more or less free from pack ice from Cape Farewell (lat. 60°) to north of Upernævik in lat. 73°. It is washed by a southerly current, which is said to carry drift timber from the Siberian rivers into its fiords, and enjoys a far milder climate, and consequently has a more luxuriant vegetation. A somewhat similar contrast is exhibited between West Greenland and the opposite shores of Baffin's Bay, against which latter the northerly arctic current from Lancaster Sound drives great masses of polar ice, derived from the regions beyond that estuary, and to which the bergs that float away from the glaciers in the Greenland fiords are also drifted. It is important to bear in mind these features of the two shores of Greenland and of Baffin's Bay and Davis Straits, because they may in some degree explain their differences of vegetation. There is also another difference between the polar islands and Greenland, inasmuch as the former are for the most part low, without mountains or extensive glaciers; while the latter is exceedingly mountainous, with valleys along the shore terminating in glacier-headed fiords, and the coast is bound by glaciers of prodigious extent from Melville Bay northwards to Smith Sound.

The isothermal lines in Greenland all follow one course, from S.W. to N.E., running more parallel to one another in this meridian than in any other. The isotherm of 32° passes through the southern extremity of the peninsula, and that of 5° through its north extreme at Smith Sound. The June isotherm of 41° skirts its east coast, and that of 32° passes north of Disco; the June temperature of Disco is hence as low as that of the north of Spitzbergen, of middle Nova Zembla, and of the extreme north of Asia, and yet Disco contains quadruple their number of plants. The autumn cold is very great; the September isotherm of 32° crossing the arctic circle on the west coast; and to this the scantiness of the flora may to some extent be attributed.

The Arctic Greenland flora contains 206 species, according to Lange's catalogue (in Rincke's ' Greenland '); or 207, according to my materials (Monocot. 67, Dicot. 140=1 : 2·1); the proportion of genera to species being 1 : 2.

Of these 207 species the following 11 alone are not European :—

| | |
|---|---|
| Anemone Richardsonii (Asiatic). | Potentilla tridentata (Labr. to Alleghanies). |
| Turritis mollis (Asiatic). | Saxifraga triscuspidata (do. to L. Superior). |
| Vesicaria arctica (American only). | Erigeron compositus (American only). |
| Draba aurea (Rocky Mountains). | Pedicularis euphrasioides (Asia). |
| Hesperis Pallasii (Asia and America). | Salix arctica (Asia). |
| Arenaria Grœnlandica (Labr. to U. S.). | |

On the other hand, no less than 57 Arctic Greenland species are absent in Arctic East America, and the following 36 Arctic Europe and Greenland species are either absent in all parts of Eastern Temperate America, or are extremely local there:—

| | |
|---|---|
| Arabis alpina (Labrador only). | Stellaria cerastioides (absent). |
| Lychnis alpina (do. and Newfoundland). | Alchemilla alpina (do.). |
| —— dioica (absent). | —— vulgaris (Labrador only). |
| Spergula nivalis (do.). | Sibbaldia procumbens (Labr. to Wht. Mts.). |
| Arenaria uliginosa (do.). | Rubus saxatilis (absent). |
| —— ciliata (do.). | Potentilla verna (Labrador only). |

Sedum villosum (absent).
Saxifraga Cotyledon (Labrador and Rocky Mountains only).
Galium saxatile (absent).
Gnaphalium sylvaticum (Labrador only).
—— supinum (do. and Wht. Mounts.).
Cassiopeia hypnoides (Labr. to W. Mts.)
Phyllodoce taxifolia (Labrador to W. Mts.).
Gentiana nivalis (Labrador only).
Thymus serpyllum (absent).
Veronica alpina (White Mountains only).
—— saxatilis (absent).
Euphrasia officinalis (N. U. S. & Canada).

Bartsia alpina (Labrador only).
Rumex acetosella (absent).
Salix arbuscula (do.).
Peristylus albidus (do.).
Carex capitata (White Mountains only).
—— microglochin (absent).
—— microstachya (do.).
—— pedata (do.).
Elyna caricina (Rocky Mountains only).
Phleum alpinum (Labrador to White Mountains).
Calamagrostis lanceolata (Labrador only).
Deschampsia alpina (absent).

When it is considered how extremely common most of these plants are throughout Europe and Northern Asia, and that some of them inhabit also N. W. America, their absence in Eastern America is even more remarkable than their presence in Greenland.

A small colony of Greenland plants has been found by Mr. Taylor in Cumberland Gulf, on the West side of Baffin's Bay, where the following Arctic Greenland plants occur, viz. :—

Arabis alpina.
Gnaphalium sylvaticum.
Cassiopeia hypnoides.

Phyllodoce taxifolia.
Euphrasia officinalis.
Kœnigia Islandica.
Luzula spicata.

Carex Hebonastes.
—— vulgaris.
Agrostis vulgaris.

Another singular feature of both Arctic and Temperate Greenland is its wanting a vast number of Arctic plants which are European, and found also in America. The following is a list of most of these, excluding about 15, which are water-plants, or species whose range is limited. The letter I. placed before a species signifies that it is Icelandic, and is introduced to show not only how many are absent from this island, but also how many are present. The letter S. indicates that the species is found in the south temperate or antarctic circle. The asterisk (*) indicates that the species is arctic both in East America and Europe:—

Anemone alpina,
—— nemorosa.
—— narcissiflora.
* Ranunculus Purshii.
* I. Caltha palustris.
* Aconitum Napellus.
Actæa spicata.
Nuphar luteum.
Nasturtium amphibium.
S. Barbarea præcox.
S. Turritis glabra.
Thlaspi montanum.

Sisymbrium Sophia.
* I. Erysimum lanceolatum.
Arabis hirsuta.
I. S. Cardamine hirsuta.
* Parrya arctica.
I. Draba muralis.
I. Subularia aquatica.
* I. Drosera rotundifolia.
I. —— longifolia.
I. Viola tricolor.
* I. Arenaria lateriflora.
* Stellaria longifolia.

I. Stellaria crassifolia.
Linum perenne.
Geranium Robertianum.
Hypericum 4-angulum.
Oxalis acetosella.
* Phaca frigida.
* Astragalus alpinus.
* —— hypoglottis.
* Oxytropis campestris.
—— Uralensis.
Lathyrus palustris.
Spiræa salicifolia.

S. Geum urbanum.
I. —— rivale.
\* Rubus arcticus.
Potentilla fructicosa.
—— Pennsylvanica.
—— argentea.
\* I.S. Fragaria vesca.
I. Sanguisorba officinalis.
Rosa cinnamomea.
—— blanda.
\* Circæa alpina.
\* I. S. Epilobium tetragonum.
\* I. S. —— alsinæfolium.
S. Lythrum salicaria.
\* Ribes rubrum.
\* —— alpinum.
\* I. Parnassia palustris.
Saxifraga Sibirica.
\* —— - hieraciifolia.
—— bronchialis.
\* Bupleurum ranunculoides.
Conioselinum Fischeri.
Cicuta virosa.
\* I. Carum carui.
Adoxa moschatellina.
Viburnum Opulus.
Lonicera cærulea.
\* Linnæa borealis.
\* I. Galium boreale.
—— rubioides.
I. —— trifidum.
S. —— aparine.
\* Valeriana capitata.
\* Nardosmia frigida.
\* Chrysanthemum arcticum.
I. Pyrethrum nodosum.
—— bipinnatum.
\* Artemisia vulgaris.
S. Bidens bipartita.
Tanacetum vulgare.
Antennaria Carpatica.
\* Senecio resedæfolius.
\* —— frigidus.
\* —— palustris.
\* —— campestris.
—— aurantiacus.
\* Solidago Virga-aurea.
\* Aster Sibiricus.
\* —— alpinus.
S. Erigeron acris.

S. Sonchus arvensis.
I. Hieracium boreale.
\* Saussurea alpina.
I. Vaccinium myrtillus.
\* Andromeda polifolia.
Cassandra calyculata.
\*\* I. Arctostaphylos alpina.
\* I. Pyrola secunda.
\* I. Gentiana amarella.
I. —— tenella.
\* Myosotis sylvatica.
—— palustris.
I. —— arvensis.
\* Scutellaria galericulata.
I. S. Prunella vulgaris.
Glechoma hederaceum.
S. Stachys palustris.
\* Gymnandra Pallasii.
\* Castilleja pallida.
I. S. Veronica officinalis.
S —— scutellata.
I. S. —— serpylifolia.
Melampyrum pratense.
—— sylvaticum.
\* I. Pedicularis palustris.
\* —— versicolor.
Scrophularia nodosa.
Utricularia vulgaris.
\* Pinguicula villosa.
Glaux maritima.
Trientalis Europæa.
\* Androsace septentrionalis.
\* —— Chamæjasme.
Naumbergia thyrsiflora.
I. S. Primula farinosa.
I. Plantago major.
—— lanceolata.
S. Chenopodium album.
I. S. Atriplex patula.
Corispermum hyssopifolium.
\* Polygonum Distorta.
I. —— amphibium.
\* Myrica Gale.
I. Betula alba.
I. —— pumila.
I. Alnus incana.
I. Salix pentandra.
I. —— myrtilloides.
I. Triglochin maritimum.
Scheuzeria palustris.

Veratrum album,
\* Lloydia serotina.
\* Allium schænoprasum.
\* Smilacina bifolia.
\* Platanthera obtusata.
\* Calypso borealis.
Godyera repens.
Cypripedium guttatum.
Calla palustris.
Typha latifolia.
Narthecium ossifragum.
Luzula maxima.
S. Juncus communis.
I. —— articulatus.
I. —— bulbosus.
—— stygius.
Carex pauciflora.
—— tenuiflora.
S. —— stellulata.
I. —— chordorrhiza.
—— teretiuscula.
—— paradoxa.
S. —— Buxbaumii.
I. —— limosa.
S. —— Magellanica.
—— ustulata.
—— livida.
I. —— pallescens.
—— maritima.
I. —— cæspitosa.
I. —— acuta.
—— stricta.
—— filiformis.
I. S. Eleocharis palustris.
S. —— acicularis.
S. Scirpus triqueter.
S. —— lacustris.
Eriophorum alpinum.
Rhynchospora alba.
Alopecurus pratensis.
I. Milium effusum.
S. Phalaris arundinacea.
I. S. Phragmites communis.
\* I Hierochloe borealis.
\* —— pauciflora.
\* I. Catabrosa aquatica.
\* I.S. Glyceria fluitans.
\* I. Atropis distans.
I. Festuca elatior.
S. Bromus ciliaris.
I. S. Triticum caninum.
S. Hordeum jubatum.

Altogether there are absent in Greenland upwards of 230 Arctic European species, which are all of them American plants. The most curious feature of this list is the absence throughout Greenland of the genera *Spiræa, Senecio, Astragalus, Trifolium, Phaca,*

*Oxytropis, Androsace, Aster, Myosotis, Rosa, Ribes, Thlaspi, Sisymbrium, Geranium,* etc., and of such ubiquitous arctic species as *Fragaria vesca, Caltha palustris\*, Barbarea præcox.* It is remarkable that *Astragalineæ* are also absent from Spitzbergen and Iceland.

Iceland possesses 432 species (Monocot., 157; Dicot., 275) amongst which I find about 120 Arctic European plants that do not enter Greenland; whereas only 50 of the European plants that inhabit Greenland are absent in Iceland. The more remarkable desiderata of Iceland are *Astragalineæ, Anemone, Aconitum, Braya, Turritis, Artemisia* and *Androsace; Alopecurus alpinus, Luzula arcuata, Hierochloe alpina, Rubus chamæmorus, Cassiopeia tetragona, Arnica montana, Antennaria dioica,* and *Chrysosplenium alternifolium.* On the other hand, Iceland contains of arctic genera absent in Greenland: *Caltha* (one of the most common plants about Icelandic dwellings), *Cakile, Geranium, Trifolium, Spiræa, Senecio,* and *Orchis.*

But perhaps the most remarkable fact of all connected with the Greenland flora is that its southern and temperate districts, which present a coast of 400 miles, extending south to lat. 60° N., do not add more than 74 species to its flora, and these are almost unexceptionably Arctic European plants; and inasmuch as these additional species increase the proportion of Monocotyledons to Dicotyledons of the whole flora, Greenland as a whole is botanically more arctic in vegetation than Arctic Greenland alone is!

The only American forms which Temperate Greenland adds to its flora are, *Ranunculus Cymbalaria, Pyrus Americana,* a very trifling variety of the European *P. Aucuparia, Viola Muhlenbergii,* a mere variety of *V. canina, Arenaria Grœnlandica,* a plant elsewhere found only on the Mountains of New England, etc., and *Parnassia Kotzebuei,* a species which is scarcely different from *P. palustris.*

The only plants which are not members of the arctic flora elsewhere, and which are confined in Greenland to the temperate zone, besides the above American plants, are *Blitum glaucum, Potamogeton marinus, Sparganium minimum,* and *Streptopus amplexifolius :* the rest will all be found in the column of the arctic plant

---

\* This is the more remarkable because it forms a conspicuous feature in Iceland, and is a frequent native of all the Arctic American coasts and islands.

catalogue devoted to Greenland, where S. signifies that the species is found only south of the arctic circle in that country.

On the other hand, Temperate Greenland adds very materially to the number of European Arctic species that do not enter Eastern America (Arctic or Temperate), amongst which the most remarkable are:—

| | | |
|---|---|---|
| Cerastium viscosum. | Galium palustre. | Betula alpestris. |
| Vicia cracca ? | Leontodon autumnale. | Juncus squarrosus. |
| Rubus saxatilis. | Hieracium murorum. | Anthoxanthum odoratum. |
| Sedum annuum. | —— alpinum. | Nardus stricta. |
| Galium uliginosum. | Gentiana aurea. | |

Another anomalous feature in the Greenland flora is the presence, on the East Arctic coast, of some species not found on the west, nor in the temperate southern end of the peninsula. These are:—

Lychnis dioica (Arctic Europe).
Saxifraga Hirculus (abundant in all extreme arctic latitudes but West Greenland).
Polemonium cæruleum (all arctic longitudes, but West Greenland).
Deschampsia cæspitosa (all arctic longitudes, but also absent in Spitzbergen).

For data connected with the Greenland flora, I am mainly indebted to the collections of the various polar voyagers in search of a north-west passage, especially to Drs. Lyall's and Sutherland's; to Lange's catalogue in Rincke's 'Greenland'; and to the notices of Vahl, Greville, Sir William Hooker, etc., on the plants collected by Sabine, Scoresby, Ross, Jameson, Graah, Vahl, etc.; to Sutherland's appendix to Penny's voyage and Durand's to Kane's voyage.

There is a curious affinity between Greenland and certain localities in America, which concerns chiefly a few of the European plants common to these countries. First, there are in Labrador, or on the Rocky Mountains, or on the Mountains of New England, etc., a certain number of European plants found nowhere else in the American continent. They are:—

| | |
|---|---|
| Ranunculus acris (Rocky Mountains). | Gentiana nivalis (Labrador). |
| Arabis alpina (Labrador). | Veronica alpina (White Mountains). |
| Lychnis alpina (do. & Newfoundland). | Bartsia alpina (Labrador). |
| Sibbaldia procumbens (do. & Wht. & Rky. | Salix herbacea (Labr. and Wht. Mts.) |
| Potentilla verna (Labrador).          [Mts.). | Luzula spicata (White Mountains). |
| Montia fontana (Labrador). | Juncus trifidus (New England Mts.). |
| Gnaphalium sylvaticum (Labrador). | Carex capitata (White Mountains). |
| —— supinum (Labr. and N. E. Mts.). | Kobresia scirpina (Rocky Mountains). |
| Cassiopeia hypnoides (Labr. & U. S. Mts.). | Phleum alpinum (Labr. to White Mounts.). |
| Phyllodoce taxifolia (Labr. to N. E. Mts.). | Calamagrostis lanceolata (Labrador). |

There are also two plants peculiar to Greenland, Labrador and the

Mountains of New England, or to Greenland and the Rocky Mountains, which have not hitherto been found elsewhere. They are :—

Draba aurea (Rocky Mountains).
Arenaria Grœnlandica (White Mountains northward to Labrador).

## IV.—ON THE ARCTIC PROPORTIONS OF SPECIES TO GENERA, ORDERS, AND CLASSES.

The observations which have hitherto been made on this subject are almost exclusively based on data collected on areas too small to yield general results. Especially in determining the influence of temperature in regulating the proportions of the great groups of flowering plants, it is of the highest importance to take comprehensive areas, both because of the wider longitudinal dispersion of some orders, especially the Monocotyledons, and the effects of local conditions, such as bog land, which determine the overwhelming preponderance of Cyperaceæ in some arctic provinces compared with others.

The proportion of genera to species in the whole arctic phænogamic flora is 323 : 762, or 1 : 2·3 (Monocot., 1 : 2·8; Dicot., 1 : 2·2); and that of orders to species 1 : 10·8; in the several provinces as follows :—

|  | Gen. | Gen. to Sp. | Orders. | Ord. to Sp. |
|---|---|---|---|---|
| Arctic Europe | 277 | 1 : 2.3 | 64 | 1 : 9.6 |
| "     Asia | 117 | 1 : 2.0 | 38 | 1 : 6.1 |
| "     West America | 172 | 1 : 2.1 | 48 | 1 : 7.6 |
| "     East America | 193 | 1 : 2.5 | 56 | 1 : 6.8 |
| "     Greenland | 104 | 1 : 2.0 | 38 | 1 : 5.5 |

Thus Europe presents the most continental character in its arctic flora, and West America the most insular; which may be attributable to the same cause in both; namely, the uniformity or variety of type. In West America we have, as in an oceanic island, a great mixture of types (Asiatic, European, East and West American) and paucity of species; in Europe the contrary. The proportions of species to orders are still more various; but here, again, Europe takes the lead decidedly.

The proportions of genera and orders to species of all Greenland differ but little from those of its arctic regions; whereas the contrast between Arctic Europe and this, together with Norway as far south as 60° N. lat., is very much greater. This is in accordance with the observation I have elsewhere made, that the whole of Greenland is comparatively poorer in species than Arctic Greenland is.

|  | Gen. Sp. Ord. Sp. |  | Gen. Sp. Ord. Sp. |
|---|---|---|---|
| Arctic Scandinavia .. | 1 : 2.3 — 1 : 9˙6 | Arctic Greenland..... | 1 : 2.0 — 1 : 5.5 |
| All Scandinavia..... | 1 : 2.8 — 1 : 11.6 | All Greenland ....... | 1 : 2.3 — 1 : 6.6 |

The proportions of Monocotyledons to Dicotyledons are:—

| Arctic Flora................. | 1 : 2.6 | Arctic East America........ | 1 : 3.1 |
|---|---|---|---|
| "    Europe .............. | 1 : 2.3 | "    Greenland ............ | 1 : 2.1 |
| "    Asia................. | 1 : 4.5 | All Greenland .............. | 1 : 2.0 |
| "    West America......... | 1 : 3.8 |  |  |

THE PROPORTION OF LARGEST ORDERS TO THE WHOLE FLORA.

| | Gram. & Cyp. | Salicin. | Polygon. | Scroph. | Eric. & Vaccin. | Comp. |
|---|---|---|---|---|---|---|
| Arctic Flora........ | 1 : 5˙6 | 1 : 30˙5 | 1 : 50˙2 | 1 : 27˙1 | 1 : 33˙1 | 1 : 10˙0 |
| "    Europe...... | 1 : 5˙2 | 1 : 38˙4 | 1 : 56˙0 | 1 : 23˙7 | 1 : 30˙8 | 1 : 12˙3 |
| "    Asia ........ | 1 : 10˙6 | 1 : 16˙6 | 1 : 23˙3 | 1 : 16˙6 | 1 : 21˙2 | 1 : 9˙6 |
| "    W. America. | 1 : 6˙7 | 1 : 24˙3 | 1 : 52˙0 | 1 : 33˙0 | 1 : 22˙7 | 1 : 9˙6 |
| "    E. America. | 1 : 5˙8 | 1 : 27˙0 | 1 : 76˙0 | 1 : 34˙5 | 1 : 23˙7 | 1 : 10˙5 |
| "    Greenland... | 1 : 3˙8 | 1 : 29˙6 | 1 : 51˙7 | 1 : 23˙0 | 1 : 17˙3 | 1 : 20˙7 |
| All Greenland....... | 1 : 3˙7 | 1 : 34˙0 | 1 : 42˙7 | 1 : 24˙9 | 1 : 21˙4 | 1 : 15˙0 |

| | Saxif. | Ros. | Leg. | Caryop. | Crucif. | Ranun. |
|---|---|---|---|---|---|---|
| Arctic Flora........ | 1 : 26˙2 | 1 : 17˙3 | 1 : 24˙6 | 1 : 15˙0 | 1 : 14˙1 | 1 : 17˙7 |
| "    Europe...... | 1 : 34˙2 | 1 : 21˙2 | 1 : 30 8 | 1 : 15˙4 | 1 : 17˙7 | 1 : 24˙6 |
| "    Asia ........ | 1 : 15˙5 | 1 : 19˙4 | 1 : 29˙1 | 1 : 14˙5 | 1 : 11˙6 | 1 : 21˙2 |
| "    W. America. | 1 : 19˙1 | 1 : 16˙6 | 1 : 28˙0 | 1 : 15˙9 | 1 : 18˙9 | 1 : 17˙3 |
| "    E. America.. | 1 : 21˙0 | 1 : 23˙7 | 1 : 27˙0 | 1 : 17˙2 | 1 : 11˙9 | 1 : 18˙9 |
| "    Greenland... | 1 : 17˙2 | 1 : 20˙7 | 0 : 207˙0 | 1 : 10˙3 | 1 : 10˙9 | 1 : 23˙0 |
| All Greenland....... | 1 : 27˙2 | 1 : 19˙8 | 1 : 149˙6 | 1 : 12˙4 | 1 : 12˙0 | 1 : 27˙2 |

The great differences between these proportions show how little confidence can be placed in conclusions drawn from local floras. Ericeæ is the only order which is more numerous proportionally to other plants in every province than in the entire arctic flora, and Cruciferæ is the only one that approaches it in this respect; and Leguminosæ is the only one which is less numerous proportionally in them all. East and West America agree most closely of any two provinces; then (excluding Leguminosæ) all Greenland and Europe; next Arctic Greenland and all Greenland.

The greatest differences are between Arctic Europe and Asia, and Arctic Asia and West America; they are less between Arctic Greenland and Asia (excluding Leguminosæ); they are great between Arctic Greenland and East America; and as great between all Greenland and Arctic America.

The proportion formerly deduced by Brown, etc., for the high arctic regions was a much smaller one; the Monocotyledons being in comparison with the Dicotyledons 1 : 5; and this still holds for some isolated, very arctic localities, as North-east Greenland; whereas Spitzbergen presents the same proportion as all the arctic regions, 1 : 2·7; the Parry Islands, 1 : 2·3; the west coast of Baffin's Bay, from Pond Bay to Home Bay, 1 : 3·3; and the extreme arctic plants mentioned at p. 333, 1 : 3. Of the prevalent arctic plants mentioned at p. 332, the proportion is 1 : 3·4.

I have dwelt more at length on these numerical proportions than their slight importance seems to require; my object being to show how little mutual dependence there is amongst the arctic florulas. Each has profited but little through contiguity with its coterminous districts, though all bear the impress of being members of one northern flora.

V.—ON GROUPING THE FORMS, VARIETIES, AND SPECIES OF ARCTIC PLANTS FOR PURPOSES OF COMPARATIVE STUDY.

Considering the limited extent of the arctic zone, the poverty of its flora, which is almost confined to 14° of latitude in the longitudes most favorable to vegetation, and to only 10° in the Asiatic area, and the number of able botanists who have studied it, it might be supposed that the preliminary task of identifying the species, and tracing their distribution within and beyond the arctic circle would have been short and simple; but this is not the case; for owing to the number of local floras, voyages, travels, and scientific periodicals that have to be consulted, to the variability of the species, and the consequent difficulty of settling their limits, and to the impossibility of reconciling the divergent opinions of my predecessors regarding them, I have found this a very tedious and unsatisfactory operation.

Of all these sources of doubt and error, the most perplexing has been the well-known variability of polar plants; and in the existing state of the controversy upon Mr. Darwin's hypothesis, it requires to be treated circumspectly. In several genera I have not only had to decide whether to unite for purposes of distribution dubious or spurious arctic species, but also how far I should go in examining and uniting cognate forms from other countries, which, if included, would materially affect the distribution of the species. These questions became in many instances so numerous and complicated, that I have often resorted to the plan of treating several very closely allied species and varieties as one aggregate or collective species. This appears at first sight to be an evasive course; but as it offered the only satisfactory method of solving the difficulty, I was obliged, after many futile attempts to find a better, to resort to it, and hence I feel called upon to enter more fully into my reasons for doing so; premising that all my attempts to treat each variety, form, and subspecies as a distinct plant involved the discussion of a multitude of details from which any generalization was hopeless; the results in every case defeated the object of this paper.

Of the plants found north of the arctic circle, very few are absolutely or almost exclusively confined to frigid latitudes (only about 50 out of 762 are so), the remainder, as far as their southern dispersion is concerned, may be referred to two classes; one consisting of plants widely diffused over the plains of Northern Europe, Asia, and America, of which there are upwards of 500; the other of plants more or less confined to the Alps of these countries, and still more southern regions, of which there are only about 200. *Glyceria fluitans, Atropis maritima,* and *Senecio campestris* are good examples of the first, as being high arctic and boreal but not alpine; while most of the species of *Saxifraga, Draba,* and *Androsace,* are examples of the second.* Both these classes abound in species, the limitation of which within the arctic circle, and the identification of whose varieties with those of plants of more southern countries, present great difficulties.

Those plants of the temperate plains which enter the arctic regions are often species of large, widely dispersed, and variable genera, most or all of whose species are very difficult of limitation; as *Ranunculus,* of which the arctic species *auricomus, aquatilis,* and *acris,* are each the centre of a nœud of allied temperate species or varieties, as to whose limits no two botanists are agreed; and the same applies to the species of *Viola, Stellaria, Arenaria,* and *Hieracium.* This has often led to the grouping of names of plants considered as synonymous by some authors, varieties by others, and good species by a third class. Furthermore, such

---

* Conversely the only arctic genus unknown in the Alps of the middle temperate zone is *Pleuropogon,* and the only alpine genera containing several species which inhabit the highest Alps of the north temperate zone, but not the polar regions, are *Soldanella* in Europe, *Swertia* in Europe and the Himalaya, etc.

genera are often represented in the temperate regions of two or
more continents (and some of them in the south temperate zone
also) by closely allied groups of intimately related species. This
always complicates matters extremely; for an arctic species, being
generally in a reduced or stunted state, may be equally similar to
alpine or reduced forms of what in two or more of these geographically sundered groups may rank as good species, and its
affinities and distribution be consequently open to doubt. Thus
under the arctic *Stellaria longipes* are included five other arctic
forms (*læta, Edwardsii, peduncularis, hebecalyx,* and *ciliatosepala*);
but amongst these forms some specimens approach closely the
American *S. Longifolio* Muhl., or slight varieties of it; while
others resemble the European *S. Fricsiana* Scr., others *S.
graminea,* others certain Tasmanian forms, and others again
Chilian. My own impression is, that some of these may prove
but slight modifications of one common, very widely dispersed
plant, between all whose varieties no constant definable characters
will eventually be found; but in the present state of science I
have abstained from including any of them, because to prove this
or disprove it, the whole genus wants a far longer and closer study
than it has yet received or than I can give it. *Arenaria verna*
and its forms offer a very parallel case, and these I have included
more largely, because I have the published opinion of many
botanists to bear me out in doing so. *Viola epipsila, palustris*
and *blanda,* are thus included, though they are more constant and
have to a considerable extent different distributions; because I
have found no differences of any moment between their normal
forms, because such as exist seem to me to be too slight to attach
specific value to, and because, though well distinguished by
Scandinavian botanists, they have not been so carefully collected
and studied in other parts of the arctic zone. *Viola canina,
Fragaria vesca,* and *Sanguisorbia officinalis,* afford other examples : all these arctic plants affect the temperate plains rather
than the mountains of the northern hemisphere.

Turning to those arctic plants that chiefly affect the Alps of the
temperate or tropical zones, their limitation is quite as difficult;
alpine plants being as proverbially variable as arctic. Many
alpine plants are now considered to be only altered forms of lowland ones; and this affects the estimated distribution of every
arctic species that is identified with an alpine one. As an example, *Saxifraga exilis* is a very slight variety of *S. cernua ;*

both are arctic and alpine plants, but *S. cernua* is considered by some botanists to be an alpine form of the lowland *S. granulata*, whose limits and distribution are very difficult to settle, because it apparently passes into several oriental forms, which have been distinguished as species. In this case I have not included *S. granulata* with *S. cernua;* because the latter is everywhere easily distinguished as a well-marked plant, having a restricted range both in area and elevation, which *S. granulata* does not share. At the same time I am in favour of a hypothesis that would give these a common origin previous to the glacial epoch.

Other reasons for adopting the system of including very closely allied species are the following :—When species have been founded in error ; this generally arises from their authors having imperfect specimens, or too limited a series of them ; various species founded by Brown on the first Arctic American collections come under this category, as do Adams's Arctic Siberians pecies ; the genera *Ranunculus, Draba, Arenaria,* and *Potentilla,* offer many examples : When the species, besides belonging to very variable genera, are apparently identical both in the herbarium and according to their descriptions, and present the same or a continuous distribution ; of this *Trientalis, Senecio, Aster, Erigeron, Mertensia, Sedum, Claytonia, Turritis,* and many others, afford examples.

It may be asked what useful scientific results can be obtained from the study of a flora whose specific limits are in so vague a condition ? the answer is, that though much is uncertain, all is not so ; and that if the species thus treated conjointly really express affinities far closer than those which exist between those treated separately, a certain amount of definite information, useful for my purpose, is obtained ; and it is a matter of secondary importance to me whether the plants in question are to be considered species or varieties. Again, if, with many botanists, we consider these closely allied varieties and species as derived by variation and natural selection from one parent form at a comparatively modern epoch, we may with advantage, for certain purposes, regard the aggregate distribution of the very closely allied species as that of one plant. When sufficient materials shall have been collected from all parts of the arctic and sub-arctic areas, we may institute afresh the inquiry into their specific identity or difference, by selecting examples from physically differ ing distant areas, and comparing them with others from inter

mediate localities. An empirical grouping of allied plants for the
purpose of distribution may thus lead to a practical solution of
difficulties in the classification and synonymy of species.

My thus grouping names must not therefore be regarded as a
committal of myself to the opinion that the plants thus grouped
are not to be held as distinct species; I simply treat of them
under one name, because for the purposes of this essay it appears
to me advisable to do so. Every reflecting botanist must acknowledge that there is no more equivalence amongst species than there
is amongst genera; and I have elsewhere* endeavoured to show
that, for all purposes of classification, species must be treated as
groups analogous to genera, differing in the number of distinguishable forms they include, and of individuals to which these forms
have given origin, and in the amount of affinity both between
forms and individuals. My main object is to show the affinities
of the polar plants, and I can best do this by keeping the specific
idea comprehensive. It is always easier to indicate differences
than to detect resemblances, and if I were to adopt extreme views
of specific difference, I should make some of the polar areas appear
to be botanically very dissimilar from others with which they are
really most intimately allied, and from which I believe them to have
derived almost all their species. A glance at my catalogue will
show that, had I ranked as different species the few Greenland
forms of European plants (called generally by the trivial name
*Grœnlandica*), I should have made that flora appear not only
more different from the European than it really is, but from the
American also; and that the differences thus introduced would be
of opposite values, and hence deceptive, in every case when the
European species (of which the *Grœnlandica* is often not even a
variety or distinct form) was not also common to America.

I wish it then to be clearly understood that the catalogue here
appended is intended to include every species hitherto found
within the arctic circle, together with those most closely allied
forms which I believe to have branched off from one common
parent within a comparatively recent geological epoch, and that
immediately previous to the glacial period or since then. Further,
I desire it to be understood that I claim no originality in bringing
these closely allied forms together; from the appended notes it will

---

* Essay on the Australian Flora; introductory to the Flora
Tasmanica,, etc.

be seen that there is scarcely one of them that has not been treated as a synonym, variety, subspecies, form, or lusus, by one or more very able and experienced botanists, some of them by many.  Furthermore, it is curious to observe how much the botanists of each country do to a considerable extent agree amongst themselves as to the specific identity or difference of the same forms—the Scandinavian agreeing with Fries, the German with Koch, and the American with Hooker's ' Flora Boreali-Americana'; also to observe, that in all these cases the authors I quote are independent observers, and not copyers or followers.  I think this fact indicates that the same plant presents a different aspect (probably obliterated in drying) in each country.  This observation is consonant with what we know of the tendency of all species to run into local varieties in isolated areas, which varieties are often appreciable to the eye or to the touch, but are not expressable by words.

Of the 762 species enumerated, I have compared arctic or boreal specimens of all but a few which I have indicated in the appended notes, and in most cases I have compared specimens from all the southern areas indicated; but I do not pretend to have made such a critical study of all the grouped species, or of all those belonging to difficult genera (as *Draba*, *Poa*, etc.), as to enable me to say that I have given all their distribution, or satisfied myself of all their affinities and differences.  There are, on the contrary, fully 60 genera out of the 323 arctic ones enumerated, each of which requires careful monographing, and months of study before the limits, systematic and geographical, of its common European species can be ascertained.  In two of the largest and most difficult of these I have been indebted to others ; namely, to Dr. Boott, who has revised my list of Carices, and to Dr. Andersson of Stockholm, who has drawn up that of the Salices: each has extensively modified the conclusions of his predecessors in arctic botany ; quite as much or more so than I have done in any genus, and I have every confidence in their judgment.  Colonel Munro has twice revised the list of grasses with a like result.  In these important genera, therefore, the groups express the opinions of these acute botanists as to the limits of the species.

With regard to the probable completeness of our knowledge of the flowering plants of the arctic zone, I think it is pretty certain that there are few or no new species to be discovered.  The collectors in the numerous voyages undertaken since 1847 in search of the Franklin expedition have not added one species to the flora of the Arctic American islands, and but one to that of Arctic Greenland. The Lapponian region is, of course, as well known as any on the globe; but further east, and especially in Arctic Siberia, much remains to be done; not perhaps in the discovery of new plants, but in ascertaining the southern limits of various Siberian ones that probably cross the arctic circle. Of Arctic Continental America the same may be said.

The method which I adopted in finally arranging the materials for geographical purposes was the following. I took Wahlenberg's 'Flora Lapponica,' Fries's 'Summa Vegetabilium Scandinaviæ,' Ledebour's 'Flora Rossica,' Hooker's 'Flora Boreali-Americani,' and Lange's 'Plants of East Greenland,' which together embrace in outline almost everything we know of arctic botany, geographical, systematic, and descriptive. I put together from these all the matter they contained, and arranged it both botanically and geographically into a 'Systema,' which I studied with an Admiralty north circumpolar chart; and by this means arrived at a general idea of the position and extent of the centres of vegetation within the polar circle. I then again went through the catalogue with the herbarium, with every work treating on arctic plants that was accessible to me, and lastly revised it, verifying the habitats, comparing specimens from each province, adding new localities from more recent floras, catalogues, and voyages; tracing the extra-arctic distribution of the species, and noting all points requiring further investigation.

(*To be continued,*)

## NOTICES OF SOME REMARKABLE GENERA OF PLANTS OF THE COAL FORMATION.

By J. W. DAWSON, LL.D., F.R.S., etc.[*]

GENUS SIGILLARIA.—The Sigillariæ, so named from the scal-like scars of fallen leaves stamped on their bark, were the most important of all the trees of the coal-swamps, and those which contributed most largely to the production of coal. Let us take as an example of them a species very common at the Joggins, and which I have named *S. Brownii*, in honour of my friend, Mr. R.

---

[*] From "Acadian Geology," 2nd edition, with specimens of the illustrations.

Brown, of Sydney. Imagine a tall cylindrical trunk spreading at the base, and marked by perpendicular rounded ribs, giving it the appearance of a clustered or fluted column. These ribs are marked by rows of spots or pits left by fallen leaves, and toward the base they disappear, and the bark becomes rough and uneven, but still retains obscure indications of the leaf-scars, widened transversely by the expansion of the stem. At the base the trunk spreads into roots, but with a regular bifurcation quite unexampled in modern trees, and the thick cylindrical roots are marked with round sunken pits or areoles, from which spread long cylindrical rootlets. These roots are the so-called Stigmariæ, at one time regarded as independent plants, and, as the reader may have already observed, remarkable for their constant presence in the underclays of the coal-beds. Casting our eyes upward, we find the pillar-like trunk, either quite simple or spreading by regular bifurcation into a few thick branches, covered with long narrow leaves looking like grass, or, more exactly, like pine leaves greatly increased in size, or, more exactly still, like the single leaflets of the leaves of Cycads. Near the top, if the plant were in fruit, we might observe long catkins of obscure flowers or strings of large nut-like seeds, borne in rings or whirls encircling the stem. If we could apply the woodman's axe to a Sigillaria, we should find it very different in structure from that of our ordinary trees, but not unlike that of the Cycads, or false sago-plants of the tropics. A lumber-man would probably regard it as a tree nearly all bark, with only a slender core of wood in the middle; and, botanically, he would be very near the truth. The outer rind or bark of the tree was very hard. Within this was a very thick inner bark, partly composed of a soft corky cellular tissue, and partly of long tough fibrous cells like those of the bark of the cedar. This occupied the greater part of the stem even in old trees four or five feet in diameter. Within this we would find a comparatively small cylinder of wood, not unlike pine in appearance, and even in its microscopic structure; and in the centre a large pith, often divided, by the tension caused in the growth of the stem, into a series of horizontal tables or partitions. Such a stem would have been of little use for timber, and of comparatively small strength. Still the central axis of wood gave it rigidity, the surrounding fibres, like cordage, gave the axis support, and the outer shell of hard bark must have contributed very materially to the strength of the whole. Growing as these trees did in swampy flats close together, and the bark of which they were chiefly composed being less susceptible of rapid decay than most kinds of wood, and too impervious to fluids to be readily penetrated by mineral matter, they were admirably fitted for the production of the raw material of coal. (Fig. 161.)

\*          \*          \*          \*

I include under Sigillariæ the remarkable fossils known as Stigmaria, being fully convinced that all the varieties of these plants known to me are merely roots of Sigillaria; I have verified this fact in a great many instances, in addition to those so well described by Mr. Binney and Mr. Brown. The different varieties or species of Stigmaria are no doubt characteristic of different species of Sigillaria, though in very few cases has it proved possible to ascertain the varieties proper to the particular species of stem. The old view, that the Stigmariæ were independent aquatic plants, still apparently maintained by Goldenberg and some other palæobotanists, evidently proceeds from imperfect information. Independently of their ascertained connexion with Sigillaria, the organs attached to the branches are not leaves, but rootlets. This was made evident long ago by the microscopic sections published by Goeppert, and I have ascertained that the structure is quite similar to that of the thick fleshy rootlets of Cycas. The lumps or tubercles on these roots have been mistaken for fructification; and the rounded tops of stumps, truncated by the falling in of the bark or the compression of the empty shell left by the decay of the wood, have been mistaken for the natural termination of the stem.\*   The only question remaining in regard to these organs is that of their precise morphological place. Their large pith and regular areoles render them unlike true roots; and hence Lesquereux has proposed to regard them as rhizomes. But they certainly radiate from a central stem, and are not known to produce any true buds or secondary stems. In short, while their function is that of roots, they may be regarded, in a morphological point of view, as a peculiar sort of underground branches. They all ramify very regularly in a dichotomous

---

\* For examples of the manner in which a natural termination may be simulated by the collapse of bark or by constriction owing to lateral pressure, see my papers, Quart. Jour. Geol. Soc., vol. x. p. 35, and vol. vii. p. 194.

Fig. 161.—*Sigillariæ.*

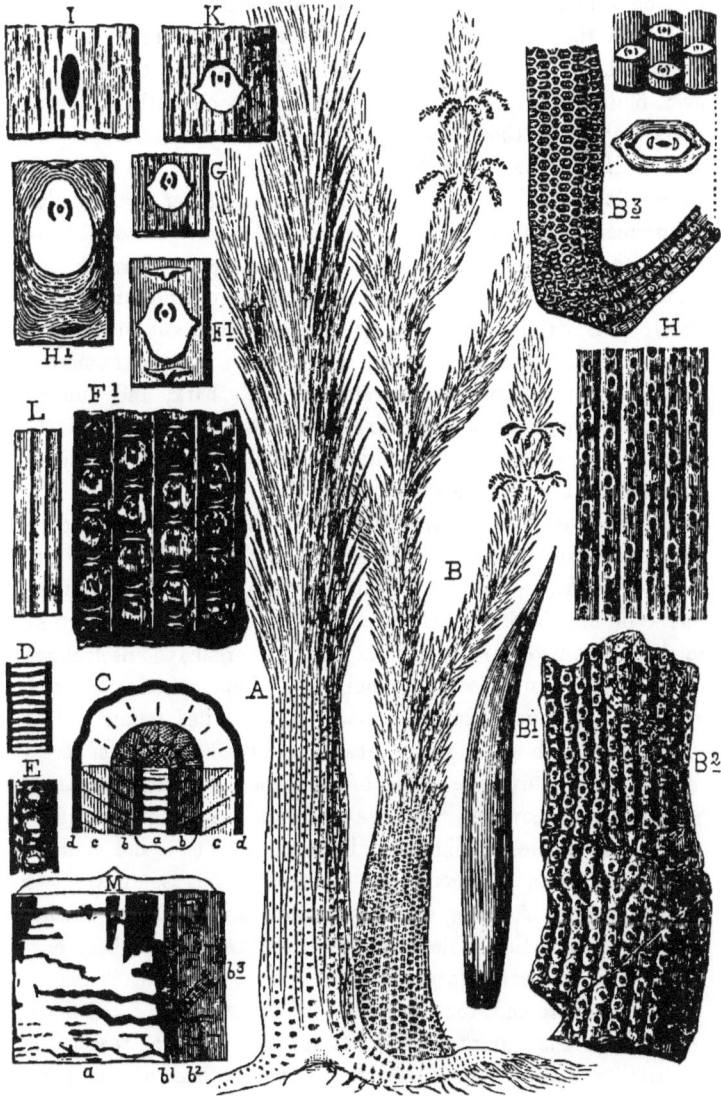

A, *Sigillaria Brownii*, restored.　　　B, *S. elegans*, restored.　　　B1, Leaf of S. elegans.
B2, Portion of decorticated stem, showing one of the transverse bands of fruit-scars.
B3, Portion of stem and branch reduced, and scars nat. size.
C,　Cross section of *Sigillaria Brownii* (?), reduced, and portion at (M) natural size. (*a*) Stern-
　　bergia pith. (*b*1) Inner cylinder of scalariform vessels. (*b*2) Outer cylinder of discigerous
　　cells, with modullary rays and bundles of scalariform vessels going to the leaves at (*b*3).
　　(*c*) Inner bark. (*d*) Outer bark.
D, Scalariform vessel magnified.　　　　H, *S. eminens*, reduced. (H1) areole, half n. size.
E, Discigerous woody fibre, magnified.　I. *S. catenoides*, half nat. size.
F, *Sigillaria Bretonensis*, ⅔. (*f*1) Areole n. size. K, *S. planicosta*, half nat. size.
G, *S. striata*, nat. size.　　　　　　　L, Portion of leaf of *S. scutellata*.

manner, and, as Mr. Brown has shown, in some species at least, give off conical tap-roots from their underside.

In all the Stigmariæ exhibiting structure which I have examined, the axis shows only scalariform vessels. Corda, however, figures a species with wood-cells, or vessels with numerous pores, quite like those found in the stems of Sigillaria proper; and, as Hooker has pointed out, the arrangement of the tissues in Stigmaria is similar to that in Sigillaria. After making due allowance for differences of preservation, I have been able to recognize eleven species or forms of Stigmaria in Nova Scotia, corresponding, as I believe, to as many species of Sigillaria.* At the Joggins, Stigmariæ are more abundant than any other fossil plants. This arises from their preservation in the numerous fossil soils or Stigmaria underclays. Their bark, and mineral charcoal derived from their axes, also abound throughout the thickness of the coal beds, indicating the continued growth of Sigillaria in the accumulation of the coal.

Our knowledge of the fructification of Sigillaria is as yet of a very uncertain character. I am aware that Goldenberg has assigned to these plants leafy strobiles containing spore-capsules : but I do not think the evidence which he adduces conclusive as to their connexion with Sigillaria ; and the organs themselves are so precisely similar to the stobiles of Lepidophloios, that I suspect they must belong to that or some allied genus. The leaves, also, with which they are associated in one of Goldenberg's figures, seem more like those of Lepidophloios than those of Sigillaria. If, however, these are really the organs of fructification of any species of Sigillaria, I think it will be found that we have included in this genus, as in the old genus Calamites, two distinct groups of plants, one cryptogamus, and the other phænogamous, or else that male strobiles bearing pollen have been mistaken for spore-bearing organs.

I cannot pretend that I have found the fruit of Sigillaria attached to the parent stem ; but I think that a reasonable probability can be established that some at least of the fruits included, somewhat vaguely, by authors under the names of Trigonocarpum and Rhabdocarpus, were really fruits of Sigillaria. These fruits are excessively abundant and of many species, and they occur not only in the sandstones, but in the fine shales and

---

* See Paper on Accumulation of Coal, Journ. Geol. Soc., vol. xxii.

coals and in the interior of erect trees, showing that they were
produced in the coal-swamps. The structures of these fruits
show that they are phœnogamous and probably gymnospermous.
Now the only plants known to us in the Coal formation, whose
structures entitle them to this rank, are the Conifers, Sigillariæ,
and Calamodendra. All the others were in structure allied to
cryptogams, and the fructification of most of them is known.
But the Conifers were too infrequent in the Carboniferous swamps
to have afforded numerous species of Carpolites; and, as I shall
presently show, the Calamodendra were very closely allied to
Sigillariæ, if not members of that family. Unless, therefore,
these fruits belonged to Sigillaria, they must have been produced
by some other trees of the coal-swamps, which, though very
abundant and of numerous species, are as yet quite unknown to
us. Some of the Trigonocarpa have been claimed for Conifers,
and their resemblance to the fruits of Salisburya gives countenance to this claim; but the Conifers of the Coal period are much
too few to afford more than a fraction of the species. One species
of Rhabdocarpus has been attributed by Geinitz to the genus
Nœggerathia; but the leaves which he assigns to it are very like
those of *Sigillaria elegans*, and may belong to some allied species.
With regard to the mode of attachment of these fruits, I have
shown that one species, *Trigonocarpum racemosum* of the
Devonian strata,[*] was borne on a rhachis in the manner of a
loose spike, and I am convinced that some of the groups of inflorescence named Antholithes are simply young Rhabdocarpi or
Trigonocarpa borne in a pinnate manner on a broad rhachis and
subtended by a few scales. Such spikes may be regarded as
corresponding to a leaf with fruits borne on the edges, in the
manner of the female flower of Cycas; and I believe with Goldenberg that these were borne in verticils at intervals on the stem.
In this case it is possible that the strobiles described by that
author may be male organs of fructification containing, not spores,
but pollen. In conclusion, I would observe that I would not
doubt the possibility that some of the fruits known as Cardiocarpa may have belonged to sigillarioid trees. I am aware that
some so-called Cardiocarpa are spore-cases of Lepidodendron;
but there are others which are manifestly winged nutlets allied to

---

[*] 'Flora of the Devonian Period,' Quart. Journ. Geol. Soc., vol. viii.
p. 324.

Trigonocarpum, and which must have belonged to phænogams. It would perhaps be unwise to insist very strongly on deductions from what may be called circumstantial evidence as to the nature of the fruit of Sigillaria; but the indications pointing to the conclusions above stated are so numerous that I have much confidence that they will be vindicated by complete specimens, should these be obtained.

All of the Joggins coals, except a few shaly beds, afford unequivocal evidence of Stigmaria in their underclays; and it was obviously the normal mode of growth of a coal-bed, that, a more or less damp soil being provided, a forest of Sigillaria should overspread this, and that the Stigmarian roots, the trunks of fallen Sigillariæ, their leaves and fruits, and the smaller plants which grew in their shade, should accumulate in a bed of vegetable matter to be subsequently converted into coal—the bark of Sigillaria and allied plants affording ' bright coal,' the wood and bast tissues mineral charcoal, and the herbaceous matter and mould dull coal. The evidence of this afforded by microscopic structure I have endeavoured to illustrate in a former paper.*

The process did not commence, as some have supposed, by the growth of Stigmaria in ponds or lakes. It was indeed precisely the reverse of this, the Sigillaria growing in a soil more or less swampy but not submerged, and the formation of coal being at last arrested by submergence. I infer this from the circumstance that remains of cyprids, fishes, and other aquatic animals, are rarely found in the underclays and lower parts of the coal-beds, but very frequently in the roofs, while it is not unusual to find mineral charcoal more abundant in the lower layers of the coal. For the formation of a bed of coal, the sinking and subsequent burial of an area previously dry seems to have been required. There are a few cases at the Joggins where Calamites and even Sigillariæ seem to have grown on areas liable to frequent inundation; but in these cases coal did not accumulate. The non-laminated, slicken-sided and bleached condition of most of the underclays indicates soils of considerable permanence.

In regard to beds destitute of Stigmarian underclays, the very few cases of this kind apply only to shaly coals filled with drifted leaves, or to accumulations of vegetable mud capable of conversion

---

* ' On the Structures in Coal,' Quart. Journ. Geol. Soc., 1859.

into impure coal.  The origin of these beds is the same with that
of the carbonaceous shales and bituminous limestones already
referred to.  It will be observed in the section that in a few
cases such beds have become sufficiently dry to constitute underclays, and that conditions of this kind have sometimes alternated
with those favourable to the formation of true coal.

There are some beds at the Joggins, holding erect trees *in situ*,
which show that Sigillariæ sometimes grew singly or in scattered
clumps, either alone or amidst brakes of Calamites.  In other
instances they must have grown close together, and with a dense
underground of ferns and Cordaites, forming an almost impene-
trable mass of vegetation.

From the structure of Sigillariæ I infer that, like Cycads, they
accumulated large quantities of starch, to be expended at intervals
in more rapid growth, or in the production of abundant fructification.  I adhere to the belief expressed in previous papers that
Brongniart is correct in regarding the Sigillariæ as botanically
allied to the Cycadaceæ, and I have recently more fully satisfied
myself on this point by comparisons of their tissues with those of
*Cycas revoluta*.  It is probable, however, that when better known
they will be found to have a wider range of structure and
affinities than we now suppose.

\*          \*          \*          \*

GENUS LEPIDODENDRON, Sternberg.—This genus is one of the
most common in the Coal formation, and especially in its lower
part.  Any one who has seen the common Ground-pine or Club-
moss of our woods, and who can imagine such a plant enlarged to
the dimensions of a great forest tree, presenting a bark marked
with rhombic or oval scars of fallen leaves, having its branches
bifurcating regularly, and covered with slender pointed leaves,
and the extremeties of the branches laden with cones or spikes of
fructification, has before him this characteristic tree of the coal
forests,—a tree remarkable as presenting a gigantic form of a
tribe of plants existing in the present world only in low and
humble species.  Had we seen it growing, we might have first
mistaken it for a pine, but the spores contained in its cones,
instead of seeds, and its dichotomous ramification, would undeceive us ; and if we cut into its trunk, we should find structures
quite unlike those of pines.  As in Sigillaria, we should perceive
a large central pith, and surrounding this a ring of woody
matter ; but instead of finding this partly of disc-bearing wood

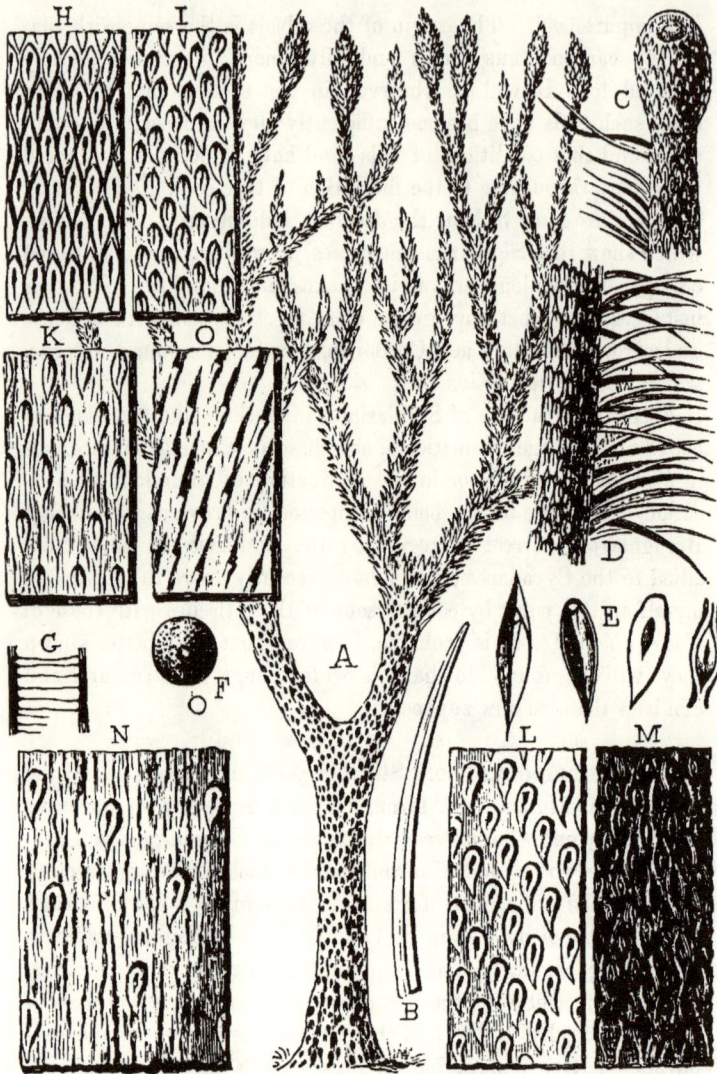

Fig. 168.—*Lepidodendron corrugatum.*

A, Restoration.
B, Leaf, nat. size.
C, Cone and branch.
D, branch and leaves.
E, Various forms of leaf areoles.

F, Sproangium.
G, Scalariform vessel, magnified.
H, I, K, L, M, Bark with leaf-scars.
N, Do. of old stem.
O, Decorticated stem (Knorria.)

cells, as in Sigillaria, and divided into regular wedges by medullary rays, we should find it a continuous cylinder of coarser and finer scalariform vessels. Outside of this, as in Sigillaria, we should have a thick bark, including many tough elongated bast fibres, and protected externally by a hard and durable outer rind. The Lepidodendra were large and graceful trees, and contributed not a little to the accumulation of coal. Several attempts have been made to divide this genus. My own views on the subject are given below.

Of this genus nineteen species have been recorded as occurring in the Carboniferous rocks of Nova Scotia. Of these six occur at the Joggins, where specimens of this genus are very much less abundant than those of Sigillaria. In the newer Coal formation, Lepidodendra are particularly rare, and *L. undulatum* is the most common species. In the middle Coal formation, *L. rimosum*, *L. dichotomum*, *L. elegans*, and *L. Pictoense* are probably the most common species; and *L. corrugatum* is the characteristic Lepidodendron of the Lower Carboniferous, in which plants of this species seem to be more abundant than any other vegetable remains whatever.

To the natural history of this well-known genus I have little to add, except in relation to the changes which take place in its trunk in the process of growth, and the study of which is important in order to prevent the undue multiplication of species. These are of three kinds. In some species the areoles, at first close together, become, in the process of the expansion of the stem, separated by intervening spaces of bark in a perfectly regular manner; so that in old stems, while widely separated, they still retain their arrangement, while in young stems they are quite close to one another. This is the case in *L. corrugatum*. In other species the leaf-scars or areoles increase in size in the old stems, still retaining their forms and their contiguity to each other. This is the case in *L. undulatum*, and generally in those Lepidodendra which have very large areoles. In these species the continued vitality of the bark is shown by the occasional production of lateral strobiles on large branches, in the manner of the modern Red Pine of America. In other species the areoles neither increase in size nor become regularly separated by growth of the intervening bark; but in old stems the bark splits into deep furrows, between which may be seen portions of bark still retaining the areoles in their original dimensions and arrangement. This is the case with *L. Pictoense*. The cracking of the bark no doubt occurs in very old trunks of the first two types, but not at all to the same extent.

*       *       *       *

GENUS LEPIDOPHLOIOS.—Under this generic name, established by Sternberg, I propose to include those Lycopodiaceous trees of the Coal measures which have thick branches, transversely elongated leaf-scars, each with three vascular points and placed on elevated or scale-like protuberances, long one-nerved leaves, and large lateral strobiles in vertical rows or spirally disposed. Their structure resembles that of Lepidodendron, consisting of a Sternbergia pith, a slender axis of large scalariform vessels, giving off from its surface bundles of smaller vessels to the leaves, a very thick cellular bark, and a thin dense outer bark, having some elongated cells or bast tissue on its inner side.

Regarding *L. Laricinum* of Sternberg as the type of the genus, and taking in connexion with this the species described by Goldenberg, and my own observations on numerous specimens found in Nova Scotia, I have no doubt that *Lomatophloios crassicaulis* of Corda and other species of that genus described by Goldenberg, *L. Ulodendron* and *L. Bothrodendron* of Lindley, *Lepidodendron ornatissimum* of Brongniart, and *Halonia punctata* of Geinitz, all belong to this genus, and differ from each other only in conditions of growth and preservation. Several of the species of Lepidostrobus and Lepidophyllum also belong to Lepidophloios.

The species of Lepidophloios are readily distinguished from Lepidodendron by the form of the areoles, and by the round scars on the stem, which usually mark the insertion of the strobiles, though in barren stems they may also have produced branches; still the fact of my finding the strobiles *in situ* in one instance, the accurate resemblance which the scars bear to those left by the cones of the Red Pine when borne on thick branches, and the actual impressions of the radiating scales in some specimens, leave no doubt in my mind that they are usually the marks of cones; and the great size of the cones of Lepidophloios accords with this conclusion.

The species of Lepidophloios are numerous, and individuals are quite abundant in the Coal formation, especially toward its upper part. Their flattened bark is frequent in the coal-beds, and their roofs, affording a thin layer of pure coal, which sometimes shows the peculiar laminated or scaly character of the bark when other

Fig. 171.—*Lepidophloios Acadianus.*

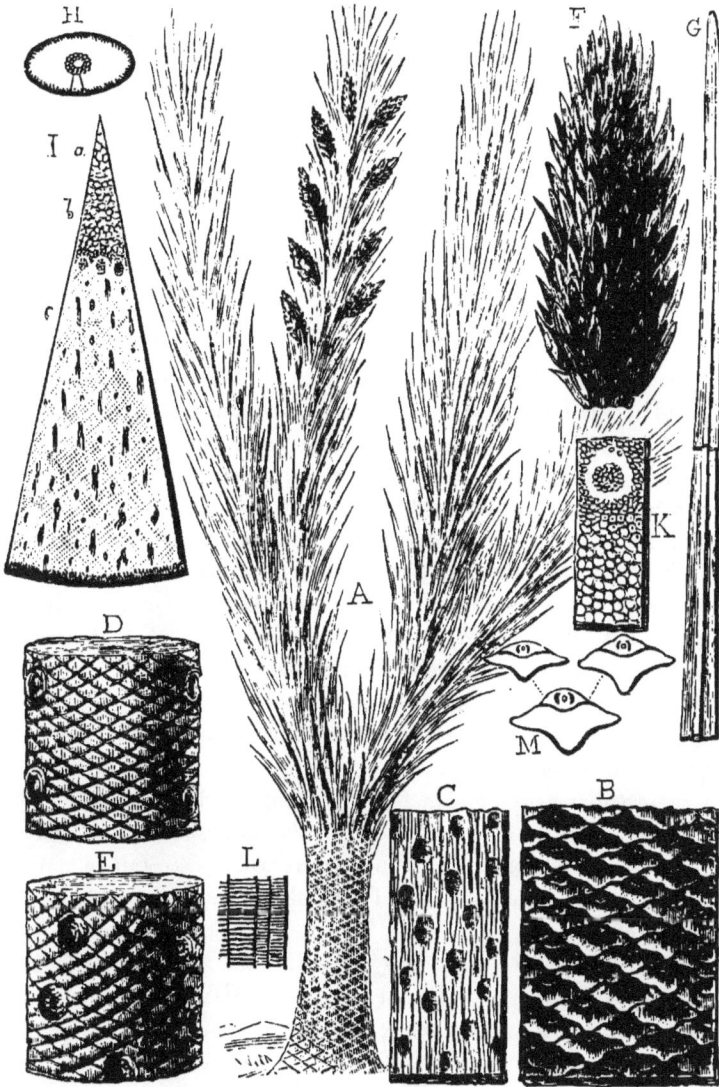

A, Restoration.
B, Portion of bark, ⅔ natural size.
C, Ligneous surface of the same.
D, Lower side of a branch, with scars of cones.
E, Upper side of the same.
F, Cone, ⅔ natural size.
G, Leaf, natural size.
H, Cross section of stem, reduced.

J, Portion of the same, nat. size, showing (a) pith, (b) cylinder of scalariform vessels, (c) inner bark.
K, Portion of woody cylinder, showing outer and inner series of vessels magnified.
L, Scalariform vessels, highly magnified.
M, Various forms of leaf scars, natural size.

characters are almost entirely obliterated. The leaves also are nearly as abundant as those of Sigillaria in the coal-shales. They can readily be distinguished by their strong angular midrib.

I figure, in illustration of the genus, all the parts known to me of *L. Acadianus.* (Fig. 171.)

---

# ON OZONE.

## A SOMERVILLE LECTURE IN 1866.

### By CHARLES SMALLWOOD, M. D., LL.D., D.C.L.

What is Ozone? Again, and perhaps, a question of greater import—more especially at the present time—What is the peculiar action and atmospheric influence, during Cholera and some other of those diseases, usually called Epidemics? This subject has engaged the attention alike of the chemist, the physician and meteorologist; to each it has presented a prolific field for investigation and research, and the subject becomes at the present time of still greater importance from the existence of cholera on the continent of Europe. As to whether cholera may visit us or not, I shall not speculate, but content myself simply to lay before you some points of interest in relation to a powerful and subtile agent, a component of our atmosphere, and which, from numerous observations, has been found to possess a wonderful influence over some diseases, and to exert some peculiar action on the lives of animals and vegetables.

The nature and composition of the atmosphere was long involved in mystery; its properties were not ascertained until chemistry and other branches of natural science were considerably advanced.

The discovery of oxygen, by Priestly, was the first-fruits of modern chemistry; and after its properties have been investigated for so many years, and in so ample and varied a manner, we are only just now beginning to find out how utterly ignorant we are of its real nature;—a substance which is the very breath of life for all created beings, both animal and vegetable, which inhabit and propagate on our globe.

In furtherance of our views on this subject, let us notice the progress of Electrical science, one which now takes its rank among the most important branches of natural philosophy, and which has made most rapid strides within the past few years; it embraces subjects curious and interesting from their close relation to almost every other branch of natural and physical investigation. It may be true that the ancients were familiar with some of its peculiar properties—that property possessed by amber, which, when smartly rubbed on a piece of linen or cloth, attracted light substances when thus excited by friction—the shock felt on touching the electric fish—and the appearance of sparks which are seen to issue from the human body under some peculiar conditions, are among the familiar and earlier examples of electrical knowledge, and it was at this period of history, and by slow degrees, that the knowledge thus acquired was reduced to something like system. That toy—the kite—which the renowned Benjamin Franklin floated under the canopy of the American firmament in June, 1752, caught from the storm-cloud the electric sparks which are now, in our day, made subservient to man, to flash our messages of commerce and daily wants along the slender pathway of a single wire.

Recent investigations have brought to light many interesting facts in connection with the sources of atmospheric electricity, which is said to have a certain bearing on the subject under our present consideration. Some of these have their origin in evaporation, which takes place constantly from the whole surface of our globe, and from the waters of the sea, lakes, and rivers; thus furnishing a constant moisture in our atmosphere, holding therein, in solution, a number of foreign substances which plants imbibe and eliminate for their own peculiar use; and it is a well ascertained fact that no electrical action takes place unless accompanied with some chemical change. Now this constant evaporation and the chemical change that is thus going on upon the surface of the earth, in the respiration of animals and plants, and the various cosmical phenomena of our globe, are supposed to be some of the sources which give rise to the generation and frequent changes of the electrical state and tension of our atmosphere. I would just allude to a theory which has a certain reference to the supposed connection between the amount of Ozone and the electrical tension of the atmosphere. It is stated that the earth is always charged with negative electricity, or that the earth is negatively electrified, and that the vapours which rise from its surface are, like itself, of a negative character; but from a constant law observed in electrical phenomena, named induction, (which is a property it possesses of producing in bodies a state opposite to its own) these particles of aqueous vapour once having left the surface of the earth, by evaporation or any other cause, become of an opposite or positive character, and are repelled in accordance with another well-known electrical law; this action of repulsion repels the positive electricity towards the upper strata of the atmosphere, carrying with it its positive character. During the night the aqueous vapour becomes condensed into dew by cold and radiation, and by the absence of the sun's rays, the amount of positive electricity in the atmosphere is diminished, and the upper vapours possess a less amount of water; the effects of heat, furnished by the rising sun, cause the dew and water to assume again its state of elastic vapour, to be again subjected to the same laws of induction and repulsion, and again placed between the negative earth and the positive celestial space. The first particles, which change from dew to the elastic state of vapour, come off the earth at a higher negative tension, which is obtained by weakening or diminishing the tension and repulsive power of the vapour they leave behind, and which has become less negative than the earth itself, thus keeping up an everchanging amount of electricity, differing both in character and tension.

It was in the year 1785, that Van Marum first called the attention of scientific men to the existence of some anomalous body, which further investigation proved to be Ozone; for he discovered, in passing the electrical spark through atmospheric air, that there was generated a peculiar and strong odour which, says he, is certainly the smell of electrical matter. For more than fifty years this fact remained forgotten or unheeded, until Schonbien, in 1839, while conducting some experiments by passing the electric current through gases, became struck with the same thought, and wrote to M. Arago, the French Astronomer Royal, that for some years he had remarked the perfect analogy that exists between the odour which is developed when ordinary electricity passes the point of a conductor into the surrounding atmosphere, and that which takes place when water is decomposed by the galvanic current.

To Schonbien, then, must be awarded the discovery of Ozone; it was he who gave it its present name, taken from a Greek verb which signifies to give out an odour, but the name reveals nothing of its real nature,

It is not my purpose to enter into a very long and argumentative chemical reasoning on the composition of Ozone. Some difference of opinion still exists as to its present character. Schonbien looked upon it as a regular constituent of our atmosphere, forming a part of, and always present in the air we breathe. I might casually mention that Cavendish, more than half a century ago, found, what he stated was nitrous-acid, present in atmospheric air, and he attributed the beautiful green colour of plants, after a thunderstorm, to a chemical combination of ammonia and nitrous-acid, making a nitrate of ammonia. This effect upon plants, after thunderstorms, is now referred to the effects of Ozone in increased quantity.

The absolute and uniform composition of Ozone has been the subject of much controversy. Schonbien claimed it as a binoxide or peroxide of hydrogen. Faraday denied this, and considered Ozone as oxygen in an isomeric state, or as a simple modification of oxygen in an allotropic condition of that body. Williamson says that, according as Ozone is produced by a galvanic battery, developed by the electric spark, or brought forth by the action of phosphorus on atmospheric air, it is a peroxide of hydrogen and azotic-acid, or a mixture of both. Berzelius opposed this idea, and went to show that Faraday was correct. De la Rive and others stated that it was only oxygen in a peculiar condition given to it by electricity. Freney and others instituted experiments to confirm their ideas, and went on to state that the presence of Ozone would not be developed unless the oxygen was electrified,—for it was shown that in the presence of oxygen alone, or electricity singly, no development of Ozone took place, but as soon as the oxygen became electrified, Ozone became manifest; they placed a strip of test paper in a glass filled with oxygen and hermetically sealed, and by means of metallic bulbs at each end, electric sparks were made to flash across, or through the volume of oxygen; the result was, the test paper immediately became blue, indicating the presence of Ozone.

Test papers have been suspended in oxygen for ten days without any apparent change, but when electrified at the end of that time, they became blue, thereby indicating the presence of Ozone. Test papers of the same quality have been placed in a vacuum, and when the electric spark has been passed through it, no change of colour in the test papers took place, but the moment oxygen gas was introduced, and the otherwise same conditions were fulfilled, the test papers showed the presence of Ozone, thus demonstrating that neither electricity nor oxygen alone, was sufficient to cause any change in the test papers. From these facts it has received the name of electrified oxygen.

Ozone can be made artificially by taking a piece of phosphorus, about half an inch long, cleaning its surface by scraping, putting it into a clean quart bottle, and adding as much water as will cover half the surface of the phosphorous; close the bottle with a loose fitting stopper, and set it aside at a temperature of about 60° Fahrenheit; Ozone will soon then begin to form in the bottle, and in five or six hours it will be abundant. Remove the phosphorus, shake a little water in the bottle, and throw this out to remove the phosphoric acid. This washing must be repeated several times; the Ozone will not be washed away but will remain with the atmospheric air in the bottle. Oil of turpentine, exposed to the sun's rays, in a bottle, partly filled, will also generate Ozone; also some other chemical combinations. The chemical agencies of magnetism and galvanism evolve Ozone, and a current of electricity passed across the surface of water produces it. It might be stated in reference to the formation of Ozone by phosphorus, that the atmospheric air in the vessel should be of the average barometrical pressure, and of a temperature not under 50° or over 90°, for Ozone is not formed in this artificial way at zero Fahrenheit. The formation becomes very rapid at 75° Fahrenheit. It is also formed by the ordinary electrical machine in rapid motion, when the electric fluid is evolved from the conductor—which fact, as before stated, led to its discovery. It may also be formed in various other ways, but enough for our present purpose. When formed by the decomposition of water by means of the galvanic pile, Ozone is always manifest at the positive pole.

I shall now proceed briefly to state the means used to ascertain its presence, and its amount in the atmosphere. The method of detecting its presence is by means of a combination of the iodide of potassium and starch. Take one part of iodide of potassium, ten parts of starch, and 100 parts of water; boil the starch with the water, allow the water to cool, and stir intimately with it the iodide of potassium; then spread the mixture on slips of good glazed paper by means of a soft brush or a sponge. My experience is that good glazed or sized paper is preferable to bibulous or blotting paper. Cream-laid post has been used by me for years; but I have since found that strips of well washed calico, after dipping them in the solution and smoothing their surfaces, answers better than paper; the calico seems more readily to absorb any moisture present, and also to retain it better than the paper, and for experiments will be found better suited for the purpose than paper slips. The exposure of these tests, free from rain, but placed in the light, causes them to become first a pale straw colour, increasing to the tint of a dried leaf, then a deep brown or dark violet approaching to black, which being wetted with pure water resolves into a blue. The decomposition which takes place in these tests is owing to the fact that the Ozone acts similarly to an acid, uniting with the potassium forming potash, and a portion of the iodine is set free, which unites to the starch, giving the peculiar blue colour just alluded to; the starch is only used to estimate the amount present by the depth of colour, and this test is sometimes called an Ozonoscope. The amount is measured from 0 to 10, the different degrees of shade indicating its amount, 10 being the deepest shade. Dr. Moffatt advises that the test papers be placed free from light, but having a free access to air; I have followed both these methods, and the results are nearly alike. Should there be a great amount of moisture in the atmosphere, the exposed test paper attains at once its blue colour, which becomes brown as it dries, but the blue colour may be again attained by moisture or re-wetting with water. Ozone is colourless, possessing a peculiar odour. resembling chlorine, and when diluted cannot be distinguished from the electrical smell; its density is said to be four times that of oxygen; it is a most powerful oxydizing agent, converting most of the metals into peroxides; it is very slightly absorbed by water after long contact; a very high temperature destroys its properties; it possesses bleaching qualities—hence its affinity to chlorine; it combines with chlorine, bromine and iodine; it is also rapidly absorbed by albumen, fibrine, blood, and other such like solutions. It is a most powerful disinfectant, and when even largely diffused in atmospheric air causes difficult respiration, acting powerfully on the mucous membrane, and in still larger quantities may prove fatal. Its presence is easily detected in the state produced in the laboratory as well as the atmosphere; its rapid production, its peculiar smell and other marked properties, render it somewhat less difficult to investigate than many other substances.

Winds influence the amount of Ozone, the amount depending upon the quarter from which they come, and in some cases on their velocity : easterly and southerly winds may be called ozonic winds, while westerly and northerly winds barely ever indicate a trace.  Rain and snow generally give indications of a large amount.  A N.E. land wind does not generally indicate Ozone; whenever there is Ozone in a N. E. wind it may be attributed to the sea-breeze passing over the land, for we have very often, in this vicinity, a dry N. E. wind with a very high barometer for some days, with no indications of Ozone.  Atmospheric temperature does not seem to influence the amount;  I have observed its presence at some 30° to 40° below zero, and at 98° above zero, Fahrenheit.

The variation in its daily amount, has been the source of some discussion.  Observations were carried on for some years at the Isle Jesus observatory, by means of a movable ozonometer, time being taken as an element ; the strips of calico were by a simple contrivance passed over an opening exposed to light and air at the rate one inch per hour.  From upwards of 3000 observations, tending to confirm this important point, it was found that the increase and decrease of the daily ozonic periods corresponded in a striking manner to the bi-daily variations of the atmospheric humidity.  There were also some slight fluctuations corresponding in a marked degree to the bi-daily variations of the barometer.  Upwards of 20,000 observations on Ozone have been taken and recorded, and I am ashamed to say, unaided, thus depriving us of any means of comparison, or confirmation ; but I can but express a wish that brighter and better days will come in the future, and that observers will not be found wanting to set at rest the important problem of the effects of the absence or presence of Ozone on the health of animals and vegetables.  Assuredly, a substance which has been found to exert an important bearing on the health of individuals, and upon the agricultural and commercial wealth of nations, demands from men of science a calm and patient investigation.  It requires, for its due prosecution, a systematic method of recording its amount ; it is for common purposes observed twice in twenty-four hours, and a mean of the two observations is recorded, and also a register of rates of disease and mortality, and a correct register of the nature of these diseases ; these of course must be simultaneous with the usual meteorological observations, of atmospheric pressure, temperature and humidity, the force and the direction of the winds, and such like conditions.

It has been stated that the higher we ascend the greater the amount of Ozone found present in the atmosphere. For many years past observations were taken at the Isle Jesus observatory, with an ozonometer hoisted nearly 80 feet high, but the observations at that altitude yielded no different results from those taken at five feet from the surface of the soil. [I might mention, the height five feet is now considered a standard one for observation ; it is, probably, at that distance, removed far enough, from the earth, to prevent the action of moisture which is emitted at the surface]. At very high altitudes, as it would appear from Glaisher's balloon experiments, a very trifling difference was apparent, much of course depending upon the wind and its direction; and if it is to be received as a general law that there is always a westerly current of wind in the higher regions of the atmosphere produced by the rotation of the earth on its axis, it is not probable that any great increase in amount would be found, as westerly are not generally known as ozonic winds.

Captain Jansen, of the Dutch Navy, in a voyage to Australia, confirms the assertion as to the ozonic winds, he says:—That in the Northern hemisphere those winds which have a southing in them are more abundant in Ozone, and that in the Southern hemisphere, those winds which have a northing in them are those more abundant in Ozone; and he further says:—That the Equatorial calm belts, with their thunder and lightning, constant rain and moisture, may well be said to be its birth-place.

So far as there is any connection between the amount of Ozone coinciding with the variations in the amount and kind of atmospheric electricity, I would beg leave to state, that from some 6000 observations taken at the Isle Jesus observatory simultaneously with the various electrometers and other apparatus connected with the investigation of atmospheric electricity, no apparent connection was evident between the amount of Ozone and the changes in the tension and kind of electricity.

In passing to the next part of the subject—its influence on some epidemics—it might be observed that epidemics generally are said to be generated by miasmata, a term used for designating a highly important class of febrific agents of a gaseous form, which act on the animal system through the medium of the atmosphere. This class of agents is generally divided into two orders: First, infectious — comprehending those febrile effluvia which are generated by the decomposition of vegetable and animal matter; Second, æriform contagious, generated by the animal system in a state of disease.  First, infection may result from the humid decomposition of vegetable and animal matter, contained in the filth of cities, in marshes, and some soils furnishing these materials, hence the designation marsh-miasma.  Second, it may result from the decomposition and natural exhalations and excretions of the human body, under ill-conditioned circumstances; to this has been applied the term idio-miasma, expressive of the personal or private character of its source.  Marsh-miasma has also received the name of malaria.  Much has been written of malaria but little of its true nature is understood, although it is supposed to be the effluvia that generates fevers, cholera, and such like diseases; many physicians of eminence have written elaborately on the subject—but after all, very little is really known of its subtile influence.

Here is a picture drawn by Dr. Macculoch:—"The fairest portions of Italy are a prey to the invisible enemy, malaria—its fragrant breezes are poison, the dews of the summer evenings are death.  The banks of its refreshing streams, its rich and flowery meadows, the borders of its glassy lakes, the luxuriant plains of its overflowing agriculture, the valleys, where its aromatic shrubs regale the eye and perfume the air, these are the chosen seats of this plague—the throne of malaria.  Death here walks hand-in-hand with the resources of life, sparing none.  The labourer reaps his harvest but to die, or he wanders amid the luxuriance of vegetation and wealth, the ghost of man, a sufferer from his cradle to his impending grave; aged even in childhood, and laying down in misery that life which was but one disease.  He is driven from some of the richest portions of this fertile, yet unhappy country; and the traveller contemplates, at a distance, deserts—but deserts of vegetable wealth—which man dares not approach, or he dies." Whatever is its composition, it may be enough for us to know that its existence in the atmosphere is incompatible with health. Now, Ozone is said to destroy this malaria; no deleterious substance is found in the atmosphere where Ozone is manifest, for one of the peculiar properties of Ozone is, its disinfecting powers; putrid meat exposed to the action of ozonized air soon becomes disinfected.  Manure heaps and foul drains, where there is decomposition going on, become quite innocuous: and it has been shown that when putrid organic matter is subjected to the action of

Ozone, the bad odour is destroyed as long as the ozonometer gives evidence of the presence of Ozone, but as soon as the ozonometer ceases its indications, the odour immediately returns. Schonbien's experiments proved that air containing one-6000th part of Ozone can disinfect 540 times its volume of air from putrid meat. Apartments are now being purified by means of Ozone; and during the visitation of cholera, last summer, in London, Ozone was extensively used as a disinfectant. Pieces of phosphorus were also suspended over the gratings of the sewers, so as to generate Ozone and neutralize the spread of the choleraic-contagion. It is here necessary to remark that the phosphorus must be luminous to produce Ozone, and the height of the barometer and the degree of temperature must be taken into account; even the direction of the wind has some influence on its development.

It is a matter of history that, in 1854, cholera visited many cities of the old world and of the new. It has been asserted, and that by numerous observers, that during this visitation, there was always indicated a deficiency of Ozone in the air; and further, that the increase or decrease of cholera coincided strictly with the development or absence of this mysterious substance.

Below is a table shewing for seven years the comparative day of precipitation (rain or snow) each year, and the amount of Ozone indicated, in quantity more than five-tenths of the scale.

| | | | | | | | | |
|---|---|---|---|---|---|---|---|---|
| 1850 | there were | 106 | days of precipitation and | 110 | days of ozone in more than $\frac{5}{10}$ |
| 1851 | do. | 123 | do. | 136 | do. |
| 1852 | do. | 136 | do. | 135 | do. |
| 1853 | do. | 156 | do. | 114 | do. |
| 1854 | do. | 133 | do. | 73 | do. |
| 1855 | do. | 140 | do. | 110 | do. |
| 1856 | do. | 144 | do. | 126 | do. |

Shewing the comparatively small amount of ozone in the year 1854, the year this cholera was prevalent.

A commission of the members of the Medical Society of Strasburgh, during the visitation of cholera in 1854, was named for testing the subject, and their united report was:—That during the days that Ozone was deficient in the atmosphere, cholera was at its greatest rate of mortality. From observations taken at Isle Jesus observatory and carefully compared with the death rates in Montreal, and the country parts visited by the epidemic in 1854, this opinion was certainly confirmed. At Newcastle, in England, during the prevalence of cholera, in 1854, Ozone was at its minimum; in London, in the same year, from the 24th of August until the 11th of September, Ozone was only present

once, and then in a minute quantity, and cholera was at its height
during that period. On the 11th of September, a southerly breeze
set in, with indications of Ozone, and from day to day the number
of cases diminished. In a paper, read by me in Montreal, before
the American Association for the Advancement of Science at their
meeting in 1857, I stated that moisture in the atmosphere
was necessary for the development of Ozone; this opinion has
been opposed by the only American observer, Captain Pope, during
some journeys that he made across the great plains in 1856 and
1857. He says:—" Ozone increases in quality, rapidly and regularly, in receding from the low lands which border the Gulf of Mexico,
and is greatest on the table lands of the interior"; he goes on further
to state that on the low lands animal and vegetable decomposition
is very rapid, and on the table lands very slow and with little
escape of offensive gases—therefore, on account of the moisture in
. the low lands, there should be more Ozone developed than in the
table lands. But another cause must, with all deference, be
brought to bear on the observations of Captain Pope, and it is a
very important one : for as already shown, there is a considerable
amount of fever and malaria in these wet, low lands, hence the
deduction that Ozone has been partially destroyed by the malaria,
consequently a less amount was indicated by the ozonometer on
the low lands than on the higher table lands. These reasons will
account for Captain Pope's observations, without in the least dis-
paraging the theory, that moisture is necessary for the development of Ozone. The fact, that a humid state of the atmosphere
better developes Ozone, is confirmed by the observation of Dr.
Moffatt, Mr. Lowe, and other Europeans, who have paid attention
to the subject. I shall read a short extract from my 1857 paper,
showing the amount of precipitation as a test for determining its
presence in the atmosphere, and the amount of Ozone corresponding to the days of precipitation; and showing, also, the
diminished quantity of Ozone during the months of July, August,
and September, 1854, which were the months of the greatest
mortality during that visitation of cholera in this neighbourhood. During the visitation of cholera, in most places there
were high readings of the barometer. In 1854, here, the mean
reading for the month of July was 29.961 ; for August, 29.910 ;
and for September, 30.201 inches—the lowest reading during the
period was 29.619. The thermometer also ranged high—the
mean temperature for July being 76.2, and for August 68.31 ; the

dryness of the atmosphere for July was .709, and for August, .714—taking saturation as 1.000—with which number at 9 P.M. on the 11th of August, the thermometer even stood at 76°. There was a haze in the atmosphere, which led to the supposition of fires in the woods being the cause; the weather was calm, and the wind north-westerly, but very light. There was a great thunderstorm at Isle Jesus on the 6th of September, from 6 to 8 P.M., and a slight frost occurred on the morning of the 11th, and snow fell at Quebec on the 21st. The ozonometer, soon after these meteorological events, indicated its usual amount. On the other hand, influenza and pulmonary diseases, when prevalent, are accompanied by a high amount of Ozone, while all gastric diseases, diarrhœa and its allies are accompanied by a decrease in the average amount. The air coming from the sea shows a high amount of Ozone, and it is presumed that it is this property that makes the sea-breeze so beneficial to health. It is a direct stimulant to animal and vegetable life, and it must be borne in mind, that a 2000th part of Ozone in the atmosphere would make it fatal to small animals, and a little more than this would be fatal to man in an atmosphere which gives the maximum number 10 in the ozonoscope or ozonometer; Ozone only exists in the proportion of 1 to 10,000 parts of atmospheric air. When considering the source of Ozone it would seem reasonable to suppose that there should be but little of this agent manifested in the atmospheres of large and crowded cities; repeated experiments have proved this to be the case. In such cities there is always a large consumption of Ozone going on; on the contrary, in the pure air of the country, and at the sea-side, Ozone is generally abundant, and the consumption is manifestly less. There is, indeed, a marked difference between the amount observed at my own residence, which is not in a crowded part of the city, and at the observatory in McGill College grounds. Ozonometers placed in the wards and halls of hospitals give no trace of Ozone, while at the exterior of these buildings a reasonable amount is indicated, shewing that the atmosphere of a city, where large numbers are dwelling together, tells largely on the consumption of this peculiar body, and it must be self-evident that any thing tending to its conservation, such as good and efficient drainage, free currents of air and plenty of ventilation, will directly contribute to the health of cities; and the removing of the causes of its consumption, if not destruction, is the paramount duty of every citizen; and it is thus to the interest of the rich to aid the poor by a cheerful submission to such taxes as may be necessary for the proper cleansing and scavengering our city. It has been beautifully put by one of England's favourite writers:—"That the universal diffusion of common means of decency and health is as much the right of the poorest of the poor, as it is indispensable to the safety of the rich, and of the State; that a few petty boards and corporate bodies —less than drops in the great ocean of humanity around them—are not for ever to let loose fever, malaria, and consumption on God's creatures at their will, or always to keep their jobbing little fiddles going, for a Dance of Death."

Chemical and physical agents produce Ozone, while the decay of vegetable and animal matter consumes it, and when the balance is destroyed between its production and consumtion, disease is the consequence. Ozone is apparent in large quantities in the pine-forests of America, and but few of the diseases arising from malaria exist in their neighbourhood, except where marshes are numerous—their exhalations, under a tropical sun, producing what is termed marsh-miasma. Ozone is generally found to exist in larger quantities in the winter than in summer—more particularly in Montreal, because there then is a much less decomposition of animal and vegetable matter.

Ozone in excess has been found to prevail when disease of the lungs and catarrh are in the ascendant; it has been frequently remarked that easterly winds aggravate these diseases. Dr. Beckel, jr., of Strasburgh, selected cases suffering from pulmonary, bronchial, and heart diseases, carefully comparing the numbers admitted into hospital through a long period of time, and by the fluctuation of the ozonometer, and the variation of the temperature, he came to the conclusion that pulmonary diseases are in adverse relation to the quantity of Ozone, and in reverse relation with the degree of temperature. When there is much Ozone with a low temperature, such diseases increase, and death often ensues; whereas, when there is little Ozone with a high temperature, the contrary occurs. Scoutetten's tables show similar results. Schonbein states, that in Berlin a diminution of atmospheric Ozone coincides with the production of gastric disorders, and that there was a complete absence of Ozone in that city, during the invasion of the cholera, and that indications of Ozone in large quantities give rise to pulmonic affections.

Persons interested in the bleaching of linen fabrics have of late directed attention to the amount of Ozone in the atmosphere, and have been induced to keep daily registers of its amount, so that it would seem that it has an important bearing upon our economic wants. Experience shows that upon days when Ozone was present in large quantities, the bleaching was better accomplished; and from experiments carried on in this department, it has been proved that our test papers rather underrate the amount of Ozone absolutely present. The bleaching properties of Ozone have been carried out, still further, for restoring books and prints that have become brown by age and exposure to the light, or have been soiled or smeared with colouring matter—a short time only being required to render them perfectly white, as if just issued from the press, and this without the slightest injury to the blackness of the printer's ink, or the lines of a pen and ink sketch or crayon drawing.

Writing ink may readily be discharged by Ozone, if the paper be subsequently treated with chlorohydric acid to remove the oxide of iron. Vegetable colouring matters are completely removed by it; but it does not act so readily on metallic colouring matters or on grease spots.

Much still remains to be said on this interesting subject. I trust the day is not far distant when it will receive from the scientific world the attention which is due to its great importance as bearing on the health and welfare of the whole community, and that observers will not be wanting to aid in carrying out the important objects embraced in its study.

---

## ON THE AZOIC AND PALÆOZOIC ROCKS OF SOUTHERN NEW BRUNSWICK.

### By F. G. MATTHEW.[*]

While exploring with my brother, Mr. R. Matthew, the Manganese district of King's County, in the summer of 1866, we made some observations on the geology of this County, having an important bearing on the subject of the article above named.

HURONIAN.—A more extended examination than had pre-

---

[*] Supplementary note to my paper in the Journal of Geol. Society of London, vol. xxi., p. 422.

viously been given to the Cambrian rocks in the Quaco Hills, led to the discovery of an important part of this series not previously recognized as sedimentary; it consists of shales, grits, and conglomerates, usually highly metamorphic, so much so, as in general to have lost all traces of stratification. In this condition they appear to be syenites, granulites and felsites, all highly coloured by the bright red felspar of which they are chiefly composed. Masses of these rocks were observed by our party, in 1864, on the Hammond River, and in the adjacent hills, but their sedimentary character was not at that time recognized.

With this addition the grand lithological features of the older supra-Laurentian rocks in the Southern Hills of New Brunswick appear to be:—

LOWER SILURIAN.—The lingula bearing flags and shales of St. John, etc., at the base of which the primordial fauna occurs.

HURONIAN.—Red sediments of comparatively small volume, perhaps not recognizable in other parts of Acadia. ( No. 5 in article on Azoic Rocks.)

Dark coloured trap-slate rocks (Nos. 2 and 4, art. cit.) of great thickness; parted about midway by a rusty-colored calcareo-arenaceous zone charged with iron and manganese. ( No. 3, art. cit.).

Red sediment, usually converted into red felspar rocks, also of great thickness, resting upon the Laurentian series ( No. 1 of article on Azoic Rocks is here included). The felsites referred to (No. 3, in my article,) may be of this lower horizon, but I have not been able to verify this point. The succession throughout this immense series of beds is greatly obscured by faults. An instance is given at page 28, of Mr. Bailey's Report.

It is noteworthy that the core of the Northern Highlands of New Brunswick consists, in a great degree, of red felspathic rocks (vide Bailey's Notes on Geology and Botany of N. B., Can. Nat.), and that these are flanked by metalliferous slates, frequently of a dark brown colour, which may be of the same age as the main portion of the Huronian in the south (Nos. 2—4) above noticed.

The resemblance of the Lower Silurian of Saint John to the gold and antimony bearing slates of the central part of the Province has been already noticed in the article cited above. Thus the Northern metamorphic region may present a full representation of the older Palæozoic series in the Southern Hills.

There is a large area of red felspar rocks in northern Cape Breton, and masses of a similar character in Charlotte Co., N. B., both of which may prove to be Lower Cambrian.

It will be seen that these views are partially at variance with conjectures offered in the last paragraph of page 428, and on page 427; the latter should be applied to the southern band of Cambrian slates (yielding gold and antimony) only.*

LOWER CARBONIFEROUS.—There is a great development of this formation in the area N. and N. E. of the Quaco Hills, drained by the Kennebeckasis and Petticodiac Rivers. The following succession (see wood-cut,) observed on the slopes of these hills, and in the lower valleys parallel to them, are beds, collectively, of very considerable thickness, but some of them vary much in bulk in other parts of this tract. They represent, as nearly as can be judged without actual measurement, the thickness of the formation in eastern Kings County.

Nos. 1 to 5 are much attenuated in the western part of this L. C. district, and have not been detected west of Hammond River valley. In this western quarter also the upper members, especially 6, 7, and 8 have a more considerable thickness than elsewhere. The first of these (6) is much reduced in bulk about the middle of the area; and 7 changes its character or disappears entirely in the east.

The limestone and gypsum beds are but a small part of this voluminous series, in which we were unable to find more than one calcareous horizon; the other outcrops of these rocks in the valley appearing to be merely repetitions of the same beds thrown up by faults.

In No. 6 the salt springs of Sussex and Upham occur. No. 4 is rich in manganese derived from the Cambrian rocks, upon and against which much of the lower carboniferous sediments of this tract rest.

Nos. 6 and 8 have complimentary characters in different parts of it; thus, the first towards the east has much bright-red sandstone, but on the Lower Kennebeckasis it is mostly chocolate coloured, and largely made up of thick shale beds, while the converse holds in regard to No. 8. The general prevalence of

---

* Observations made for the Canadian Survey during the past summer indicate that much of the slate country of the interior may be of Upper Silurian or Lower Devonian age.—Oct. 1868.

chocolate coloured rocks appears to be due to the presence of oxides of iron and manganese, derived from the Huronian system in the adjacent hills.

### LOWER CARBONIFEROUS SERIES IN KINGS CO.

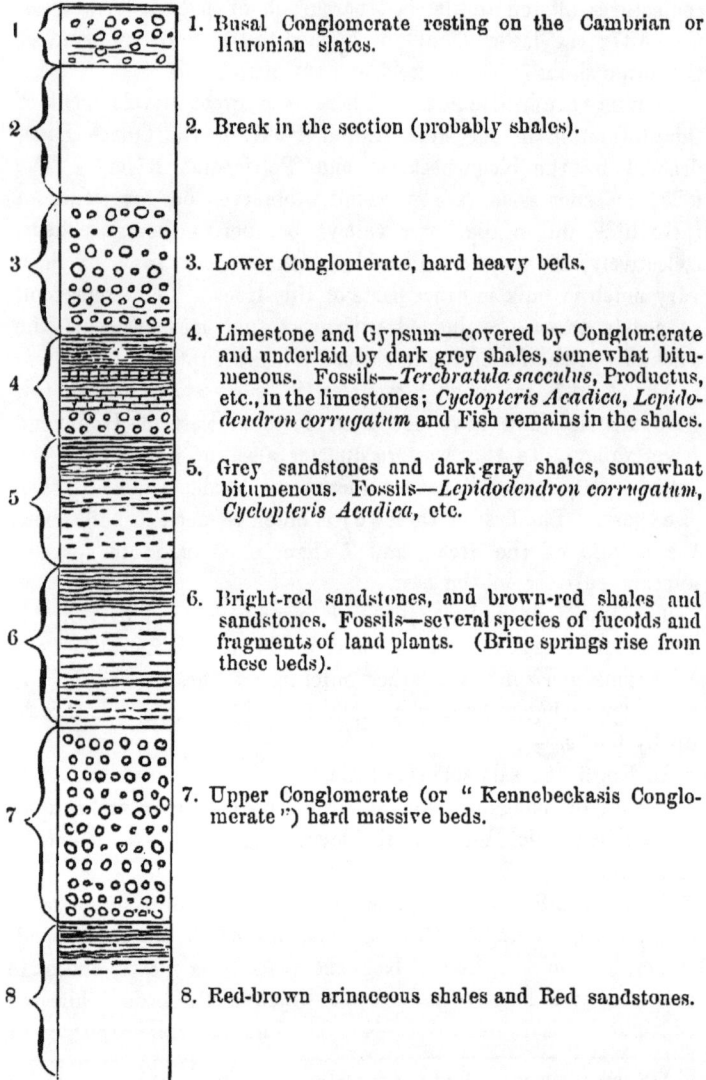

1. Basal Conglomerate resting on the Cambrian or Huronian slates.

2. Break in the section (probably shales).

3. Lower Conglomerate, hard heavy beds.

4. Limestone and Gypsum—covered by Conglomerate and underlaid by dark grey shales, somewhat bituminous. Fossils—*Terebratula sacculus*, Productus, etc., in the limestones; *Cyclopteris Acadica, Lepidodendron corrugatum* and Fish remains in the shales.

5. Grey sandstones and dark-gray shales, somewhat bitumenous. Fossils—*Lepidodendron corrugatum, Cyclopteris Acadica*, etc.

6. Bright-red sandstones, and brown-red shales and sandstones. Fossils—several species of fucoids and fragments of land plants. (Brine springs rise from these beds).

7. Upper Conglomerate (or " Kennebeckasis Conglomerate ") hard massive beds.

8. Red-brown arinaceous shales and Red sandstones.

Nos. 4, 5, and 6, which are comparatively soft, are frequently seen on the slopes and at the bottom of valleys of erosion, formed between the hard conglomerates of Nos. 3 and 7. These softer members also yield the elements of the fertile loamy soils, for which the valleys of Kings County are famous.

Along the margin of the great central coalfield, these " Lower Coal-measures" * are much reduced in bulk; and volcanic outbursts have left traces of their presence in that quarter, at epochs corresponding to those marked by the spread of conglomerate beds (Nos. 3 and 7) among the Southern Hills.     See Prof. Bailey's Report, page 98.

The following changes in that part of my article which relate to this formation, will bring it into accord with the preceding remarks† :—

Page 431, line 11, for "which may represent" read " of later origin than"

"     "   29, for "at or near" read " not far from".

---

SEA-WEEDS IN MEDICINE.—The genus *Laminaria* consists chiefly of large plants growing abundantly in deep water. They are very rich in iodine, chlorine, sulphur, silica, lime, potash, and soda. They are burnt in large quantities on the French shores of the British Channel and Atlantic, and produce the best raw soda from which iodine is afterwards extracted. There are three species :—*Laminaria digitata, L. saccharina,* and *L. bulbosa ;* and these almost exclusively yield the 70,000 kilogr. of iodine annually brought into the market. There are also other algæ such as *Fucus vesiculosus, F. nodosus, F. serratus,* etc., which generally yield bromine. The inhabitants of the Cordilleras of the Andes were in the habit of using the decoctions of sea-weeds, in cases of scrofula, wens, and lymphatic tendencies. These liquids are, however, very unpalatable; to avoid which M. Moride proceeds as follows:—The plants are slightly rinsed in fresh water, then dried and exposed to the sun, whereby they lose their smell and taste of wrack; after which they are pounded in a mortar and macerated in strongly alcoholized water at a somewhat high temperature. The iodized tincture thus obtained is found useful in all affections for which iodine is prescribed.—*Ex.*

---

* Dawson.—Synopsis of the Flora of the Carboniferous period in Nova Scotia.

† Journal of Geological Society of London, Vol. xxi.

## NATURAL HISTORY SOCIETY.

### REOPRT OF THE COUNCIL TO THE ANNUAL MEETING OF THE NATURAL HISTORY SOCIETY, MAY 18, 1867.

The Council begs to congratulate the members on the more hopeful condition of the Society in many of its aspects.

#### MEMBERSHIP.

During the last year, twenty additional ordinary members have been elected; but as ten of these have been proposed as life members, the real addition from this source only amounts to ten.

In order to meet the increased expenses of the Society, it has been agreed, after mature and frequent deliberation, to raise the subscription from four dollars to five dollars per annum. It will be an important branch of the labours of the incoming Council to endeavour to increase the list of ordinary members, as the working revenue of the Society depends principally on this source.

Two new life members have been added to the Society; but they regret to record the decease of one, Mr. W. H. A. Davies, who was also a Vice-President. The number of life members is now forty-one, which will shortly be increased by ten of the ordinary members, as above noted. The payments received from life members will now be $50 instead of $40 as before.

A new bye-law has lately been passed admitting ladies to the privileges of the Society as Associate Members, on payment of two dollars per annum. Thirty names have already been proposed; and if members will exert themselves to add to this good beginning, the income will not only be increased, but the attendance at the meetings, the visits to the Museum, and the general interest felt in the concerns of the Society will receive a very healthy augmentation. It is hoped that this new source of income may more than counterbalance the loss incurred by the transference of many names from the list of ordinary to that of life members,—a change which otherwise would be of questionable benefit to the Society.

#### FINANCE.

The present income from ordinary and associate members may be stated at $800. The Society is still under great obligations to Mr. Ferrier for his valuable services as Treasurer. The financial position during the past year is set forth in the balance sheet herewith presented.

Dr.    THE NATURAL HISTORY SOCIETY OF MONTREAL, IN ACCOUNT WITH JAMES FERRIER, JR., TREASURER.    Cr.

| | | |
|---|---|---|
| 1866, May 1. | | |
| To Balance due the Treasurer................................ | $188 | 51 |
| 1867. | | |
| Cash paid, J. F. Whiteaves, salary........................ | 300 | 00 |
| "   Wm. Hunter,   do. ........................................ | 200 | 00 |
| "   Commission for collecting............................... | 31 | 70 |
| "   Interest.......................................................... | 243 | 75 |
| "   for Wood and Coal......................................... | 166 | 86 |
| "   Gas Accounts................................................ | 42 | 72 |
| "   Water Rent.................................................... | 40 | 65 |
| "   City Taxes..................................................... | 40 | 00 |
| "   Insurance...................................................... | 39 | 00 |
| "   Repairs, and petty expenses........................... | 107 | 12 |
| "   Post Office Accounts...................................... | 14 | 50 |
| "   Printing......................................................... | 17 | 15 |
| "   Advertising.................................................... | 11 | 58 |
| "   for Specimen Cases........................................ | 110 | 00 |
| "   Specimens purchased by Mr. Whiteaves........... | 24 | 75 |
| May 1. | | |
| To Balance in Treasurer's hands......................... | 128 | 97 |
| | $1707 | 26 |

| | | |
|---|---|---|
| 1867. | | |
| By Cash, Government Grant............................... | $750 | 00 |
| "   Donations towards liquidation of debt:— | | |
|        Wm. Molson, Esq.....$100 | | |
|        Thos. Rimmer, Esq....  50 | | |
|                                              | 150 | 00 |
| "   Members' yearly subscriptions....................... | 640 | 00 |
| "   Museum entrance fees.................................. | 18 | 00 |
| "   Rent of Lecture Room.................................. | 52 | 50 |
| "   Proceeds of Conversazione........................... | 96 | 76 |
| | $1707 | 26 |

JAMES FERRIER, Jr., Treasurer, N. H. S.

## STATEMENT OF LIABILITIES, MAY 1ST, 1867.

| | | |
|---|---|---|
| Open Accounts.................................................. | $ 159 | 45 |
| Mortgage on Society's Building, favour Royal Institution.......... | 2000 | 00 |
| Do   do   do   Wm. Watson, Esq........................ | 400 | 00 |
| | $2559 | 45 |

E. & O. E.,
Montreal, May 1, 1867.

After considerable discussion on the liabilities of the Society, it was determined, by an appeal to the public, to raise a special fund to defray the debt incurred by the Building Committee, now amounting to $2,400; and, if possible, to increase the Library and Museum. The object was announced by the President at the Conversazione; the appeal has been printed and circulated, and a special Collecting Committee appointed. It was decided that all subscribers of $50 or upwards to this fund should be recommended as life members, or be able to nominate a friend if they were themselves qualified. The Council earnestly recommend that this most important committee be re-appointed. The subscriptions already promised amount to $1,430, of which the following is a list :—

SUBSCRIPTIONS for the Liquidation of debt owed by the Natural History Society of Montreal, and thereafter for the improvement of its Museum and Library.

| | | | |
|---|---|---|---|
| Mr. John Frothingam | $100 | Mr. T. Macfarlane | $50 |
| — William Molson | 100 | — Champion Brown | 50 |
| — Thomas Workman, M.P. | 50 | — John Swanston | 50 |
| — William Workman | 50 | — Alexander McGibbon | 50 |
| — Thomas Morland | 50 | — Jas. Ferrier, Jr. | 50 |
| — Peter Redpath | 50 | — T. J. Claxton | 50 |
| — John W. Molson | 50 | — F. J. Claxton | 50 |
| — George Barnston | 50 | Rev. A. DeSola, LL.D. | 50 |
| — J. Henry Joseph | 50 | Rev. Canon Balch, D.D. | 20 |
| — Thomas Rimmer | 50 | Sir W. E. Logan, LL.D., F.R.S. | 50 |
| — G. A. Drummond | 50 | Mr. E. Billings, F.G.S. | 25 |
| — William Muir | 50 | — J. F. Whiteaves | 25 |
| — William Ewan | 50 | — A. S. Ritchie | 20 |
| — John Leeming | 50 | — Jas. Ewan | 20 |
| — W. Fred. Kay | 50 | — Jno. Lovell | 20 |

### PUBLIC LECTURES.

The yearly course of the Somerville Free Lectures was delivered last winter as follows :

1.—General Sketch of the Gasteropodous Mollusks. By P. P. Carpenter, B.A., Ph. D.

2.—On the Chemistry of the Stars. By J. B. Edwards, Ph. D., F.C.S. •

3.—On the Origin of Continents. By the President.

4.—On the Anatomy of the Common Sea-Urchin. By Principal Dawson, LL.D., F.R.S.

5.—From Granite to Basalt. By Mr. T. Macfarlane.

6.—On Coleoptera. By G. P. Girdwood, M.D.

In consequence of the interest excited by the very beautiful experiments made by Dr. Edwards in illustration of the second lecture, he kindly consented to deliver a supplementary lecture

on Artificial Auroras, illustrated by Rumkorff's Induction Coil
and Geissler's vacua Tubes. A small charge was made for
admission to this, to defray the expenses of illustration.

## CONVERSAZIONE.

The annual Conversazione was held at the Museum on Feb.
18th, and was numerously attended. After an address by the
President, a series of interesting experiments on Force was made
by Dr. Edwards. Objects were exhibited in microscopes lent
by Messrs. Ferrier, Watt, Muir, Clarke, Ritchie, Murphy, Baillie,
and others. A binocular microscope was lent by Mr. F.
Cundill. Principal Dawson exhibited a collection of Fossils and
Canadian Pearls; Mr. Rimmer a series of Fossils; Mr. Chap
man of Fictile Ivories; Mr. Stanley Bagg of Coins and Medals;
Mr. Reynolds of Illustrated Works and Roman Antiquities.
The rooms were tastefully ornamented by a committee of ladies;
and a choice collection of flowers was exhibited from the con-
servatory of Mr. Donald Ross. The band of the Rifle Brigade
enlivened the meeting with beautiful music. A novel feature on
this occasion was the execution of permanent decorations, designed by Mr. M'Cord, which recall to mind the names of the leaders
in different departments of science, emblazoned with mottoes and
emblems, in a very attractive manner. It is hoped that every
year permanent additions will be made of the same character.

## MUSEUM.

The Council has pleasure in again expressing their high
appreciation of the services of Mr. Whiteaves, whose special
report to the Council has fully set forth the labours and acquisi-
tions of the year. They have renewed the previous engagement
with him, subject to due notice being given on either side. In
consequence of the great additions to the collection, and especially those generously presented by the University of Oxford at the
instance of Mr. Whiteaves, it has been found necessary to erect
three new glass cases. A sub-committee was appointed to assist
the Curator in this and other changes in the Museum. Special
donations to the fund for cases were made by Mr. Rimmer of
$40, and Mr. Reynolds of $45.

The two extremities of the Museum room being now fitted with
permanent cases, the much greater work of fitting up the two
sides on the same plan ought to be proceeded with without delay.
The existing cases are not only unsightly, but they afford no

room for additional specimens. Upwards of a hundred new species of birds (of which ninety-two specimens were presented by the University of Oxford), several fine and rare mammals, and specimens in every other branch of natural history, make the new cases urgently called for.

Perhaps the most important alteration introduced this year has been the throwing the Museum open to the public gratuitously on Saturdays, from 1 to 4 p.m. in winter, and from 2 to 6 p.m. in summer. This step, which was not taken without full deliberation and some difference of opinion among the members, has at any rate proved their desire to spread the knowledge and pleasure to be derived from their collections as widely as possible among the inhabitants of Montreal and the strangers visiting the city. At first considerable damage was done to the property of the Society; but, an appeal having been made to the Mayor, two policemen have been regularly placed in attendance, and the conduct of visitors has been such as to warrant the Council in recommending the present as a permanent arrangement. The visitors have varied from 30 to 130 on these occasions,—a small number for so large a population.

During the past summer one of the Vice-Presidents, Mr. Leeming, kindly made arrangements to send the Cabinet-keeper on a collecting excursion to the coast of Maine. This was not only an agreeable change from his ordinary employments, but Mr. Leeming (who defrayed all the expenses of the expedition,) generously presented the specimens obtained to the Museum. Many other places might be visited with great advantage if other gentlemen are disposed to follow this excellent example.

The Scientific Curator reports as follows:—

MAMMALIA.—Fourteen specimens of North American mammals, mostly Californian species, and six specimens of Australian marsupials, have been procured. These additions made it necessary to re-group and re-arrange the whole of this part of the collection. The collection of antlers has been taken down, cleaned, re-arranged, and conspicuously labelled.

BIRDS.—The collection of birds has largely increased, especially in the department of British and exotic species. Ninety-two specimens have been presented by the authorities of the University of Oxford and by the late Rev. F. W. Hope, through Prof. Westwood. Mr. Angas has given an Australian eagle, Mr. Jno. Molson a specimen of the "black-headed plover" of the

Nile, and six exotic species have been purchased, making a total of one hundred specimens. Twenty new Canadian birds have been added, some of which are new to our series. The new British and exotic species have been named as far as possible, and have been arranged in a temporary manner until proper cases are provided for their reception.

REPTILES.—Dr. Gunther, of the British Museum, has kindly given thirty-five species of exotic reptiles; seven have been acquired by purchase; and Mr. Vining has given two Geckos from Jamaica. This portion of the collection has more than doubled during the past year. With the exception of about half the exotic snakes, all the specimens have been labelled and arranged.

FISHES.—Mr. Leeming's donation above referred to consisted of twelve species from the Portland coast; Mr. Morland gave the head of a Tunny caught at Gaspé; Dr. Gunther seven species of exotic fishes; and other donors six specimens of Canadian fresh-water fishes. A specimen of the rare Port Jackson Shark, and four species from the Pacific Ocean have been purchased.

INVERTEBRATES.—Thirty species of shells, principally fine cones, have been presented by Mr. B. M. Wright. A collection of beetles and butterflies from Jamaica was given by Mr. Vining, and some of the rarer Canadian moths by Mr. Fowler. The insect cabinet has been re-arranged. Seventeen species of crustacea (from Dr. Dawson and Mr. Wright), three of corals and five of Echinodermata, have been received during the year.

BOTANY.—In this department a set of specimens of the woods of New Zealand has been presented by Mr. Wright, and a beautiful specimen of the fibre of the lace-bark tree of Jamaica by Mr. Vining. In the Aquarian room a space has been set apart for the illustration of structural botany and botanical economics after the plan adopted by the British Museum.

GEOLOGY.—About one hundred and thirty species of fossils have been added during the past session, mainly through the kindness of Mr. Henry Woodward, Mr. Wright and Mr. Mason. These have been mounted on tablets, labelled, and arranged in their respective places in the Museum. Sixty-six fine specimens of rare exotic minerals have been presented by Mr. Wright; these are named, and have been provisionally placed in one of the cases in the gallery.

MISCELLANEOUS.—The ethnological and miscellaneous objects in the cases and on the walls of the gallery have been re-grouped and, as far as possible, labelled. A new case has been put up in the gallery for the reception of objects of antiquarian and of general ethnological interest. A collection of medals and medallions given some years ago by Dr. Gibb, has been arranged and labelled. Want of cases has prevented the formation of a collection to illustrate the comparative anatomy of our Canadian vertebrates; still, a beginning has been made, and the few specimens we have, have been collected together and some of them cleaned.

The Council desire to renew their expression of satisfaction at the manner in which the varied duties of Janitor, Taxidermist and Cabinet-keeper have been performed by Mr. Hunter,—whose labours have been necessarily increased by the opening of the Museum to the public.

### LIBRARY.

The Council regret that no funds have been at their disposal to increase the Library, or even to bind the periodicals, which at present are almost useless for reference. It is recommended that during the forthcoming year the Council take steps to render this department more attractive to members, and that gentlemen be invited to contribute books and periodicals thereto.

### ORIGINAL PAPERS.

The following are among the communications laid before the Society :—

On the Mineralogy of Crystalline Limestones. By the President.

On the Classification of the genus ATHYRIS M'Coy, as determined by the laws of Zoological Nomenclature. By E. Billings, F.G.S.

On certain discoveries in regard to *Eozoon Canadense;* On Insects from the Devonian and Carboniferous Formations; and On Canadian Pearls. By Principal Dawson, LL.D., F.R.S.,

On the Distribution of Plants in Canada, as related to its physical and geological conditions. By A. T. Drummond, B.A.

On Some Mammals and Birds recently added to the Society's Museum. By the Scientific Curator.

On certain peculiarities in the Shell-structure of Chitonidæ; and on the Vital Statistics of Montreal. By P. P. Carpenter, B.A., Ph. D.

The last paper belongs rather to the unnatural than to the natural history of our species, and might therefore be regarded as somewhat foreign to the objects of the Society. As, however, it is impossible at present to organize a Society in this city for the prosecution of every branch of scientific knowledge, it is to be hoped that the subject on which it treats, which is confessedly of the greatest importance, will be fully discussed from time to time at the monthly meetings.

### MISCELLANEOUS.

In consequence of the unnecessary labour caused by the appointment of sub-committees for separate but connected objects, a bye-law has been passed providing that a committee should be nominated by the Council and elected by the Society at the meetings in October, to make the necessary arrangements for both the Conversazione and the Somerville Lectures.

A new bye-law has also been adopted, changing the date of the Council meetings from the Thursday to the Tuesday preceding the monthly meetings, in order to allow more time for the issuing of the necessary circulars.

It is recommended that steps be taken by the Council now to be appointed, to codify and print these and all other new bye-laws of the Society which have been passed since 1859.

In conclusion, the Council beg to recommend that the Silver Medal of the Society be awarded to Mr. Billings. It was owing to his exertions that the *Canadian Naturalist*, which has become so valuable an organ for the Society's operations, was first established. His contributions to scientific literature and to the geology of Canada, although unobtrusive, and of a nature not to attract the general attention, have been singularly careful and exact, and have won the praises of all on this continent and in Europe, who are competent to pass judgment on their merits. And at the present time there is a special reason why this mark of appreciative respect should be no longer delayed,—the Council wishing to bear testimony to the singular ability which Mr. Billings has displayed in the volume on the Palæozoic Fossils of Canada and other publications, which have been issued by the Geological Survey during the last year.

Respectfully submitted by

PHILLIP P. CARPENTER,
Chairman.

# BOOK NOTICES.

### ACADIAN GEOLOGY.*

Canada has been upon the whole liberal to science. Not so liberal, it is true, as the neighbouring State of New York, whose splendid series of quartos are known the world over ; not perhaps so liberal as some even of our sister colonies, who have cheerfully contributed their share of the expense necessary to publish the series of works known as the Colonial Floras, while Canada has hitherto refused hers. Yet, withal, she has been, in her own way, liberal. She has for many years back spent something like $20,000 per annum on literary and scientific societies. It might have been better if this money had been given to these several societies for some specific object—for research into some defined branch of literature or science (excluding geology), to be pursued from year to year, and the results published ; nevertheless, though probably there may not be much to show for it, this money has, doubtless, been upon the whole well spent. Canada's greatest benefaction to science is, however, in the maintenance of her Geological Survey, which, under the direction of its eminent chief, has been continued for some sixteen or eighteen years, with plenty of good work to shew for the sums expended on it. Personally we are of opinion that this Survey has been too restricted,—all has been devoted to the fossil, almost nothing to the living. Had Sir William been provided with means to extend his survey so as to report on the natural productions of a district as well as on its geology, the country might have been saved the thousands it has spent in making so-called colonization roads through uncolonizable territory, and in surveying lots unfitted for settlement.

Until very recently, the Lower Provinces have not enjoyed the benefits of organized Geological Surveys, but our author, Dr. Dawson, aided to some extent by other zealous explorers, animated by a love of science for its own sake has, and that to no mean extent, in great part made up for this deficiency, though of course devoting himself to those points most likely to yield important scientific results, leaving the drudgery of details to those who

---

* ACADIAN GEOLOGY.—The Geological Structure, Organic Remains, and Mineral Resources of Nova Scotia, New Brunswick and Prince Edward Island, by John William Dawson, M.A., LL.D., F.R.S. Second edition, with a Geological map and numerous illustrations. London : MacMillan & Co. Montreal : Dawson Brothers. 8vo, pp. xxviii, 694.

might be officially entrusted with that part of the work. Not content with mere survey, Dr. Dawson has from time to time, and at his own expense, generously published his 'Reports of Progress,' the last and most complete of which is now before us, and in a form well entitled to take rank with official reports, while it is much more attractive to the general reader.

We do not propose giving any lengthy review of this work, as it is within easy reach of all our readers, and moreover we shall hereafter have opportunities of enriching our pages with copious extracts, one of which is given in the present issue of this journal. The following paragraphs are from the preface :—

"While the progress made in the Geology of Acadia since the publication of the first edition of this work is most satisfactory, it also suggests the fact that the present edition, probably the last which the author will be permitted to issue, merely marks a stage in that progress; and that the time will soon arrive when its imperfections will be revealed by the discovery of new facts, when many things now uncertain may have become plain, and when some things now held as certain will be proved to have been errors. When that time shall come, I trust that those who may build on the foundations which I have laid, if they shall find it necessary to remove some misplaced stone or decaying beam, will make due allowance for the difficulties of the work, and the circumstances under which it was executed."     *     *     *

"The lovers of the lighter kind of scientific literature may be disappointed in not finding in this work any incidents of travel or illustrations of the aspects of social life in Acadia. I have been obliged by the pressure of graver and more important matter to resist all temptation to dwell on these; but may perhaps find some future occasion to introduce the public to the incidents and adventures of my geological excursions.     *     *     *

"For myself, I confess that at an earlier period of my life it was a cherished object of ambition with me, that it might be my lot to work out in a public capacity the completion of some, at least, of the departments of geological investigation opened up to me in my native province; but it has been otherwise decreed; and however I may regret the want of that extraneous aid, which would have enabled me to devote myself more completely to original researches, by which my own reputation and the interests of my country might have been advanced, I am yet thankful that I have been enabled to do so much by my own unaided resources, and that I have also been able to assist and encourage others, who may now carry on the work more effectually in connection with an organized Geological Survey."     D. A. W.

### FILICES CANADENSES.

Under this title the undersigned has issued, for distribution among his foreign correspondents, a collection of our native ferns (filices exsiccatæ). The following is his catalogue : it includes all the species hitherto detected in Upper and Lower Canada, and were the Maritime Provinces included in the limits, the list would have been extended by only one species, (and that of very doubtful occurrence,) namely, *Asplenium marinum*, of which Sir William Hooker says in the Species Filicum, iii. p. 96, " I possess specimens from New Brunswick, Nova Scotia, from Capt. Kendal"—which contradictory note is corrected in the more recent Synopsis Filicum, so as to read "from Nova Scotia," while the Flora Bor. Am., had it from "New Brunswick, E. N. Kendal, Esq."—but its occurrence in either of those Provinces has not otherwise been authenticated. Three other species probably occur on the Canadian shores of Lake Superior, namely, *Cryptogramme crispa* (*acrostichoides* R. Br.), *Dryopteris Filix-mas*, and *Woodsia Oregana*, but have not been found there, the region being probably not yet botanized. The name attached to each species and variety is, in all cases, that of the author of the same ; when it is placed within brackets, that author put the plant in a different genus (or in the same genus differently named) from that here assigned to it. It is noteworthy that out of forty-two species twenty-nine belong to Linnæus, and five (or if *P. gracilis* be included, six) to Michaux.

### FILICES CANADENSES.

COLLECTÆ DISTRIBUTÆQUE CURA D. A. WATT.

| | |
|---|---|
| POLYPODIUM (Linn.) Mett. | |
| P. vulgare *Linn.* ; | No. 1. |
| PELLÆA, Link. | |
| 1. P. Stelleri (*Gmel.*) | |
|     sub *P. gracilis* (Michx.) ; | No. 2. |
| 2. P. atropurpurea (*Linn.*) ; | No. 3. |
| PTERIS, Linn. | |
| 1. P. aquilina *Linn.* ; | No. 4. |
| ADIANTUM, Linn. | |
| 1. A. pedatum *Linn.* ; | No. 5. |
| WOODWARDIA, Smith. | |
| 1. W. Virginica (*Linn.*) ; | No. 6. |
| SCOLOPENDRIUM (Smith) Hook. | |
| 1. S. vulgare *Smith*, | |
|   [*Aspl. Scolopendrium* Linn.] ; | No. 7. |

| | |
|---|---|
| 2. S. rhizophyllum (*Linn.*) ; | No. 8. |
| ASPLENIUM, Linn. | |
| 1. A. viride *Hudson* ; | No. 9. |
| 2. A. Trichomanes *Linn.* ; | No. 10. |
| 3. A. ebeneum *Aiton* ; | No. 11. |
| 4. A. angustifolium *Michx.* ; | No. 12. |
| ATHYRIUM, Roth. | |
| 1. A. thelypteroides (*Michx.*) ; | No. 13. |
| 2. A. Filix-fœmina (*Linn.*) ; | No. 14. |
| PHEGOPTERIS, Feé. | |
| 1. P. Dryopteris (*Linn.*) ; | No. 15. |
| 2. P. connectile (*Michx.*), | |
|   [*Polyp. Phegopteris* Linn.] ; | No. 16. |
| 3. P. hexagonoptera (*Michx.*) ; | No. 17. |

| | | | |
|---|---|---|---|
| DRYOPTERIS (Adans.) Schott. | | 2. W. glabella *R. Brown* ; | No. 33. |
| 1. D. Thelypteris (*Linn*.) ; | No. 18. | ONOCLEA, Linn. | |
| 2. D. nov-Eboracensis (*Linn.*) ; | No. 19. | 1. O. sensibilis *Linn.* ; | No. 34. |
| 3. D. spinulosa (*Müll.*) ; | No. 20. | 2. O. Struthiopteris (*Linn.*) ; | No. 35. |
|    *b.* dilatata ( *Wahl.*) ; | No. 21. | DICKSONIA, L'Herit. | |
| 4. D. cristata (*Linn.*) ; | No. 22. | 1. D. punctilobula (*Michx.*) ; | No. 36. |
| 5. D. Goldiana (*Hook.*) ; | No. 23. | OSMUNDA, Linn. | |
| 6. D. marginale (*Linn.*); | No. 24. | 1. O. regalis *B. Linn.* ; | No. 37. |
| POLYSTICHUM (Roth) Schott. | | 2. O. Claytoniana *Linn.* ; | No. 38. |
| 1. P. fragrans (*Linn.*) ; | No. 25. | 3. O. cinnamomea *Linn.* ; | No. 39. |
| 2. P. aculeatum (*Linn.*). . | | BOTRYCHIUM, Swartz. | |
|    *a.* Braunii (*Koch*) ; | No. 26. | 1. B. Lunaria (*Linn.*) ; | No. 40. |
| 3. P. Lonchitis (*Linn.*) ; | No. 27. |    *b.* simplex ; | No. 41. |
| 4. P. acrostichoides (*Michx.*) ; | No. 28. | 2. B. matricariæfolium *A. Braun* ; | No. 42. |
| CYSTEA, Smith. | |    *b.* lanceolatum ; | No. 43. |
| 1. C. bulbifera (*Linn.*) ; | No. 29. | 3. B. ternatum ( *Thunb.*). | |
| 2. C. fragilis (*Linn.*) ; | No. 30. |    *a.* lunarioides *Milde* ; | No. 44. |
| WOODSIA, R. Brown. | |    *b.* obliquum *Milde* ; | No. 45. |
| 1. W. Ilvensis (*Linn.*) ; | No. 31. | 4. B. Virginianum (*Linn.*) ; | No. 46. |
|    *b.* alpina | | OPHIOGLOSSUM, Linn. | |
|    sub *W. hyperborea* R. Br. ; | No. 32. | 1. O. vulgatum *Linn* ; | No. 47. |

The following supplementary species (of fern allies) are intended to be included in the collection :—

| | | | |
|---|---|---|---|
| Lycopodium apodum *Linn.* ; | No 48. | L. lucidulum *Michx.* ; | No. 51. |
| L. rupestre *Linn.* ; | No. 49. | Equisetum robustum *A. Braun*; | No. 52. |
| L. dendroideum *Michx.* ; | No. 50. | Eq. scirpoides *Michx.* ; | No. 53. |

A complete set will be deposited in the Herbarium of the Society. D. A. W.

---

## ARCHIVES DES SCIENCES PHYSIQUES.

Prof. Oswald Heer, of Zurich, has continued his researches into the Miocene Flora of Greenland, and has published the results, and his inferences therefrom, in the above named periodical. By these researches our knowledge of the distribution of vegetation in an era long prior to the present is increased, In Prof. Heer's details we find that the Arctic Fossil Flora, so far as known, now comprises 162 species, among which are eighteen cryptogams, nine being tall, handsome ferns, that probably covered the soil of forests, while on some of the others a growth of minute fungi can be detected, as in analogus species of our own day. Of phanerogams 31 species are conifers, 14 are monocotyledons, and 99 dicotyledons; and judging of these by the existing Flora, 78 were trees and 50 shrubs, which gives a total of 128 species of woody vegetables formerly distributed over the polar regions. The pines and firs come near to those now growing in America, particularly the *Pinus Maculrii*, which closely resembles the *Pinus alba* of Canada. Cones of this tree were brought from Banks Land by Capt. Maclure, who saw the stem of the tree in the hills of fossil wood in that country.  And, remarkable enough, that extinct Arctic Flora includes four species of the largest trees in the world, of which two only survive—the *Sequoia sempervirens* and *S. gigantea* of California.  These prodigious trees played an important part in the forests of the miocene period; they are found fossilized in Europe, Asia, and America, as well as in the polar regions.

Prof. Heer distinguishes three kinds of cypress Taxodium, Thujopsis, and Glyptostrobus, of which the last two are still living in Japan.  The elegant twigs of the Thujopsis are identical with those sometimes found embedded in amber.

Among the deciduous trees are a number which resemble the beech and chestnut of the present day.  The *Fagus Deucalinois*, which flourished beyond the 70th degree of north latitude, nearly resembles our common beech—*Fagus sylvatica*—the leaves being of the same forms and dimensions and the same venation, that, were they not toothed at the extremity, it would not be easy to describe the difference.  The tree appears to have been widely spread in the north, for its remains are found in Iceland and Spitzbergen as well as in Greenland.  There is even more variety among the oaks; eight species have been discovered in Greenland alone, most of them with large, beautifully-formed leaves.  One example (*Quercus Olafsoni*,), which can be traced from the north of Canada to Greenland and Spitzbergen, is the analogue of the *Q. Prinus* of the United States.  The plane and poplar were also largely represented.  The willow, on the contrary, is very rare; a surprising fact, when we remember that in the present day the willow forms one-fourth of the woody vegetation of the Arctic zone.  The birch was abundant in Iceland; where, also, a maple and a tulip-tree have been found.  The magnolia, the walnut, a species of plum and two species of vine grew in Greenland; a large-leafed lime and an alder in Spitzbergen.  In short, Prof. Heer, with all the interesting fossils before him, sees in imagination the polar regions of the miocene period covered with great forests of various trees, leafy and resinous, the leaves in some instances extraordinarily large, where veins and ivy interlaced their wandering branches, while numerous shrubs and handsome ferns grew beneath their shade; and these forests extended to the lands bordering on the Pole, if not to the very Pole itself.—*The Athenæum.*

Published, Montreal, 31st December, 1868.

THE

# CANADIAN NATURALIST.

SECOND SERIES.

## THE REMOVAL AND RESTORATION OF FORESTS.

By J. W. DAWSON, LL D., F.R.S., &c.*

The woods perish by the axe and by fire, either purposely applied for their destruction, or accidental. Forest fires have not been confined to the period of European occupation. The traditions of the Indians tell of extensive ancient conflagrations ; and it is believed that some of the aboriginal names of places in Nova Scotia (for example, *Chebucto*, *Chedabucto*, *Pictou*) originated in these events. In later times, however, fires have been more numerous and destructive. In clearing land, the trees when cut down are always burned, and that this may be effected as completely as possible, the driest weather is frequently selected, although the fire is then much more likely to spread into the surrounding woods. It frequently happens that the woods contain large quantities of dry branches and tops of trees, left by cutters of timber and firewood, who rarely consider any part of the tree except the trunk worthy of their attention. Even without this preparation, however, the woods may in dry weather be easily inflamed ; for, although the trunks and foliage of growing trees are not very combustible, the mossy vegetable soil, much resembling peat, burns easily and rapidly. Upon this mossy soil depends, in a great measure, the propagation of fires, the only exception being when the burning of groves of the resinous coniferous trees is assisted by winds, causing the flame to stream through their tops more rapidly than it can pass along

---

*From 'Acadian Geology,' second edition.

the ground.  In such cases some of the grandest appearances
ever shown by forest fires occur.  The fire, spreading for a time
along the ground, suddenly rushes up the tall resinous trees with
a loud crashing report, and streams far beyond their summits, in
columns and streamers of lurid flame.  It frequently happens,
however, that in wet or swampy ground, where the fire cannot
spread around their roots, even the resinous trees refuse to burn ;
and thus swampy tracts are comparatively secure from fire.  In
addition to the causes of the progress of fires above referred to, it
is probable that at a certain stage of the growth of forests, when
the trees have attained to great ages, and are beginning to decay,
they are more readily destroyed by accidental conflagrations.  In
this condition the trees are often much moss-grown, and have
much dead and dry wood; and it is probable that we should
regard fires arising from natural or accidental causes as the
ordinary and appropriate agents for the removal of such worn-out
forests.

Where circumstances are favourable to their progress, forest
fires may extend over great areas.  The great fire which occurred
in 1825, in the neighbourhood of the Miramichi river, in New
Brunswick, devastated a region 100 miles in length and 50 miles
in breadth.  One hundred and sixty persons, and more than 800
cattle, besides innumerable wild animals, are said to have perished
in this conflagration.  In this case, a remarkably dry summer, a
light soil easily affected by drought, and a forest composed of
full-grown pine trees, concurred, with other causes, in producing
a conflagration of unusual extent.

When the fire has passed through a portion of forest, if this
consist principally of hardwood trees, they are usually merely
scorched,—to such a degree, however, as in most cases to cause
their death ; some trees, such as the birches, probably from the
more inflammable nature of their outer bark, being more easily
killed than others.  Where the woods consist of softwood or
coniferous trees, the fire often leaves nothing but bare trunks and
branches, or at most a little foliage, scorched to a rusty-brown
colour.  In either case, a vast quantity of wood remains uncon-
sumed, and soon becomes sufficiently dry to furnish food for a
new conflagration ; so that the same portion of forest is liable to
be repeatedly burned, until it becomes a bare and desolate
' barren,' with only a few charred and wasted trunks towering
above the blackened surface.  This has been the fate of large

districts in Nova Scotia and the neighbouring colonies; and as these burned tracts could not be immediately occupied for agritural purposes, and are diminished in value by the loss of their timber, they have been left to the unaided efforts of nature to restore their original verdure. Before proceeding to consider more particularly the mode in which this restoration is effected, and the appearances by which it is accompanied, I may quote, from a paper by the late Mr. Titus Smith of Halifax, a few statements on this subject, which, as the results of long and careful observation, are entitled to much respect, and may form the groundwork for the remarks which are to follow.

" If an acre or two be cut down in the midst of a forest, and then neglected, it will soon be occupied by a growth similar to that which was cut down; but when all the timber on tracts of great size is killed by fires, except certain parts of swamps, a very different growth springs up; at first, a great number of herbs and shrubs, which did not grow on the land when covered by living wood. The turfy coat, filled with the decaying fibres of the roots of the trees and plants of the forest, now all killed by the fire, becomes a kind of hot-bed, and seeds which had lain dormant for centuries, spring up and flourish in the mellow soil. On the most barren portions, the blueberry appears almost everywhere; great fields of red raspberries and fire-weed or French willow spring up along the edges of the beech and hemlock land, and abundance of red-berried elder and wild red-cherry appears soon after; but in a few years the raspberries and most of the herbage disappear, and are followed by a growth of firs, white and yellow birch, and poplar. When a succession of fires has occurred, small shrubs occupy the barren, the Kalmia or sheep-poison being the most abundant; and, in the course of ten or twelve years, form so much turf, that a thicket of small alder begins to grow, under the shelter of which fir, spruce, hackmatack (*Larix Americana*) and white birch spring up. When the ground is thoroughly shaded by a thicket twenty feet high, the species which originally occupied the ground begins to prevail, and suffocate the wood which sheltered it; and within sixty years, the land will generally be covered with a young growth of the same kind that it produced of old." Assuming the above statements to be a correct summary of the principal modes in which forests are reproduced, we may proceed to consider them more in detail.

1st. Where the forest trees are merely cut down and not

burned, the same description of wood is immediately reproduced.
This may be easily accounted for. The soil contains abundance
of the seeds of these trees, there are even numerous young plants
ready to take the place of those which have been destroyed; and
if the trees have been cut in winter, their stumps produce young
shoots. Even in cases of this kind, however, a number of shrubs
and herbaceous plants, not formerly growing in the place, spring
up; the cause of this may be more properly noticed when describing cases of another kind. This simplest mode of the destruction
of the forest may assume another aspect. If the original wood
has been of kinds requiring a fertile soil, such as maple or
beech, and if this wood be removed, for example, for firewood, it
may happen that the quantity of inorganic matter thus removed
from the soil may incapacitate it, at least for a long time, from
producing the same description of timber. In this case, some
species requiring a less fertile soil may occupy the ground. For
this reason, forests of beech growing on light soils, when removed
for firewood, are sometimes succeeded by spruce and fir. I have
observed instances of this kind both in Nova Scotia and Prince
Edward Island.

2nd. When the trees are burned, without the destruction of
the whole of the vegetable soil, the woods are reproduced by a
more complicated process, which may occupy a number of years.
In its first stage, the burned ground bears a luxuriant crop of
herbs and shrubs, which, if it be fertile and not of very great
extent, may nearly cover its surface in the summer succeeding
the fire. This first growth may comprise a considerable variety
of species, which we may divide into three groups. The first of
these consists of those herbaceous plants which have their roots so
deeply buried in the soil as to escape the effects of the fire. Of
this kind are the various species of Trillium, whose tubers are
deeply embedded in the black mould of the woods, and whose
flowers may sometimes be seen thickly spread over the black
surface of woodland, very recently burned. Some species of ferns
also, in this way, occasionally survive forest fires. A second
group is composed of plants whose seeds are readily transported
by the wind. Pre-eminent among these is the species of Epilobium, known in Nova Scotia as the fire-weed or French willow,
(*E. angustifolium*), whose feathered seeds are admirably adapted
for flying to great distances, and which often covers large tracts
of burned ground so completely, that its purple flowers com-

municate their own colour to the whole surface, when viewed from a distance. This plant appears to prefer the less fertile soils, and the name of fire-weed has been given to it in conse-quence of its occupying these when their wood has been destroyed by fire. Various species of Senecio, Solidago and Aster, and Equiseta, Ferns and Mosses, are also among the first occupants of burned ground; and their presence may be explained in the same way with that of the Epilobium, their seeds and spores being easily scattered over the surface of the barren by wind. A third group of species, found abundantly on burned ground, consists of plants bearing edible fruits. The seeds of these are scattered over the barren by birds which feed on the fruits, and, finding a rich and congenial soil, soon bear abundantly and attract more birds, bringing with them the seeds of other species. In this way, it sometimes happens that a patch of burned ground, only a few acres in extent, may, in a few years, contain specimens of nearly all the fruit-bearing shrubs and herbs indigenous in the country. Among the most common plants which overspread the burned ground in this manner, are the raspberry, which, in good soils, is one of the first to make its appearance; the species of Vaccineæ or whortleberries, and blueberries; the tea-berry or wintergreen (*Gaultheria procumbens*); the pigeon-berry (*Cornus canadensis*); and the wild strawberry. It is not denied that some plants may be found in recently burned districts whose presence may not be explicable in the above modes; but no person acquainted with the facts can deny that nearly all the plants which appear in any considerable quantity within a few years after the occurrence of a fire, may readily be included in the groups which have been mentioned. By the simple means which have been described, a clothing of vegetation is speedily furnished to the burned district; the unsightliness of its appear-ance is thus removed, abundant supplies of food are furnished to a great variety of animals, and the fertility of the soil is preserved, until a new forest has time to overspread it.

With the smaller plants which first cover a burned district, great numbers of seedling trees spring up, and these, though for a few years not very conspicuous, eventually overtop and, if numerous, suffocate the humbler vegetation. Many of these young trees are of the species which composed the original wood, but the majority are usually different from the former occupants of the soil. The original forest may have consisted of white or

red pine; black, white, or hemlock spruce; maple, beech, black
or yellow birch, or of other trees of large dimensions, and capable
of attaining to a great age.  The 'second growth' which succeeds these usually consists of poplar, white or poplar birch, wild
cherry, balsam fir, scrub pine, alder, and other trees of small
stature, and usually of rapid growth, which, in good soils, prepare
the way for the larger forest trees, and occupy permanently only
the less fertile soils.  A few examples will show the contrast
which thus appears between the primeval forest and that which
succeeds it after a fire.  Near the town of Pictou, woods chiefly
consisting of beech, maple, and hemlock, have been succeeded by
white birch and firs.  A clearing in woods of maple and beech in
New Annan, at one time under cultivation, was, after thirty
years, observed to be thickly covered with poplars thirty feet in
height, presenting a striking contrast to the surrounding woods.
In Prince Edward Island, fine hardwood forests have been succeeded by fir and spruce.  The pine woods of Miramichi, destroyed
by the great fire above referred to, have been followed by a second
growth, principally composed of white birch, larch, poplar, and
wild cherry.  When I visited this place, twenty years after the
great fire, the second growth had attained to nearly half the
height of the dead trunks of the ancient pines, which were still
standing in great numbers; and in 1866 I found that the burnt
woods were replaced by a dense and luxuriant forest principally of
white birch and hackmatack, and I was informed that some of
these trees were already sufficiently large to be used in ship-
building.  This is an instructive illustration of the fact, that
after a great forest fire an extensive region may in less than half
a century be re-clothed with different species from those by which
it was originally covered.

As already stated, the second growth almost always includes
many trees similar to those which preceded it, and when the
smaller trees have attained their full height, these, and other
trees capable of attaining a great magnitude, overtop them, and
finally cause their death.  The forest has then attained its last
stage, that of perfect renovation.  The cause of the last part of
the process evidently is, that in an old forest, trees of the largest
size and longest life have a tendency to prevail, to the exclusion
of others.  For reasons which will be afterwards stated, this last
stage is rarely attained by the burned forests in countries beginning to be occupied by civilized man, and it is evident that many

circumstances may occur which will prevent this restoration of the primeval forest.

In accounting for the presence of the seeds necessary for the production of the second growth, we may refer to the same causes which supply the seeds of the smaller plants appearing immediately after the fire. The seeds of many forest trees, especially the poplar, the birch, and the firs and spruces, are furnished with ample means for their conveyance through the air. The cottony pappus of the poplar seems especially to adapt it for this purpose. The seeds of the wild cherry, another species of frequent occurrence in woods of the second growth, are dispersed by birds, which are fond of the fruit; the same remark applies to some other fruit-bearing species of less frequent occurrence. When the seeds that are dispersed in these ways fall in the growing woods, they cannot vegetate; but when they are deposited on the comparatively bare surface of a barren, they readily grow; and if the soil is suited to them, the young plants increase in size with great rapidity.

It is possible, however, that the seeds of the trees of the second growth may be already in the soil. It has been already stated, that deeply-buried tubers sometimes escape the effects of fire; and, in the same manner, seeds embedded in the vegetable mould, or buried in cradle hills, may retain their vitality, and, being supplied by the ashes which cover the ground with alkaline solutions well fitted to promote their vegetation, may spring up before a supply of seed could be furnished from any extraneous source. It is even probable that many of the old forests may already have passed through a rotation similar to that above detailed, and that the seeds deposited by former preparatory growths may retain their vitality, and be called into life by the favourable conditions existing after a fire.

If, as already suggested, forest fires, in the uncultivated state of the country, be a provision for removing old and decaying forests, then such changes as those above detailed must have an important use in the economy of nature, since by their means different portions of the country would succeed each other in assuming the state of ‘barrens,’ producing abundance of herbs and wild fruits suitable for the sustenance of animals which could not subsist in the old woods; and these gradually becoming wooded, would keep up a succession of young and vigorous forests.

3rd. The process of restoration may be interrupted by successive fires. These are most likely to occur soon after the first burning, but may happen at any subsequent stage. The resources of nature are not, however, easily exhausted. When fires pass through young woods, some trees always escape ; and so long as any vegetable soil remains, young plants continue to spring up, though not so plentifully as at first. Repeated fires, however, greatly impoverish the soil, since the most valuable part of the ashes is readily removed by rains, and the vegetable mould is entirely consumed. In this case, if the ground be not of great natural fertility, it becomes incapable of supporting a vigorous crop of young trees. It is then permanently occupied by shrubs and herbaceous plants; at least these remain in exclusive possession of the soil for a long period. In this state the burned ground is usually considered a permanent ' barren,'—a name which does not, however, well express its character ; for though it may appear bleak and desolate when viewed from a distance, it is a perfect garden of flowering and fruit-bearing plants, and of beautiful mosses and lichens. There are few persons born in the American colonies who cannot recall the memory of happy youthful days spent in gathering flowers and berries in the burnt barrens. Most of the plants already referred to, as appearing soon after fires, continue to grow in these more permanent barrens. In addition to these, however, a great variety of other plants gradually appear, especially the *Kalmia angustifolia*, or sheep laurel, which often becomes the predominant plant over large tracts. Cattle straying into the barrens deposit the seeds of cultivated plants, as the grasses and clovers, as well as of many exotic weeds, which often grow as luxuriantly as any of the native plants.

4th. When the ground is permanently occupied for agricultural purposes, the reproduction of the forest is of course entirely prevented. In this case, the greater number of the smaller plants found in the barrens disappear. Some species, as the Solidagos and Asters, and the Canada thistle, as well as a few smaller plants, remain in the fields, and sometimes become troublesome weeds. The most injurious weeds found in the cultivated ground are not, however, native plants, but foreign species, which have been introduced with the cultivated grains and grasses; the ox-eye daisy or white-weed, and the crowfoot or buttercup, are two of the most abundant of these.

When a district has undergone this last change,—when the sombre woods and the shade-loving plants that grow beneath them have given place to open fields, clothed with cultivated plants,—the metamorphosis which has taken place extends in its effects to the indigenous animals; and in this department its effects are nearly as conspicuous and important as in relation to vegetation. Some wild animals are incapable of accommodating themselves to the change of circumstances; others at once adapt themselves to new modes of life, and increase greatly in numbers. It was before stated that the barrens, when clothed with shrubs, young trees, and herbaceous plants, were in a condition highly favourable to the support of wild animals; and perhaps there are few species which could not subsist more easily in a country at least partially in this state. For this reason, the transition of a country from the forest state to that of burned barrens is temporarily favourable to many species, which disappear before the progress of cultivation; and this would be more evident than it is, if European colonization did not tend to produce a more destructive warfare against such species than could be carried on by the aborigines. The ruffed grouse, a truly woodland bird, becomes, when unmolested, more numerous on the margins of barrens and clearings than in other parts of the woods. The hare multiplies exceedingly in young second growths of birch. The wild pigeon has its favourite resort in the barrens during a great part of the summer. The moose and cariboo, in summer, find better supplies of food in second growth and barrens than in the old forests. The large quantities of decaying wood, left by fires and wood cutters, afford more abundant means of subsistence to the tribe of woodpeckers. Many of the fly-catchers, warblers, thrushes, and sparrows, greatly prefer the barrens to most other places. Carnivorous birds and quadrupeds are found in such places in numbers proportioned to the supplies of food which they afford. The number of instances of this kind might be increased to a great extent if necessary; enough has, however, been stated to illustrate the fact.

Nearly all the animals above noticed, and many others, disappear when the country becomes cultivated. There are, however, other species which increase in numbers, and at once adapt themselves to the new conditions introduced by man. The robin (*Turdus migratorius*) resorts to and derives its subsistence from fields, and greatly multiplies, though much persecuted by sports-

men.  The *Junco hyemalis*, a summer bird in Nova Scotia,
becomes very familiar, building in outhouses, and frequenting
barns in search of food.  The song sparrow and Savannah finch
swarm in the cultivated ground.  The yellow bird (*Sylvia æstiva*)
becomes very familiar, often building in gardens.  The golden-
winged woodpecker resorts to the cultivated fields, picking grubs
and worms from the ground.  The cliff-swallow exchanges the
faces of rocks for the eaves of barns and houses, and the barn
and chimney swallows are everywhere ready to avail themselves
of the accommodation afforded by buildings.  The Acadian or
little owl makes its abode in barns during winter.  The boblin-
coln, the king-bird, the wax-wing or cherry bird, and the hum-
ming-bird, are among the species which profit by the progress of
cultivation.  The larger quadrupeds disappear, but the fox and
ermine still prowl about the cultivated grounds, and the field-
mouse (*Arvicola Pennsylvanica*), which is very abundant in some
parts of the woods, is equally so in the fields.  Many insects are
vastly increased in numbers in consequence of the clearing of the
forests.  Of this kind are the grasshoppers and locusts, which, in
dry seasons, are very destructive to grass and grain; the frog-
spittle insects (*Cercopis*), of which several species are found in
the fields and gardens, and are very injurious to vegetation;
and the Lepidoptera, nearly the whole of which find greater
abundance of food and more favourable conditions in the
burned barrens and cultivated fields than in the growing
woods.

It thus appears that, in the course of between two and three
centuries, large areas of the Acadian provinces have passed
through two or more of the following conditions:—i. that of
primitive forest; ii. that of second-growth forest; iii. that of
the burned barren; iv. that of cultivated fields.  Each of these
changes is accompanied with modifications of the animal population; and in primitive states of society each would imply a
change in the habits of the people; and, if very extensive, might
even cause migrations of tribes and important changes of population.  In the old world, most countries have passed through
these vicissitudes in very early times, and have subsequently
reached a more stable condition, with more slow and gradual
changes; and in extensive regions it has usually happened that
the destruction and removal of forests have been effected piece-
meal, so as to extend only over limited areas at one time.  The

case of Denmark would seem to have been an exception to this.* At a very early pre-historic time it seems to have been covered by forests of Scotch fir. These were destroyed, probably by a great fire like that of Miramichi. The people perished or were driven from the country, and were replaced by another race, while the forests grew up again, but were now composed of oak. Still more recently the oak forests were replaced by beech. The stages of unrecorded human history connected in Denmark with these successive forests, are thus summed up by Steenstrup and Morlot :—" 1st. A *stone period*, when the inhabitants were small-sized men, brachykephalous or short-headed, like the modern Lapps, using stone implements, and subsisting by hunting; then the country, or a considerable part of it, was covered by forests of Scotch fir (*Pinus sylvestris*). 2nd. A *bronze period*, in which implements of bronze as well as of stone were used, and the skulls of the people were larger and longer than in the previous period; while the country seems to have been covered with forests of oak (*Quercus robur*). 3rd. An *iron period*, which lasted to the historic times, and in which beech forests replaced those of oak." All of these remains are geologically recent; and, except the changes in the forests, and of some indigenous animals in consequence, and probably a slight elevation of some parts of Denmark, no material changes in organic or inorganic nature have occurred.

The Danish antiquaries have attempted to calculate the age of the oldest of these deposits by considerations based on the growth of peat, and the succession of trees; but these calculations are obviously unreliable. The first forest of pines would, when it attained maturity, naturally be destroyed, as usually happens in America, by forest conflagrations. It might perish in this way in a single summer. The second growth which succeeded would, in America, be birch, poplar, and similar trees, which would form a new and tall forest in half a century; and in two or three centuries would probably be succeeded by a second permanent forest, which in the present case seems to have been of oak. This would be of longer continuance, and would, independently of human agency, only be replaced by beech, if, in the course of ages, the latter tree proved itself more suitable to the soil, climate, and other conditions. Both oak and beech are of slow

* Lyell, "Antiquity of Man "; Lubbock, in Nat. Hist. Review.

extension, their seeds not being carried by the winds, and only to a limited degree by birds. On the other hand, the changes of forests cannot have been absolute or universal. There must have been oak and beech groves even in the pine woods; and the growing and increasing beech woods would be contemporary with the older and decaying oak forest, as this last would probably perish, not by fire, but by decay, and by the competition of the beeches. The growth of peat has also been appealed to in connexion with the succession of forests as affording a mark of time; but this is very variable even in the same locality. It goes on very rapidly when moisture and other conditions are favourable, and especially when it is aided by wind-falls, drift-wood, or beaver-dams, impeding drainage and contributing to the accumulation of vegetable matter. It is retarded and finally terminated by the rise of the surface above the drainage level, by the clearing of the country, or by the establishment of natural or artificial drainage. On the one hand, all the changes observed in Denmark may have taken place within a minimum time of two thousand years. On the other hand, no one can affirm that either of the three successive forests may not have flourished for that length of time. A chronology measured by years, and based on such data, is evidently worthless; but it is interesting in connexion with our present subject to observe, that the remains preserved in the shell-heaps or 'Kjökkenmödding' of the stone age in Denmark indicate a wonderful similarity of habits and customs with those of primitive America, except that the people seem to have borne a closer resemblance to the Esquimaux than to the ordinary American Indian.

On the whole, nothing can be more striking to any one acquainted with the American Indian than the entire similarity of the traces of pre-historic man in Europe to those which remain of the primitive condition of the American aborigines, whether we consider their food, their implements and weapons, or their modes of sepulture; and it seems evident that if these pre-historic remains are ever to be correctly interpreted by European antiquaries, they must avail themselves of American light for their guidance. Much of this light has already been thrown on this subject by my friend Professor Wilson, in his " Pre-historic Man;" but one can scarcely open any European book on this subject, or glance at any of the numerous articles and papers on this fertile theme in scientific journals, without wishing that those who discuss pre-historic man in Europe knew a little more of his analogue in America. The subject is a tempting one, but I must close this notice, already too long for the space I should devote to it, by remarking, that the relations in America of the short-headed and long-headed races of men are by no means dissimilar from those of the two similar races in Europe; while it is also evident that some pre-historic skulls, supposed to be of vast antiquity, as, for instance, that of Engis, bear a very close resemblance to those of the Algonquin and Iroquois Indians.

# ON THE RESPIRATORY SYSTEM OF INSECTS.

By S. H. PARKES, Birmingham, England.

The subject of the present paper is The Respiratory System of Insects, and its direct relation to their nervous, nutritive and muscular functions, and as I trust this will only be the first of a series of papers on the structure of this remarkable and interesting class in the animal kingdom, I may perhaps be permitted to make a few introductory observations.

To some minds the discussion of insect physiology may appear a well nigh threadbare and exhausted subject, so much having been said and written on the structure, habits, and economy of these creatures. But, like other branches in the great domain of scientific research, this one has still hidden wonders, which will repay the labour of diligent and persevering inquiry.

No one ever thinks of asking, "What is a Bird?" or "What is a Fish?" but the question has yet to be answered satisfactorily and scientifically, "What is an Insect?" Nor need we wonder at the difficulty which naturalists have felt, when striving to find a distinctive name for these creatures; for of all the living things which this wondrous world presents to our view, there is no one class which contains such a strange diversity as that usually designated Insects.

There are insects with wings, and without wings; with jaws, and without jaws; with two eyes, and with many thousand eyes; some as large as humming birds, and others so small that the aid of a microscope is required to enable us to see them. Some insects, with dainty appetite, sip honey from the nectaries of flowers; while others, furnished with a pair of terrible jaws, grind down the root, bark or trunks of stately forest trees. All sorts of food is devoured by them in all sorts of ways. There are honey sippers, blood suckers, cabbage eaters, insect cannibals, and even, we regret to say, men eaters!

Insects too, have all sorts of odd ways for getting on in the world. There are creepers, runners, jumpers, fliers, swimmers and divers. Some take it into their heads to walk heels upwards; while others, with as strange a fancy, swim head downwards in the water. Very queer, too, are the occupations and habits of these strange little creatures. Some, like hermits, live alone in the wilderness; while others form themselves into well ordered communities, having a queen, government, soldiery and laws.

And what fantastic shapes do they assume! what a variety of dresses do they wear! Beasts, fishes, birds, reptiles, and even plants, have all their mimic representatives in the insect world. There are black insects, and white; blue insects, and grey; insects with smooth skins, hairy skins, horny skins, and feathery skins. Some strut about in a bright coat of armour, and others are decked from "top to toe" with sparkling gems, more brilliant and dazzling than those of an eastern prince. Some few there are that encircle themselves with a beautiful halo of light, moving about like fairy sprites, in the darkness of night.

All sorts of trades and occupations are likewise pursued by these busy little mortals. There are carpenters, builders, miners, stone-masons, paper-makers, silk-weavers, sugar-refiners, upholsterers, net-makers, fishermen, scavengers, nurses, and even slaveholders! with a few tribes of lazy epicures, who seem to think (like some of their human brethren) that life was given only for eating, drinking, sleeping and enjoyment. Without insects we should neither have honey nor wax, scarlet dye nor lac. The poor silk-weaver would have to look out for another occupation, and queens, princesses, and aristocratic ladies, would be obliged to doff their shining robes and satisfy themselves with dresses of cotton, linen and wool. Fevers and other fearful diseases would make their appearances in many places for lack of the same useful tribe of busy little scavengers, and the doctor would shake his head sorrowfully for want of some potent remedy which some insects supply. In short, the world could not wag on as comfortably as it does, if even a single tribe of these much despised creatures were wanting. And no wonder, for the great Architect has made no useless thing amid the million curiosities of earth, however idle or blind man may be in seeking to understand the sublime plan !

As, however, it is not my purpose in this paper to offer a new designation for these strange and diversified animals, but rather to describe an important and essential peculiarity in the anatomy and physiology of the entire class, (which, by the way, might perhaps form a very scientific groundwork for their classification,) I will now proceed to the discussion of my subject.

A careful study of the structure and functions of organs, as developed in the lower animals, has long been considered by comparative physiologists, an important and instructive pursuit. We may thus see functions performed by the simplest possible structural arrangements, and may learn what are the essentials of such organs. Dr. Goadby (the once English but now American professor of comparative physiology,) remarks in his beautifully illustrated work on this subject, " that in this class (Insecta) the most important problem—the ultimate structure of glands— may be studied with great ease. In the higher animals, these organs are veiled by a parenchyma, which renders investigation difficult ; but in insects we find them already analyzed—existing as simple tubuli, and offering every facility for the most minute examination of them. When the like organisms in man and the higher animals have been successfully treated and reduced to their elemental conditions, lo ! they too, are simple tubes !" Now with regard to the special function of respiration, I think some important truths may be elicited, by a careful study of the very beautiful and elaborate arrangement by which it is effected in the insect race. It will scarcely be needful to observe—even in the most casual way—what an important part is played by this function in the economy of all organized beings. Most animals can exist for a considerable period without food ; although this is an essential condition to the continuance of their life. But if the function of respiration be suspended, even for a very limited period, death is the speedy and inevitable result. Now the necessity for respiration in all animals — whether aquatic, terrestrial or aerial—results from the fact, that a continual decay takes place during every moment of such an animal's existence. Waste and renewal form one of the prominent peculiarities of organic life. And one of the peculiar phases of this physiological law is, that activity and waste bear a definite relation to each other. The more active any organ, or set of organs may be, the more rapidly does waste occur, and the greater necessity is there for rapid renewal. One of the results of waste in the animal economy is, the liberation of carbonic acid ; which carbonic acid is produced by the union of the broken down carbonaceous particles of the old body with a portion of oxygen still existing in the blood. Unless this poisonous carbonic (when thus formed) be speedily removed, death is the inevitable consequence.

Thus arises the paramount necessity for the exercise of this function of respiration—which consists essentially in the removal of carbonic acid from the fluids of an animal's body, and in the interchange for this of an equivalent amount of oxygen. The mode by which this is effected, is wonderfully varied in different classes of animals ; the respiratory apparatus of each great division being beautifully adapted to the peculiar mode of such animal's existence, and to the general plan of its structure. But in all cases, however complicated may be the structural arrangements this function is performed, it depends essentially on the effective action of a most exquisitely simple law, usually expressed as that of ' the diffusion of gases.' Thus :—if a bladder containing pure oxygen gas be hung up in a room or vessel containing common atmospheric air, although no distinguishable pores may exist in the membraneous bag thus containing the gas, still, after a while, an interchange will have taken place between the internal and external gases; and the bladder will ultimately be found to contain nothing but common air ! This interchange will take place between other dissimilar gases under the same conditions ; and thus, the beautifully simple arrangement is provided for the carrying on of this all important function of respiration. For it matters not whether an animal may exist in the water or on the land ; whenever or however the blood (which may have become overcharged with carbonic acid by its passage through the body) is brought, through the intervention of an enclosing membrane, in contact with oxygen, contained either in water or in the air, this interchange—of which we are speaking—instantly takes place, and respiration, or the revivifying of the blood, is the result. It would have been interesting to trace the various structural arrangements by which this is effected in different grades of animal life; but this would lead us too far away from the special subject under consideration. It will, however, be necessary to make a passing reference to the respiratory apparatus of other animals ; in order to show clearly the totally distinct, and very

unique means by which it is effected in the insect race. In all
other animals, whether low or high in the scale of being,
wherever there is a circulation of the blood, or nutritive fluid,
and as a consequence, some organ of propulsion termed a heart,
this blood is sent continually to some special region of the body,
where an apparatus is set apart for its constant renewal,
termed lungs, in reptiles, birds and mammals, and gills, in
fishes.   Thus, all the blood in the body of a fish is brought suc-
cesssively, through a delicate net work of vessels which spread
over the gills, into direct contact with the water which bathes
every portion of such gills ; and thus the interchange of gases we
have referred to, takes place.   In the various terrestrial animals,
however, lungs of different kinds are provided, and to these the
blood is constantly sent, to receive the necessary aeration.

Perhaps we should also remark still further that, according to
the peculiar habits of each class of animals, according to
the slowness or activity of their movements and the feebleness or
vigour of their vascular system, so are their lungs or respiratory
organs modified.   For instance, in the cold-blooded and slow
moving Reptile class, the lung is little more than a simple bag,
with a few air chambers lining its interior; and thus the blood,
which flows through the vascular net-work lining these chambers,
is somewhat slowly brought in contact with the air which is
inspired.

On the other hand, in the case of birds and mammals whose
muscular system is called into active and vigorous play, we find a
most effective and elaborate arrangement, consisting of an almost
innumerable aggregation of elastic air cells, over the walls of
which is spread an immense surface of capillary net-work ; so that,
at every fresh inspiration, a considerable portion of the animal's
blood is exposed to atmospheric influence.

Now of all the diversified grades of animals, that add variety,
beauty, vivacity and utility to the wondrous planet on which we
live, there is no one class which exhibits such marvellous evidence
of muscular force, and untiring activity as the class Insecta.   We
might therefore—reasoning from analogy—have expected to find
a most elaborate system of arteries and veins, conveying their
blood to and from an equally elaborate and vigorous respiratory
organ.   Instead of this, however, we find a sudden and startling
break, in what appeared to be the uniform and universal organic
arrangement, ordained for the performance of this function ; a

complete turning upside down of the general plan. Here, in the Insect body, we have blood, it is true, and a pulsating organ (termed the dorsal vessel), which appears to give this blood a somewhat definite and uniform motion through different parts of the animal's frame. But no blood-vessels are any where to be seen, nor can we discover any one organ set apart for the special aeration of the vital fluid. But we do find something no less wonderful and interesting; nay, I would rather say, immeasurably more interesting and instructive, because illustrative of the limitless resources of that Infinite Mind which thus condenses and concentrates within the small dimensions of a point, such an exquisitely perfect and marvelously elaborate vital mechanism !

What is there then, in the anatomy of an insect, which claims the special and careful attention of a modern physiologist ? Not only (I humbly think) the mere structural difference, which I will now briefly describe, but the physiological inference which may possibly be deduced therefrom, as to the true nature and immense importance of the respiratory function in the animal economy. As this paper will be accompanied by a series of microscopic preparations, illustrative of some of the structural peculiarities here alluded to, it will not be necessary to give any lengthened verbal description. I will merely remark, therefore, that instead of the blood (which flows in grooved channels or canals through the body of an insect) being forced to one spot to receive oxygenation, the air is conveyed to it, by means of a most elaborately arranged system of external breathing mouths, termed spiracles, and internal air tubes, termed trachea. Although the plan of respiration is the same essentially in all insects, the modifications of these breathing organs is as wonderfully varied as the external appearance and peculiar habits of the creatures themselves. When it is remembered that insects pass through a series of metamorphosis, some living in water at one period of their existence, and then assuming an aerial life ; others burying in the earth, during their early days, and then coming forth to roam abroad amid the forest trees ; and when we recollect that almost all exist under very different external conditions, at different periods of their changeful history, and that in each of these states respiration is an indispensible function, we need not be surprised to find striking and important modifications in the physical structure of their breathing organs, suited in each case to the peculiar exigencies of the individual. It will be impossible, therefore, in this paper to do more than indicate the prevailing structure. And first, with regard to the spiracles, or external breathing organs of these creatures. If you will examine the body of almost any insect, you will perceive, arranged along each side of the abdomen and thorax, a series of openings, each bounded by a dark colored ring. The office of these, is to admit air to the interior of the animal's body, and to regulate its admission and expulsion according to existing circumstances. The essentials of these spiracles appear to be, 1st, a marginal ring of horny or cartilaginous substance, capable of being opened and closed by an arrangement of muscles, (thus forming the framework of the spiracle, and serving as a support to the delicate tubes within); and 2nd, a variously arranged membrane, or fringe, or system of horny plates, placed within this horny ring, for the purpose of preventing the entrance of dust or other matter, which might stop up the air passage within, and thus cause the death of the animal. The number of these spiracles, possessed by different insects, varies of from two to eighteen; the number frequently differing in the same insect, according as it is in its larval or perfect state. In every order (as before observed) there is some peculiar modification in the structure of this important organ; and even striking variations in different members of the same order, as will be seen in the specimens sent to illustrate this paper. It is supposed by some entomologists, that some of these spiracles, (namely, the abdominal ones,) are specially concerned in the inspiration of air; and that those situated in the thorax are designed for its expulsion. The point most worthy of notice and admiration, however, in the structure of these organs is, the perfect and exquisitely beautiful manner in which provision is made for the protection of the elaborate system of vessels to which they lead. In some beetles, peculiarly liable to be infested by parasites, (which parasites attach themselves to the softer parts of the body where the spiracles are placed,) there is a membranous covering with a narrow opening, thickly studded with sharp spines. In others, whose habits are of a burrowing character, we find the entrance guarded by an admirable arrangement of horny or cartilaginous plates, while in many of the dipterous and neuropterous insects, there is an elegant arrangement of fringed processes, which, for beauty as microscopic objects, can scarcely be surpassed. Some writers have supposed that the humming or buzzing noises made by many insects, when on the wing, is pro-

duced by these spiracle appendages, during the rapid ingress and
egress of the air; an effect similar to that which is produced by
the sweeping of the air over the strings of an eolian harp.
The most important vital purpose, however, is doubtless
that to which I have already alluded—the protection afforded to
the air vessels within.　There is also another important end
which they may serve, and one which, I think, has not been
observed by any writer on the subject.　It is this : the modification of the temperature of the air, as it enters the trachea, and
the preservation of that within the body, at the normal standard
of heat, usually existing in the different members of this class.
For this purpose these fringes and plates and membranous folds,
would be admirably adapted, and would act in precisely the same
way as the metallic framework of a respirator does when
worn by consumptive persons.　A question might here naturally
arise, as to the production and maintenance of animal heat in the
insect economy.　But the full discussion of this subject would
demand more time than we have at disposal.　Many interesting
observations have been made, which show that the temperature of
different insects varies greatly, especially those living in societies
(as the hive bee) whose normal standard of heat is very much
higher than that of other classes.　There has been a prevailing
notion that the temperature of insects is altogether regulated by
that of the external atmosphere in which they live, but this opinion
is, I think, at variance with the common principles of animal
physiology; and it is, moreover, contradicted by a variety of
experiments, bearing on this question.　There can be little doubt,
I think, that the standard of heat, in different species of insects, is
regulated very much by the degree of muscular activity manifested by them; for this would involve a more rapid and vigorous
respiration, and a greater consequent evolution of heat.　Without
pursuing this question farther, however, I would remark finally
respecting the spiracles of insects, that however beautiful and
elaborate they may be in their structure, and however perfectly
adapted to the habits and peculiarities of the creatures possessing
them, they are but the portals to an inner sanctuary of wonders,
unspeakably transcending all human contrivances in execution,
and surpassing human thought, even in conception.　The fact
that insects breathe, and that their respiration is carried on by
means of an elaborate system of air tubes, which ramify extensively through the interior of their body, has long been known, and

has been described by writers on this subject. But very few, I believe, until lately, have been able to show, by actual demonstration, to what an almost infinite extent these wonderful air channels divide and sub-divide, and how they spread over and penetrate, almost every membrane and fibre of an insect's body. The principle published accounts of the Respiratory System of insects, have been descriptive chiefly of the larger species of lepidopterous caterpillars ; also of coleoptera, neuroptera and diptera. Preparations of these are of course more easily made and displayed, than the demonstration of the same system in the smaller tribes. As the microscope, however, has gradually been improved, and as microscopic manipulation has also kept honorable pace in the same onward march, so have the more minute marvels of this wondrous material world been gradually unfolded ; and a restless and insatiate craving has been awakened in the minds of physical philosophers, which has prompted them to *see* and to *touch*, not only the most minute organs, of the most minute organism, but even the very molecules of which those natural substances are composed. The great cry of the physiological microscopist now is, More magnifying power— more light. Well, suppose he could obtain both, what would he then want ? Why, most assuredly—I verily believe—something which he does not now possess : more mental power ; and a far more steady and delicate touch, to enable him to handle and separate such infinitesimal forms of matters. And even then, he would still " see through a glass darkly," for he would certainly never touch that invisible essence, which gives vitality to the visible form ! But this is a digression—for my purpose, in this paper, has been, not to speak of what is impossible and unattainable, but to show what marvellous results have been attained by patient microscopic research, and by persevering practical manipulation. As an illustration of this, I have had prepared for examination, not only the larger tracheal system, dissected from the body of a large caterpillar, but the same system of respiratory tubes taken from the body of a human flea. In another slide containing a specimen of Pediculus, the body of the creature has been rendered transparent, and so mounted, as to show the entire respiratory system *in situ*. Preparations will also accompany this paper, showing the minute ramifications of air vessels over the stomach of the house fly, and of the honey bee, also over the nerve ganglia of a caterpillar. In another slide containing the

contents of the head of the honey bee, may be seen the singular
and somewhat puzzling connection between these air vessels
distended by their peculiar spiral fibres, and the salivary glands
of this insect.  In this preparation it will be seen that, instead of
a large spiral vessel, dividing and sub-dividing into extremely
fine tubes, and these tubes ramifying over the part requiring
aeration (as in other cases), these tubes appear to be modified
and converted into the very gland structures themselves ?  And
in another slide, may be traced the connection of these wonderful air tubes, with the muscles, the ovaries, and the gizzard of a
flea.  Perhaps I should remark by the way, that the existence of
this last mentioned organ, a flea's gizzard, was, some time since,
warmly discussed by a number of microscopists.  It is well known
that insects, possessing a suctorial apparatus, are not usually
furnished with a gizzard, of which is essentially a grinding or
triturating organ.  But the late Professor Quckett (whom it was
the writer's great privilege to know) asserted in spite of all
opposition, and contrary to analogy, that the flea possessed this
organ; and so it turns out !  For the clever little Frenchman
who made this flea preparation for me, has managed to demonstrate the fact; and to mount the minute dissection (thus made
with an amazing amount of patient persevering skill) in a
most exquisitely beautiful manner.

But what of these air tubes, about which so much has already
been said ?  On examining the preparations which accompany
this paper, you will observe that they consist of two membranous
tubes—one inside the other—and that between these delicate
membranes, there is coiled a spiral fibre which tapers down
smaller and smaller, as the tubes subdivide; and which continues
its course down to the most minute vessel that the microscope
can reveal.  The purpose which this spiral fibre serves, affords a
striking and beautiful illustration of that marvellous design and
adaption, which is exemplified in the whole of the great Creator's
works.  As these tubes contain only air, they would be liable to
collapse by the constant pressure of surrounding organs, and still
more by the violent contortions of the animal when moving about
were it not for these spiral fibres, which combine lightness,
firmness, elasticity, and every other needful requisite.  So
admirably do they fulfil their intended purpose, that the human
inventor has copied them, to strengthen his elastic india rubber
gas pipes and other tubes of similar character.

But what of the termination,—the ultimate distribution of these elaborately constructed tubes? And what of the purpose they are intended to subserve? With regard to their distribution; no one, perhaps, has gone so far in demonstrating their universality and extreme fineness, as Dr. Beale, with his 25th-inch object glass, and with this, which gives a magnifying power of nearly 3000 diameters, he has traced both air tubes and nerve fibres interlacing and spreading over the sarcolemma of muscular fibre, taken from the larva of the blow fly, a single fibre of this insect's muscle being completely encased in a net work of these inconceiveable minute and wondrous air tubes, whose very existence requires a power of 3000 diameters to reveal?

And not only do they thus intertwine about the fibres of an insect's muscles, but they penetrate the very substance of the nerve ganglia of the body; entering the head, and spreading over that optic nerve which receives impressions through ten thousand compound eye lenses; penetrating the wings, and giving lightness and energy to those untiring organs of flight; spreading over the stomach and other abdominal viscera; and aerating every particle of that blood which bathes and surrounds all the internal organs! I know not, gentlemen, what your feelings may be when you examine with your microscopes such unspeakably wonderful and complicated organisms, condensed and crowded within an almost invisible point of space; and this mechanism vitalized, directed and controlled during the period of its existence by an individual will, and by an unerring instinct. I know not, I say, what you may think and feel about the origin and design of such manifestations of constructive wisdom and skill; but for myself, I can say, it produces in my mind the most profound emotions of humility and awe; nay, rather, I would say, of adoring gratitude to that Infinite Being, who, while he displays to my astonished sight a spectacle so grand and glorious, as I look through my telescope at a starry universe, has also stooped so low, as to lay at my very feet the same incontestible proofs of His own " Infinite power and Godhead."

But what of the Physiological necessity for such a complicated mechanism? Can we suppose that the mere general aeration of the blood, such as is supposed to take place in the pulmonary respiration of higher animals, calls for this excessive elaboration and minute sub-division of air tubes in the insect economy. These tubes penetrate and twine about the interior of organs,

which cannot possibly be bathed as other parts are, by the
nutritive fluid. What is this atmospheric air ?—this component
fluid which all animals must breathe, but which to insects appears
to be pre-eminently " the breath of life." Does it contain something more than oxygen, carbonic acid and nitrogen ? Is there
not ammonia, and that wonderful substance ozone ? And is it
not the carrier of that still more wonderful something, which we
call electricity ? It may yet appear, as science advances, that in
our respiration, there is something more effected than the mere
interchange of oxygen and carbonic acid, with one or two subordinate results ; and that the character of the air we breathe,
and the air we live in, is a question of no mean importance to
individuals and to communities. Not only do we, like all other
terrestrial beings, draw this atmospheric air within our bodies,
during the process of respiration, but, like a great ocean, it
encompasses us about on every side. And like that deep and
dark blue ocean of waters, whose restless vicissitude of storm and
calm, is changing our land marks, and modifying our climates ;
so this great ocean of air, carries in its bosom the same wonder-
ous law of mutation. For, the electrical changes which are
constantly taking place in its upper strata, producing sometimes
very sudden hygrometric and thermometric changes in the lower.
regions, must and do affect the conditions of animal health, to
a very great extent. The effect produced by physical alterations
in the atmosphere upon the nervous system of animals, and the
peculiar influence of atmospheric air upon the bodies of animals
(especially upon man) externally, when freely exposed to its
action, have not, we think, had that attention from the scientific
men that the subject deserves.

I must not, however, go further with this subject, but will
conclude by quoting the eloquent language of Dr. Williams;
which langugage he also puts into the form of interrogation.
" What can be the meaning of these incomparable pneumatic
plexuses, which embrace immediately the very ultimate elements
of the solid organs of the body ?—those minute microscopic air-
tubes, which carry oxygen in its gaseous form, unfluidified by
any intervening liquid, to the very seats of the fixed solids
which constitute the fabric of the organism ? The intense
electrical and chemical effects, developed by the immediate
presence of oxygen at the actual scene of all the nutritive
operations of the body, fluid and solid, give to the insect its vivid

and brilliant life, its matchless nervous activity, its extreme muscularity, its voluntary power to augment animal heat. Such contrivances, subtle and unexampled, reconciles the paradox of a being, microscopic in corporeal dimensions and remarkable for the minuteness of the bulk of its blood, sustaining a frame, graceful in its littleness, yet capable of prodigious mechanical results."

# SOME STATISTICAL FEATURES OF THE FLORA OF ONTARIO AND QUEBEC,

## AND A COMPARISON WITH THOSE OF THE UNITED STATES FLORA.

### By A. T. DRUMMOND.

The recent issues by Prof. Gray of a fifth edition of his Manual of Botany of the Northern United States and by Mr. Horace Mann of a Catalogue of the Phænogamous Plants of the United States east of the Mississippi, have suggested the thought that with the materials for a flora of Ontario and Quebec, which have been for some years accumulating, the prominent statistical characteristics of our local vegetation might now be indicated with reasonable certainty, and a fair comparison instituted between them and those of the flora of the United States. That any statistics given will, in coming years, be altered in consequence of additions made to our flora, is certain. There is reason to believe that a considerable number of phænogamous and filicoid plants not at present known to occur within our geographical limits, will yet be detected there. Whilst, however, these statistics are not invested with absolute certainty, they can, I think, be regarded as fair general conclusions.

The works of Michaux, Pursh, Hooker, Torrey and Gray, etc., afford much information regarding the flora of this part of the continent, but since their publication our knowledge of it has been greatly extended. Foreign as well as provincial scientific journals have within the past few years contained valuable papers on the subject of Canadian botany. The institution of a society, whose special aim was the promotion of botanical research in our midst, infused for a time much interest in the study, and resulted in the accumulation of considerable material for a provincial flora. Some of the papers and catalogues were published in the society's 'Annals,' but many are still in manuscript. To these

latter, as well as to other catalogues in the hands of the editor
of this journal, I have been permitted to have access, and from
them have derived much aid in arriving at the results given
hereafter.*

Endeavours have already been made to bring the flora of
Ontario and Quebec into one connected view. The work of
the Abbè Provancher, in the French language, which was
published some years since, is upon an ample scale, and contains
descriptions of the plants referred to in it, whilst the more recent
brochure of the late Prof. Hubbert is simply an arranged catalogue, which was intended as the precursor of his contemplated
Hand-book of the Canadian Flora. Prof. Hubbert's list, in
addition to the results of his own collections, as well as of those
of his correspondents, probably contains all previously published
information bearing on the subject.

The views of authors, of course, vary considerably with regard
to orders, genera and species; however, for the purposes of
comparison with the flora of the United States, those of Prof.
Gray, as expressed in the recent edition of his Manual of
Botany, are here adopted. Further, it should be premised that
only flowering and filicoid plants are referred to in this paper,
our knowledge of the lower cryptogams being as yet too limited;
and it should be added that when speaking of the Northern
States and the United States or Union, no more extended
geographical limits are intended than are kept in view in the
Manual on the one hand and Mr. Mann's catalogue on the
other.

The prominent features in the distribution of the plants of
Quebec and Ontario have been indicated in another place. With
regard to the nature of the flora of the United States, it may be,
in a general way, said that in the eastern and central portions of
the Northern States the vegetation embraces a mountain
and a woodland flora, which, excluding the more southern

---

* In addition to the catalogues cited in the foot note to p. 406, vol. i.
(new series) of this journal, I have had access to those of Dr. Thomas, of
the Rivière-du-Loup flora, and Dr. J. Bell, of the Maintoulin Island
flora; to the notes of Prof. Hincks on Toronto plants (through Prof.
Hubbert), and to the elaborate lists of Dr. McLaggan and Mr. John
Macoun, the former of whom collected in different sections of the provinces, but chiefly in the western peninsula, and the latter in the vicinity
of Belleville.

forms, is similar to that of Ontario and Quebec; that as the Mississippi is approached there is a transition to a prairie flora in some districts, and in others to the flora of the western plains and wooded country; that along the Atlantic coast there is a maritime flora, some former members of which now occur in special inland localities; that the line of distribution of many of the United States plants has a north-westward trend; and that the Southern States have their semi-tropical species, many of which do not range as far as, whilst others extend within, the geographical limits of the Northern States. All these circumstances largely affect the number and character of the species in each region.

In our two Provinces there are representatives of one hundred and fourteen natural orders. Of these Magnoliaceæ, Melastomaceæ, Dipsaceæ, Bignoniaceæ, Phytolaccaceæ, Lauraceæ, Ceratophyllaceæ, Platanaceæ, Amaryllidaceæ, Commelynaceæ, and Xyridaceæ, are, as far as known, confined to Ontario. No order is, however, peculiarly provincial; all have their representatives in the Northern States among the one hundred and thirty-two orders which embrace the flora of that section of the Union. It is nevertheless a not uninteresting circumstance that, although there are eighteen of these Northern States orders which have no place in our Provincial flora, they comprise only thirty-five species, most of which are Southern States forms.

The genera which have representatives in Ontario and Quebec number 575, of which 428 are dicotyledenous, 124 are monocotyledenous, and 23 comprise the filicoid plants.

Of indigenous genera five are unknown south of the Great Lakes. These are Cochlearia, Crepis, Armeria, Pleurogyne, and Elæagnus, each of which comprises a single species. Crepis and Elæagnus are, with us, only found along the upper lakes, and are probably entirely western in their distribution, whilst the remaining three are of semi-arctic range. In addition to the above there are some introduced genera, as Scabiosa, Tragopogon, Ajuga, and Borago, which apparently have not been noticed in the United States. Within the geographical limits of Prof. Gray's work are 834 genera, 631 of which are dicotyledonous, 175 monocotyledonous, and 28 are filicoid. There are thus 263 genera in the Northern States which are without either indigenous or introduced representatives in either Ontario or Quebec.

The relative numerical proportion of monocotyledonous and

dicotyledonous genera decreases from our section of the continent
southward. Thus, in Ontario and Quebec monocotyledons are to
dicotyledons as 1:3.46; in the Northern States as 1:3.61, and in
the whole of the States east of the Mississippi as 1:4.13. The
numerical relations of filicoid to phænogamous genera present
much more marked differences. In the Provinces the proportion
is as 1:24, whilst in the Northern States it is as 1:28.9.

The relative positions of the orders with respect to the number
of genera in them vary to some, though not to any considerable,
extent in the two countries. In the Northern States and the
whole Union these relative positions are not much different.
Compositæ and Graminæ, however, assume the precedence there
in each case as well as here. Arranging the large orders represented in each country according to priority in point of number
of included genera, the following results are presented:

*In Ontario and Quebec.*

| | | |
|---|---|---|
| Compositæ | 56 | Filices, Liliaceæ and Umbelliferæ, each 19 |
| Graminæ | 47 | Cruciferæ and Rosaceæ, ............." 17 |
| Labiatæ | 24 | Ranunculaceæ and Scrophulariaceæ, " 15 |
| Ericaceæ | 22 | Orchidaceæ ........................ 14 |
| Leguminosæ | 21 | Caryophyllaceæ ........................ 12 |

*In Northern States.*

| | | |
|---|---|---|
| Compositæ | 86 | Umbelliferaeæ ........................ 27 |
| Graminæ | 67 | Scrophulariaceæ ..................... 25 |
| Leguminosæ | 39 | Filices.............................. 22 |
| Labiatecleæ | 33 | Ranunculaceæ and Cruciferæ, each.... 20 |
| Liliaceæ and Ericaceæ, each | 28 | Rosaceæ ............................. 18 |

Of the 576 genera in the two Provinces, 291 or rather more
than one-half, are referable to the twelve orders which take
precedence in the first of these lists. The aggregate of the
genera in the second list barely attains the half of the whole
number of genera which have representatives in these States.

The largest interest is of course invested in the species which
occur within our geographical limits, and in the numerical
relations of the orders and genera with regard to the species
which they embrace. The details given with respect to them
will be less wearisome.

Recent discoveries have confirmed the occurrence in Canada of
several species whose previous claims to a place in our flora rested
solely on the authority of Michaux or Pursh. I have therefore
experienced a reluctance to exclude any of their species—unless
the occurrence of the plant is very improbable—on the mere
ground that it has not been noticed by subsequent observers.
This reluctance is increased by the circumstance that the Lake
Superior and lower St. Lawrence districts, where many, if not

most, of these species are supposed to occur, have received but a limited exploration. Though *Sabbatia gracilis, Utricularia subulata,* and *Ilex glabra* are probably errors, I have had no hesitation in admitting *Rhododendron maximum, Phlox maculata, Trichostema dichotomum, Andromeda tetragona,* and even *Gnaphalium sylvaticum,* which occurs in Labrador and may very well be found within our extreme north-eastern limits. The same course in admitting or rejecting species has been adopted with regard to other authors.

Special reference will hereafter be made to introduced plants. Here, in order to exhibit the mass of the vegetation of each country and the relative proportions which classes, orders and genera bear to one another with regard to the entire number of species which they include, both indigenous and introduced plants are, without distinction, embraced in the statistics of species now given.

As far as considerable care can extend the catalogue, there are 1,676 flowering and filicoid plants in Ontario and Quebec. Of these, 1,161 are referable to dicotyledonous, 450 to monocotyledonous, and 65 to filicoid species. Monocotyledons are thus to dicotyledons as 1:2.5, and to phænogams as 1:3.5. In the Northern States the relative numerical proportions are almost identical, and the extension of the comparison to the whole Union does not much alter them. The large number of monocotyledonous species is very remarkable, and evinces a climate and physical conditions very favourable to these plants. Again, filicoid plants are to phænogams in the Provinces as 1 to 25, whilst in the Northern States they are as 1 to 28.7.

Some facts of considerable interest are presented by the relations which the different orders bear to one another, and to flowering plants, with respect to the number of included species. In ten natural orders are grouped nearly one-half of our indigenous and introduced species, and eighteen orders represent about two-thirds of them. Another interesting feature which appears quite as conspicuous in the United States flora, is that Cyperaceæ, Graminæ, Orchidaceæ, and Liliaceæ embrace the greater portion of our endogenous plants. Again, in the United States, east of the Mississippi, the Compositæ number 1-7th, and the Cyperaceæ 1-11th of the entire phænogamous flora ; whilst in the Provinces the same orders comprise nearly 1-9th and 1-11th, and in the Northern States 1-8th and 1-10th respectively. The

grasses bear very nearly the same relations to flowering plants—
1-12th to 1-13th—in the three divisions of country mentioned.
Among other orders there are some marked differences in the
proportions as they are exhibited in the different geographical
regions;—in some the species proportionably increase from Canada
southward ; in others, the reverse of this is the feature. The five
examples cited below will illustrate these particulars :—

|  | Ontario and Quebec. | Northern States. | United States. |
|---|---|---|---|
| Leguminoseæ ..................... ................. | 1-29th | 1-21st | 1-18th |
| Euphorbiaceæ ........................ .... | 1-95th | 1-72nd | 1-58th |
| Rosaceæ...., .... ...... ....... ....... ......... | 1-25th | 1-32nd | 1-40th |
| Cruciferæ ..................... ...... ......... | 1-31st | 1-39th | 1-49th |
| Ericaceæ................................... ...... | 1-34th | 1-38th | 1-43rd |

Among the smaller orders there are instances quite as marked.
Convolvulaceæ increases from eight species within our limits to
twenty-four in the Northern States, and forty-one in the whole
Union ; and the Malvaceæ are similarly augmented from eight to
twenty-two and forty-four ; whilst in Cupuliferæ the species, in
which are sixteen, twenty-three, and thirty-one, respectively, the
numbers proportionally diminish.    These circumstances tend, of
course, to indicate the well-known facts, that, whilst some of the
orders mentioned are semi-tropical and southern temperate, others
are more abundant in the northern temperate regions of America.

The number of species occurring within our limits in each of
the large orders is indicated below.    To admit of a comparison
being more easily made, the numbers in the same orders in the
United States are placed in parallel columns.

|  | Ontario and Quebec. | Northern States. | United States. |
|---|---|---|---|
| Compositæ......... ......... ......... | 194 | 324 | 481 |
| Cyperaceæ.................... .... ......... | 155 | 248 | 336 |
| Graminæ............ ..... .. ........ | 124 | 212 | 287 |
| Rosaceæ ..................... ......... | 65 | 81 | 92 |
| Leguminosæ......................... | 55 | 120 | 199 |
| Cruciferæ..... ... ................... | 51 | 65 | 74 |
| Ericaceæ........................... | 47 | 68 | 84 |
| Labiatæ.......... ...... ............. | 47 | 76 | 108 |
| Orchidaceæ ..................... ..... | 46 | 57 | 71 |
| Scrophulariaceæ............... ..... | 44 | 66 | 94 |
| Filices.... ......................... | 44 | 57 | 76 |
| Liliaceæ....... ... .................. | 42 | 62 | 78 |
| Caryophllaceæ.............. ......... | 34 | 33 | 70 |
| Polygonaceæ.......... ............. | 34 | 38 | 54 |
| Umbelliferæ.......... ....... .......... | 28 | 45 | 58 |

To somewhat complete the parallel drawn, it will be useful to
bring to view the number of species in the more important
genera of Ontario and Quebec and of the Northern States.    To
extend the comparison to the flora of the Southern States may
diminish its interest, as many of the conspicuous genera there are

but scantily or not at all represented north of the Great Lakes or in the valley of the St. Lawrence. The carices, it will be observed, constitute nearly 1-14th of our flowering plants. The asters comprise thirty-one and the solidagos twenty-six species—the larger number in each case being in Ontario—and together form 1-28th of phænogams. The maximum development of these two genera is probably in the Northern States, but they do not there form so conspicuous a relation to the entire vegetation as, though they comprise seventy-eight species, they constitute but 1-33rd of the flowering plants. Along the northern banks of the lower St. Lawrence and among the Laurentide hills to the northward, the same genera are, in both number of species and individuals of each species, poorly represented; and in the effect which they elsewhere have upon the aspect of the shubby and herbaceous vegetation, they are replaced by *Cornus Canadensis* and *Vac-cciniums*

| Ontario and Quebec. | | Northern States. | |
|---|---:|---|---:|
| Carex | 118 | Carex | 153 |
| Aster | 31 | Aster | 41 |
| Solidago | 26 | Solidago | 37 |
| Polygonum | 19 | Juncus | 26 |
| Ranunculus and Juncus, each | 18 | Potamogeton and Euphorbia, each | 23 |
| Salix | 17 | Polygonum | 22 |
| Viola | 16 | Cyperus and Scirpus, each | 21 |
| Euphorbia and Habenaria, each | 15 | Panicum and Helianthus, each | 20 |
| Panicum | 14 | Desmodium and Ranunculus, each | 19 |
| Potamogeton and Rumex, each | 13 | Habenaria | 18 |
| Poa | 12 | Quercus, Viola and Eleocharis, each | 17 |
| Vaccinium | 11 | | |

Common to Ontario and Quebec on the one hand, and to the Northern United States on the other, there are no less than 1,591 flowering and filicoid plants. Of these, 1,089 are dicotyledonous, 440 monocotyledonous, and 62 filicoid species. There are thus eighty-five species which are without representatives across the border. Of these, however, it should be specially observed nineteen are manifestly introduced, and there are therefore only sixty-six indigenous plants which, as between the two Provinces and the Northern States, are peculiar to the former. There is thus a very marked similarity between the floras of these two sections of country. The indigenous species referred to include the following:—

Anemone narcissiflora, L.
Thalictrum alpinum. L.
Ranunculus affinis, R. Br.
R. cardiophyllus, Hook.
Caltha natans, Pallas.
Aquilegia vulgaris. L.
Arabis patula, Graham sp.
A. brachycarpa, Torr. & Gray sp.
A. retrofracta. Graham.
Erysimum lanceolatum, R. Br.

Vesicaria arctica, Richn.
Draba hirta, L.
D. muralis, L.
Thlaspi montanum, L.
Cochlearia tridactylites, DC.
Arenaria arctica, Steven.
Linum perenne, L.
Astragalus Labradoricus, DC.
Dryas octopetala, L.
D. Drummondii, Hook.

Geum geniculatum, Michx.
Rubus arcticus L.
Rosa stricta, Lindl.
Epilobium tetragonum, L.
Ribes oxyacanthoides, L.
Saxifraga Grœnlandica, Hook.
S. nivalis, L.
Angelica lucida, L.
Sium latifolium, L.
Cornus suecica, L.
Nardosmia frigida, Hook.
Aster Lamarckianus, Nees.
A. cornuti, Nees.
Matricaria inodora, L.
Gnaphalium sylvaticum, L.
Antennaria Carpathica, R. Br.
Senecio canus, Hook.
Hieracium vulgatum, Fries.
Crepis runcinata, T. & G.
Andromeda tetragona, L.
Ledum palustre, L.
Armeria vulgaris, L.
Penstemon gracilis, Nutt.

Pedicularis palustris, L.
Melampyrum pratense, L.
Mertensia Sibirica, Don.
M. pilosa, Don.
Gentiana acuta Mx. v. stricta, Hook.
Pleurogyne rotata, L.
Rumex acetosa, L.
R. domesticus, Hartm.
Elæagnus argentea, Ph.
Salix reticulata, L. var. vestita.
Alisma natans, Ph.
Echinodorus subulatus, Engel.
Iris tridentata, Ph.
Eriophorum capitatum, Host.
E. russeolum, Fries.
Carex Macounii, Dew.
Carex bicolor, Allioni.
C. ovata, Rudge.
Elymus Europæus, L.
Triticum Macounii, Dew.
Asplenium viride, Hudson.
Woodsia hyperborea, R. Br. *
Equisetum littorale, Kuhl.

A critical examination of the above catalogue suggests some remarks. *Ranunculus affinis* and *R. cardiophyllus* will by some authors be referred to *R. auricomus* Linn., which, however, is a known British-American plant, and is absent from the United States flora; *Geum geniculatum*, *Angelica lucida*, and *Aster cornuti* are species of which not much is known; *Carex Macounii* and *Triticum Macounii* were only discovered about two years since, and, when their range is more fully known, may be found to occur south of the lakes; *Sium latifolium* Prof. Gray rejects from his manual as erroneously applied to the broad-leaved form of *S. lineare* Michx., and here a similar mistake may probably have been made; and *Equisetum littorale* perhaps requires confirmation. Prof. Gray, again, in the manual, takes no notice of *Arabis brachycarpa*, which Torrey and Gray locate at Fort Gratiot, Michigan; of *Nardosmia frigida* (to which *N. sagittata* Hook. is referred) which, on Pursh's authority, occurs on the highest mountains of Vermont and New Hampshire; of *Ledum palustre*, whose occurence in Vermont and Pennsylvania is mentioned by Beck; or of *Penstemon gracilis*, to which Wood gives a place in his flora, with Chicago as a locality. It should be further observed that *Matricaria inodora* is adventive though not native in Maine. *Aster borealis*, Prov., if a good species, and not a variety of *A. æstivus*, must be added to the list. If the twelve

---

* EDITOR'S NOTE.—My esteemed correspondent, the late Mr. Horace Mann, sent me specimens of this fern, collected by himself on Willough-by Mountain, Vermont. *Lycopodium alpinum*, long known as a New-foundland plant, may be added to this list; it occurs on the north shore east of Point de Monts, and probably elsewhere.     D. A. W.

species referred to be rejected from the catalogue, there still remains fifty-four species unrepresented in the Northern States.

In connection with the non-occurrence of these plants in the Northern States, their range becomes a subject of considerable interest. Speaking generally, some are of semi-arctic and boreal types, and only occur in the more northern or otherwise suitable stations; others are entirely western in their distribution; whilst there are a few which are sparingly distributed in the Provinces, or with whose range we have but a limited acquaintance. *Ranunculus affinis, Thalictrum alpinum, Vesicaria arctica, Cochlearia tridactylites, Saxifraga Grœnlandica* and *S. nivalis* are peculiar to the arctic climate, and, with the exception of the Ranunculus and Cochlearia, are also denizens of the coasts of Greenland. *Arenaria arctica,* an interesting discovery of which was not long since made at Muskoka Lake, Ontario, by one of Prof. Hincks's students ; *Dryas Drummondii,* a pretty species in the Gaspé collections of Dr. Bell ; *Astragalus Labradoricus, Rubus arcticus* and *Pleurogyne rotata* are examples of a less arctic type, though the little Arenaria penetrates the polar regions beyond Whale Sound on the West Greenland coast. *Ribes oxyacanthoides* is said by Torrey and Gray to occur throughout Canada ; and *Caltha natans, Aquilegia vulgaris, Linum perenne, Rosa stricta, Matricaria inodora* and *Elæagnus argentea* are probably limited to the north western parts of Ontario, and may be looked for in the neighbouring districts of the Northern States.

---

## ON LESKIA MIRABILIS (GRAY).

By Prof. S. LOVÉN.

*Communicated by Dr. CHRISTIAN LUTKEN, Assistant Zoologist in the Museum of the University, Copenhagen.

This little paper, inserted in the Proceedings of the Royal Swedish Academy for 1867, well deserves the attention of palæontologists, though its principal aim is to redescribe a little-known *recent* Sea-Urchin from the Eastern Seas, because this animal throws a peculiar light on certain important points in the morphology of Cystidea. It is, moreover, distinguished by all the ingenuity, accuracy, and profound knowledge which is peculiar to the works of the celebrated Scandinavian zoologist.

---

* From the Geological Magazine, vol. v., p. 179.

The genus *Leskia* is described, in 1851, by Dr. J. E. Gray, in the " Annals," and subsequently, in 1855, in the Catalogue of Recent Echinida, from specimens from Lugard, in Mr. Cummings's collection. It is most intimately allied to the Spatangidæ, of which it has the general stamp, but is distinguished from them, and therefore the type of a peculiar family (*Leskiadæ* Gray) or tribe (*Palæostomata* Lovén) by the peristome and periproct being closed up with a few " triangular converging valves," those of the vent with some small " spicula" in the centre. Dr. Gray has already remarked that " in the form of the mouth and vent it has considerable affinity with the fossil Cystidea, especially the genus Echinosphærites." The detailed description given by Prof. Lovén quite confirms this remarkable combination of features ; the characters assigned to the Palæostomata are as follows: " *testa oviformis, peristomium non labiatum, pentago-num, æquilaterale, ore quinqueralis, anus intra periproctium centralis, valvis clausur quinque octo ; aperturæ genitales binæ ; semita unica peripetala.*" Leskia is a true Spatangoid, *save the mouth and the vent ;* the latter, instead of being surrounded by a threefold circle of minute plates, the greater and outermost, has only five, seven, or eight great triangular outer plates, and an equal num-ber of minute inner papillæ. The peristome is not bilabiate with a prominent under-lip, nor is it formed principally by the ambula-cral plates ; it is pentagonal, and bordered almost exclusively by the interambulacralia ; there is no buccal membrane covered with three to five series of irregular plates, decreasing inwards, but the mouth is closed up by five equal triangular plates, inserted on the five sides of the peristome. " No living Echinid has such a mouth ;" but the author thinks that the genus Toxaster of the ' Neocomien Inférieur,' whose peristome was pentangular, not labiate, might possibly—though the configuration of its mouth somewhat more approaches to that of the true Spatangidæ—have had a similar organization.

In the Silurian Cystidea again, we find precisely the same structure as in the recent East Indian Sea-urchin, viz., in the commonly so-termed ' ovarian pyramid,' which, after the opinions of Gyllenhal, Wahlenberg, Pander, Hisinger, de Koninck, and Billings, is really the mouth, whilst Von Buch, with some incon-sistence, makes it the mouth of Caryocrinus, but the genital outlet in the other Cystidea, and Joh. Muller and Volborth sought the mouth in the centre of the converging ambulacral furrows. The

remarkable observations on *Sphæronites pomum* and *Echinospha-
rites aurantium*, by means of which Prof. Lovén draws the conclusion that Leskia is a Spatangoid *with the mouth of a Cysti-
dean*, we will give with his own words. (See figures on page 443.)

"Good specimens of *Sphæronites pomum* Gyll., collected by
Prof. Angelin, show its organization more distinctly than usual.
He had observed that this animal had no stalk, but adhered immediately to rocks or other objects through a part of its lower
surface, which is without pores, and surrounded by a ridge formed of the somewhat thickened, free, smooth border of the under-
most plates. This surface of attachment is of a very variable
form and extension in different specimens,—round and but little
excavated in some, oblong and deep in others,—depending upon
the nature of the object to which it adhered. On the point
opposite to this basal surface lies the apex with the ambulacral
apparatus. In the middle of a somewhat deepened area *d*, through
which five delicate but distinct ambulacral furrows pass towards
five arms, whose bases form a circle, which however is broken at *f*,
one-fifth of its circumference. Where the furrows reach the arms,
they will be seen to pass into an oblong hole *e*, which is the lumen
of the broken furrow of the lost arm: in every remaining arm-
base you will see an indication of the branching of the arms and
of the central channels of the branches. Close up to the ambulacral circle lies the 'pyramid' or mouth *a*, closed by its five valves
of unequal dimensions; two of them are emarginate on one side
in order to give space to the two adjoining outermost arms,
which are less than the others, and, as it were, crippled, the right
by its vicinity to an oral valve, the left by an apparatus *b*, that
cannot be interpreted otherwise than as an external genital organ.
When it is tolerably well preserved, it is conical, with a rounded
apex, without any terminal aperture; for vestiges of valves I have
sought in vain, but in two specimens I found the two pores
indicated in the figure. From this organ a ridge *c* runs towards
the next arm, suggesting the idea of the possible existence of a
'madreporite.' The centre of the brachial apparatus forms
with the genital organ and the oral orifice a compressed but only
slightly inequilateral triangle. In *Echinosphærites aurantium*
the relative position of these parts is the same, but the triangle
which they form with each other is much larger, longer, and
more inequilateral, because the distances are greater, especially
that of the mouth from the ambulacral apparatus, which is cor-

rectly described and delineated by Volborth and Joh. Muller.
Close to this is seen the other ' orifice,' viz., the external genital
organ. All specimens that I have examined have this so-termed
' orifice' in such a condition that it most likely is the remnant
of a prominent broken part, and it must be assumed that in this
species also it had a conical form, but remained mainly in the
surrounding stone-matrix. Volborth's figure (Ueber die Russis-
chen Sphæroniten, x. ix. f. 9) appears to be correct, but gives no
complete evidence as to the presence of the three valves. That
the ' pyramid,' which in Leskia is the armature and covering of
the mouth, is the same thing in Cystidea, is now quite certain ;
in the last-named group it was, doubtless, also the vent. The
mouth does not lie where J. Muller and Volborth sought for it,
viz., in the centre of the ambulacral furrows ; and the organ, interpreted as the vent by Volborth and Von Buch, is more correctly
regarded as an external sexual organ."

It is not my intention to criticise the various interpretations of
the morphology of Cystidea given by different authors, or to
trespass on the space here allowed me by a detailed examination
of all the questions entangled with them. But should I venture
to express any humble opinion of my own on this important point
in the morphology of Echinodermata, I must first confess that
hitherto I have been very sceptical as to the theory advocated so
very ingeniously by Mr. Billings and now upheld by Mr. Lovén.
The concordance between these two authorities is nevertheless
not so great as would be supposed—that the ' pyramid' was the
mouth of the Cystidea, and that this orifice accordingly would
lie elsewhere than in the centre of the ambulacral system, where
it lies in all living Echinoderms and (I may add, where it did lie,
I have no doubt, also in the Palæozoic Crinoids, where no superficial ambulacral channels are to be seen, but where they pursued
their way on the inferior surface of the 'vault' through the
' ambulacral orifices' at the base of the arms,—as shown by
Mr. Billings, with those researches (see Decades Geol. Survey of
Canada) I was, I regret, unacquainted when I wrote my paper
on Pentacrinus, etc.) I know no other exception to this rule;
and would it not be a dangerous thing—not to be done without very
strong arguments—to give up the leading principle of Palæontology, viz., that only from the organization of the *living* form can
we learn to understand that of the *extinct ?* Might we not thus
too often run the risk of giving up ourselves to the delusions of

fancy! When we remember how minute and concealed the mouth
often is in *recent* Crinoids, we should not be puzzled at its being
almost or quite invisible in *fossils;* and if we should search for
the interpretation of an orifice, closed by a definite low number of
triangular valves, will not several recent Echinidæ (*Echinocidaris,
Echinometra arbacia, Leskia* itself,) give us the answer, that
such an aperture could (at least) be a vent? Nor can I well
conceive that an aperture should altogether fail to exist in the
centre of the ambulacral system of Cystidea. How otherwise
could the ambulacral vessels communicate with the interior?
And if such an orifice *must* be assumed (though it be often
obliterated and hidden in the fossils), why should not this apical
or ambulacral orifice be also the mouth as in Asteridæ and
recent Crinoids, and the valvular orifice be the vent, analogous
to the proboscis of the Palæolithic Crinoids, or the oral tube
of the living?* The superiority of size of the presumed mouth is
not, as Mr. Billings thinks, a very good argument. Has not
the anal tube in many of our recent Crinoids (*Antedon, Actinome-
tra, Pentacrinus*) the same preponderance over the minute buccal
orifice? Nor has the repeated revision of the published descriptions of other Cystidea, accessible to me, convinced me of the
correctness of a theory, according to which the mouth would, in
many instances, lie very far from the arms, sometimes nearer to
the base (the stalk or point of attachment) than to the apex of
the calyx. The argument deduced in later times from the
presumed existence of five similar peristomatic valves in the
recent Pentacrini, I have elsewhere had the opportunity of
refuting;† no such hard "clapets" are to be seen in *P. Mulleri,*
and until their existence is *proved* in other recent Pentacrini, I
must doubt, or rather deny, their existence at all!‡ On the other
hand, I must confess that matters are considerably altered by
these highly valuable investigations of Prof. Lovén, who, for the
first time, supports this theory with strong (perhaps convincing)

---

* The analogy between the valvular aperture of Caryocrinus and the
'proboscis' of Crinoids is also argued by Mr. Billings (Dec. No. 3, p. 22).

† Om Vestindiens Pentacrinen, p. 205 (Vidempel. Meddel. f. d. Natur-
hist Forneing, 1864).

‡ Prof. Lovén told me himself that during his last stay in Paris he
succeeded in getting access to the original specimen of Mr. Duchassaing,
in the collection of the late Mr. Michelin. It did not show the five
valves, because it had no peristome at all!

arguments.  It is now no longer a mere hypothetical supposition
—hitherto it was in reality no more—but a real scientific explanation, borne out by well established facts and undeniable analogies from living forms.*   To Dr. Gray we certainly owe the first
intimation of this analogy between Leskia and Cystidea, but
while the knowledge of that genus rested on a single examination,
there might still linger some doubt whether its importance in this
respect had not possibly been overrated.  Science, therefore, must
be highly indebted to Prof. Lovén for his small but valuable
memoir, and for the excellent observations laid down in it.  The
absolute denying of the existence of an apical orifice in that place
where, in other Cystidea at least, such an orifice was also believed
to exist, is particularly recommended to the attention of future
investigators of Cystidea, as bearing upon the very heart of the
question.  *Adhuc sub judice lissit !*

### NOTE BY E. BILLINGS, F.G.S.

Professor Lütken is certainly mistaken when he makes use of
the expression, "It is *now* no longer a mere hypothetical supposition,
hitherto it was in reality no more," etc.   The earlier Palæontologists, Gyllenhal, Wahlenberg, Pander and Hisinger, described
the valvular orifice of the Cystidea as the mouth, but they never
proved it to be so.  Indeed they could not do so, for the data, *i. e.*,
the structure and functions of the arms of the Crinoids living in
the sea at the present time, were not known.   In 1845 Leopold
von Buch pronounced the aperture in question to be an ovarian
orifice, and the small one in the apex the mouth.  His views
were adopted by Prof. E. Forbes, in his beautiful memoir on the
British Cystidea and by Prof. J. Hall in the Palæontology of
New York.  In my first attempt at describing fossils, in 1854, I
followed these three last named distinguished Naturalists, in a
paper on the Cystidea of the Trenton Limestone at Ottawa,
published in the Canadian Journal.  But in 1858, while re-
investigating the subject for my Decade, (No. 3,) I saw that they
were wrong, and proved it according to the ordinary rules of comparative anatomy.  If any organ of an extinct animal is the exact
homologue of an organ possessed by an existing species (of the
the same zoological group), its function must have been the same.

---

* To these analogies might be added, that between the valves of
Cystideæ and those of the young (larval) Antedon.

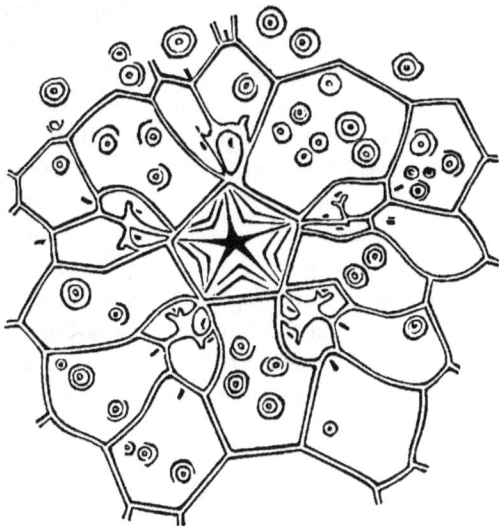

Fig. 1.

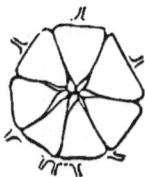

Fig. 2.

Fig. 3.

Fig. 4.

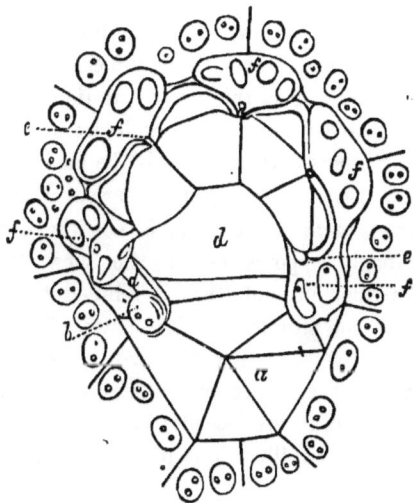

Fig. 5.

Fig. 1. Mouth and adjoining parts of *Leskia mirabilis* Gray. Fig. 2. Vent of the same. Figs. 3 and 4. The mouth of *Echinosphærites auran-tium* Gyll. Fig. 5. The apex of *Sphæronites pomum* Gyll. (*a*.) The mouth. (*b.*) The genital process. (*c.*) Its ridge. (*d.*) The ambulacral area with its furrows. (*e.*) The lumen of the furrows. (*f.*) The base of the five arms.

The principal office of the arms of the existing Crinoids is the maturing of the ova. On comparing the arms of the extinct Crinoids with those of the species living at the present day, we find that both have the same anatomical structure and, consequently, they are all the homologues of each other. The small apertures, at the bases of the arms of the ancient species, are the passages through which the ovarian tubes and the vessels of the ambulacral system gained access to the grooves and pinnulæ. Their functions were first pointed out in my Decade. The arms of the Cystidea are the homologues of those of the Crinoids. This at once proves that, in the Cystidea, the orifice at the apex, which in all cases opens out into the grooves of the arms, is the ovarian aperture. The large lateral orifice is undoubtedly the exact homologue of the valvular opening in the summit of Caryocrinus which is admitted by all to be the mouth. I proved all this in my Decade, and consequently in 1858, the date of the publication of that work, the theory that the lateral aperture of the Cystidea is the mouth, ceased to be a mere hypothetical supposition as Dr. Lütken calls it.

The Cystideans are rare fossils; few Palæontologists have occasion to examine them, and consequently only a few have given their opinion on this vexed question since 1858. J. W. Salter, the celebrated English Palæontologist says : " I strongly suspect Mr. E. Billings is right ; this is the anal, not the ovarian Pyramid,"* thus partly adopting my views. Prof. Wyville Thompson also agrees with me that it is not the ovarian orifice, but then he strongly opposes me in the view that it is the mouth on the same ground, that is alluded to by Dr. Lütken, i.e., that it is not situated in the centre of the radial system.† Prof. J. D. Dana has recognised it as the homologue of the oral and anal aperture of the Criniods, which is exactly the opinion advocated in my Decade‡ ; and now it gives me much satisfaction to add the illustrious name of Prof. S. Lovén to this short list.

With regard to the grounds taken by Prof. Wyville Thompson and Dr. Lütken, I freely admit that if it is impossible for an Echinoderm to have the mouth situated anywhere except in the

---

* Memoirs of the Geological Survey of England, vol. iii, p. 286.

† Edinburgh New. Phil. Jour. vol. xiii p. 112.

‡ Manual of Geology p. 162.

ambulacral centre, then my theory falls to the ground. But all experience in Palæontology has proved over and over again, that although we can show that the extinct animals, whose remains we find buried in the earlier formations, possessed organs identical in their functions with those of the existing races, yet they were not always combined together in the same manner. As an example we have only to refer to the Crinoidea. In the few species known to live in the seas of the present day, the mouth and the vent are separate orifices; but in the palæozoic species they were combined into one. Why, then, is it impossible that the mouth and radial centre, which are now united, could not be separate in the earlier ages? This question, however, can be decided without argument. I have specimens lying before me, in which we can see the mouth and also the radial centre, and at the same time see that they are not in the same place. A long train of reasoning is not necessary,—only simple inspection.

---

# A FEW POINTS OF INTEREST IN THE STUDY OF NATURAL HISTORY.

### THE PRESIDENT'S ADDRESS BY THE REV. A. DE SOLA, LL.D.

LADIES AND GENTLEMEN,—The study of Natural History, if merely considered in its aspect of a branch of human knowledge, has a claim on every one's attention. It is a knowledge which is not merely power, but pleasure; and has claims great and peculiar on both the theoretical and practical man. The theoretical will find in it almost boundless scope for absorbing and interesting cogitation in such inquiries as the origin of species, spontaneous generation, the animal or vegetable character of certain obscure forms of life, the correlation of physical forces, mutual relations of the physical and vital forces, and similar modern engagements of human thought. The other great class, the practical, who have been taught by the books of their earliest youth to appreciate the difference between ' eyes and no eyes,' will also be prepared fully to admit with the student of Natural History that, merely to see an object, or to remember its name, is not to know it; and that if thoroughness of knowledge be essential or desirable in all the practical engagements of life, it must be equally so in our study of the countless objects of nature's universal domain—objects that are inseparably connected with the supply of all human necessities and comforts. But this knowledge is not merely useful, it is also elevating and interesting in the highest possible degree; and this I will proceed to show as far as I can in the brief limits to which I must confine myself, by seeking in the three great kingdoms of nature some practical illustrations of the truth of these assertions.

The animal world, from which we may take our first illustration, presents, from its lowest to its highest forms, a series of organic structures progressing with almost imperceptible gradation in perfection of development and complexity of organization. Amongst the simplest of its representatives are the Protozoa, the great majority of which are too small to be distinguished without the aid of the microscope. They are graphically described by Dr. Wm. B. Carpenter as consisting of " seemingly structureless jelly." They perform those vital operations which we are accustomed to see carried on by an elaborate apparatus without any special instruments whatever; a little particle of apparently homogeneous jelly changing itself into a greater variety of forms than the fabled Proteus,—laying hold of its food without members, swallowing without a mouth, digesting without a stomach, appropriating its nutritious material without absorbent vessels or a circulatory system, moving from place to place without muscles, feeling (if it has any power to do so) without nerves, multiplying itself without eggs, and not only this, but, in many instances, forming shelly coverings of a symmetry and complexity not surpassed by those of any molluscous animal. And yet these creatures have performed, and are still performing, one of the chief parts in the history of this globe. With them, we arrive at that mysterious border-land which divides, and yet seemingly blends, the organic and inorganic world; where we find arising the simplest vegetable and animal structures scarcely distinguishable from each other, and beyond which we cannot proceed in our search for the beginning of life. Yet the earnest student when examining them feels with more than ordinary intensity the profound mystery of life, and will continue to investigate the phenomena they present in eager hope of new revelations. But the Protozoa have not ungenerously left without reward the researches made in their behalf. They have presented to man's astonished sight objects of marvellous beauty in the form and structure of the microscopic shells of many of them. They have also enabled him to obtain enlarged conceptions

respecting the nature of species and the laws of organic life, and
have taught him to recognize in these minute organisms some of
the chief builders of the earth's crust,—many of its component
rocks being the stupendous monuments of their labors, and in
which they lie entombed.

Not without interest, also, will be found the study of the shell-
fish, long considered the most inert and stupid of all animals.
" Les mollusques," wrote Virey, even within our own time, "sont
les pauvres et les affligés, parmi les êtres de la création; ils
semblent solliciter la pitié des autres animaux." On the other
hand, Lorenz Oken exclaims, " Surely a snail is an exalted symbol
of mind slumbering deeply within itself!" Shakespeare's fool hit
the happy medium between extremists, when he told King Lear
that the reason why the snail has a house, was " to put his head
in, not to give it away to his daughters, and have his horns
without a case." Lucian ridiculed the philosophers who spent
their lives inquiring into the soul of an oyster; but a modern
writer is yet more severe on the conchologists when he says
" Lucian's wiseacres were respectable when compared with their
brethren, who care for neither an oyster's soul nor body, but con-
centrate their faculties in the contemplation of its shell." But
this writer may have forgotten that the conchologist—reversing
the procedure of the lawyer of the fable, who gave to his clients
the shells and kept the oyster to himself—may be as much war-
ranted in examining the waves, scales, and ribs of the shell, as is
another to anatomize the contained creature, which, says Lentitius,
" animal est aspectu et horridum et nauseosam, sive ad spectes in
sua concha clausum," etc. Without claiming too much for the
shell-fish, we may assert that the student will find them possessing
quite a sufficiency of acuteness and sensibility, and their in-
stinctive proceedings are often very surprising. Some of these
proceedings of mollusks, it is true, we are not always inclined to
admire; for instance, those of the Teredo, or ship-worm, that
terrible destroyer of ships, landing-piers, and dockyards; though,
perhaps, he may consider he is only offering just retaliation for
man's unceasing warfare against his cousins—the oysters. I
may not stay to take a more particular view of the mollusks, but
will proceed to notice a few points of interest in the study of the
vegetable kingdom.

About a century and a quarter ago, Linnæus declared the
number of the different kinds of plants to be 5,938. Half a

century afterwards the estimate had increased five-fold. In 1847 it was announced as 92,920 ; and now, Meyers and others calculate the entire vegetation of our planet to consist of some 200,000 species. The aborigines of New Zealand have learned to distinguish by name some 700 species of the trees and plants produced on their own island, a number considerably greater than that described by Theophrastus in the first history of plants ever given to the world. But besides those plants which the pious and philosophic Ray says " are by the wise disposition of Providence proper and convenient for the meat and medicine of men and animals"—besides those which enable the botanist, like his prototype in Milton's Comus, to

> " Ope' his leathern scrip
> And show simples of a thousand names,
> Telling their strange and vigorous faculties,"

we find vegetable life in its most simple form and development represented by the mere primary cell; and of the one-celled plants the most interesting order is the Diatomaceæ. The yellow-dust, which falls like rain on the Atlantic, near the Cape-de-Verde Islands, and occasionally drifts even to Italy and Central Europe, was found by Ehrenberg ·to consist of myriads of silicious-shelled microscopic plants. Darwin discovered that a cloud of dust, drifting through the air from America to Africa, and coming in contact with the rigging of the ship in which he was sailing, consisted of the shelly coverings of diatoms. The naturalists of the Antarctic Expedition constantly found them adhering to the lead, after sounding depths in the ocean which would have engulphed the loftiest peaks of the Andes. Humboldt, on the other hand, has shown that they float in the upper currents of the atmosphere perhaps for years, until brought down to the earth by vertical currents. But, turning from these —and the almost equally interesting family of the Fungi, which are so destructive to our bread, fruits, and other objects of domestic economy,—I would now, on the Solomonian principle of ascending from the hyssop to the cedar, say a few words respecting some of the giants of vegetation. I take, as an illustration, the celebrated big-trees of California. This group of huge conifers (placed botanically between the pine and the juniper) was discovered in 1850, by some hunters when pushing their way through a hitherto unexplored forest in the Calaveras country, about 240 miles from San Francisco. It is deeply to be regretted that cupidity and vandalism have led men to hew down the largest of the group, for the purpose of making a show of it. One measured ninety-six feet in circumference, and afforded ample space for thirty-two persons to dance on; theatrical performances were given on it in 1835 ; it measured three hundred and two feet as it lay on the ground. The so-called 'Mother of the Forest' is ninety feet in circumference, and three hundred and twenty-seven high. The largest, called the 'Father of the Forest,' is forty-two feet in circumference and four hundred and fifty high—only a few feet lower than the Pyramids of Egypt. As a set-off to this barbarity—which, be it said, no where called forth greater indignation than in the United States, —the Wellingtonia, * as these trees were called by the English (Washingtonia by the Americans), have become acclimated in England and Scotland, where their growth, first recorded in inches, is now annually reported in feet. The propagation of these trees lead us to examine, as points of interest in the vegetable kingdom, the more general subjects of the propagation of plants by nature's wondrous provisions, their fertility and preservation.

Recurring for an instant to the Diatomaceæ, I may here remark that the existence of these minute uni-cellular organisms may lead the uninitiated to doubt whether they could well answer that apparently easy question, What is a plant? Further investigation would show that it is difficult for the greatest adept to do so, and that when it is attempted to draw a line of demarcation between the primary conditions and forms of animal and vegetable life, no problem in the science of nature is more obscure ; and the difficulty increases too with our knowledge. Perhaps this may be sufficiently shown by those familiar objects, the sensitive plant and the sponge. It was always held by naturalists that the property or character distinguishing animals from plants is feeling, which is evinced in the lower forms of animal life by their shrinking from the touch. But when we try vegetables as well as animals by this rule, we find many plants (one example is the *Mimosa pudica*, or sensitive plant) endowed with a far higher degree of susceptibility to external impressions than is evinced by some of the lower races of animals under the

---

* Dr. Torry has shewn conclusively that these trees belong to the genus Sequoia.—ED.

operation of tests which, if applied to the higher races, would amount to torture. Thus, the art of ingeniously tormenting has been exhausted in vain upon the imperturbable sponge, which is so endowed with vital powers as to render its animal nature unquestionable;—lacerated with forceps, bored with hot irons and saturated with the fiercest acids of the chemist, it has never once given any symptom of suffering or sensibility. These facts may be sufficient to show that no difference of a physical or chemical nature can be established between plants and animals in that low part of the organic world where these two great divergent branches have their source, and that any attempt to separate them must be arbitrary and artificial. Here, then, the student of Natural History learns the great lesson of a fundamental unity prevailing throughout organic nature; he sees exhibited to him a sequence without interruption in the working out of the divine idea of creation from man spiritual and immortal, in whose wonderful organization meet and culminate the structural perfections of all the animals, down to the primary cell in which both vegetable and animal life exhibits its simplest form of development.

Turning now to the third of nature's great kingdoms, I would remark that no one has ever questioned the utility of that study which directs and guides us in our search within the bowels of the earth for the ores and other substances that are at once the sources of national wealth and the supply of human wants and comforts. But while the utility of the study of mineralogy is everywhere conceded, geological research, which is inseparably connected with it, has been regarded not without much suspicion and disfavor. Irrespective of the fact that all quarrying and mining undertakings must be properly based on and directed by, geological knowledge, how different the aspect which a section of country exhibits to the eye of a geologist and of the uninformed spectator. Whether it present sand, gravel or alluvial soil, and in its form, hill or valley, solid rock or detached boulders—all add to the interest and pleasure of the scientific observer. The stone turned up by the ploughman, and which would not interrupt his whistle, or call forth the slightest interest in the stolid wielder of pick and mattock, has, for the geologist, sermons and histories, exhibiting to him mighty changes and wondrous revolutions, that have completely changed the surface of the globe he lives on. The careless laborer breaks the stones that have no other interest in his eye than that they are intended to mend roads; and the quarryman cuts out his slabs, the highest utility of which he deems their appropriation to building or ornamental purposes. Both crush or cut to pieces, in all the blindness of ignorance, the fossil forms of unknown organisms contained in them, but from which the geologist learns the botany and zoology of former ages of the world, and which enable him to predict the great changes to take place in the future. The achievements of geology are, however, too numerous and important even to be glanced at within my limits, but I would venture to say something respecting one of its sub-divisions—Ichnology, or the study of fossil footsteps—revealing to us wonders of the past such as the imagination of even a Milton or a Danté could never conceive.

Possibly Robinson Crusoe himself was not so much astonished at the footprints on the sands of his desolate island, as the naturalist who first saw the footmarks of birds on a slab of sandstone which was turned up by the plough of an American boy in 1802, at South Hadley, in the valley of the Connecticut River. From this valley, the tide of conjecture flowed over other continents, until it seemed finally to settle down into the theory that the Noachic flood had rolled over those sandstone slopes, the surface of which, when the waters subsided, was so soft as to readily receive the imprints of a bird's foot. The traces, then, were those by which the raven of Noah had written the historical fact of his standing on the earth itself; and so the foot-prints were finally set down as those of Noah's raven. For another quarter of a century or more, this dictum of popular ignorance remained uncontroverted, men of science paying but little attention to it, until a Scotch clergyman, Dr. Henry Duncan of Ruthwell, in 1828, called attention to fossil tracks in connection with the sandstones of Corncocklemuir. Dean Buckland, by means of his Bridgewater Treatise, gave wide circulation to Duncan's discoveries, showing that these impressions were found through a depth of forty-five feet of rock, not on a single stratum only, but on many successive strata, thus demonstrating that they had been made at successive intervals. The sandstones of Dumfrieshire are supposed to have been wide-spread expanses of sand of a littoral character, visited and covered by the ancient tides, some of their surfaces, recording atmospheric conditions, being sometimes pitted with hollows, the results of a pelting shower, and these pittings have occasionally such a well-defined and distinct course, that one can ascertain the direction of the wind, which bore the rain clouds along with it. The sandstones of Cheshire, again, exhibit sufficient evidences of solar influence. We find here the sun-dried surfaces of the clayey strata associated with the sandstone, over which animals formerly crawled, cracked and shrunk by the solar beams. Sometimes they present beautiful sand ripples, the result of a gentle breeze breaking the stiff surface of a shallow pool of sea water on these sandy shores. There may also be found instances of the evaporation of salt-water, and the crystallization of sea-salt, from the natural salt pans of the ancient beaches. Another noticeable fact is the almost constant and uniform direction of the impressions. They nearly all indicate that the animals, which Sir William Jardine shows must have belonged to some forms of tortoise, walked from the west towards the east. Further discoveries of fossil foot-steps were made in the United States in 1835; the impressions resembled the feet of birds, and were found in the sandstone rocks near Greenfield. Dr. Hitchcock, President of Amherst College, showed that they were actually produced by the feet of living birds, and that one of the tracks had been made by a pair of feet, each leaving a print twenty inches in length. Says the eminent Owen: "Under the term *Ornithichnites gigan-teus*, Dr. Hitchcock did not shrink from announcing to the geological world the fact of the existence, during the period of the deposition of the red sandstone of the valley of the Connecticut, of a bird which must have been at least four times larger than the ostrich." Says Hugh Miller, "I have already referred to flying dragons, real existences of the Oolitic period, that were quite as extraordinary of type, if not altogether so huge of bulk, as those with which the Seven Champions of Christendom used to do battle; and here we are introduced to birds that were scarcely less gigantic than the roc of Sinbad the sailor." I might add to Miller's remarks, that the Bar Yuchné, that enormous bird of the Talmudic legend, seems to find identification here.

But I must hasten to conclude these remarks, already too long. They must necessarily convey but a very faint idea of the bound-less field of interesting and pleasurable inquiry awaiting the student of Natural History; still, I trust, they will not be without effect in leading into this field, some of those who have not hitherto entered at all. To such my concluding words would be in the accents of caution and advice. I would say, You must needs fearlessly concede to modern science all that is claimed for it, to this extent, that in its dealings with the great physical powers or elementary forces which pervade and govern the material world, it has been led or even forced into a bolder form and method of inquiry,—that inductions of a higher class have been reached, and generalizations attained, going far beyond those subordinate laws in which science was formerly satisfied to rest,— that the precision and refinements of modern experimental research strikingly distinguish it from that of any anterior time,— that physical researches generally in our own day have a larger scope and more connected aim, experiment being no longer tentative merely, but suggested by views which stretch beyond the immediate result, and hold in constant prospect those general laws which work in the universe at large. But, let it be ever remembered that there is also exhibited in our own day, a marked fondness for what is new and difficult and unintelligible in philosophy,—a spirit that takes pleasure in stigmatizing as hindrances to truth in physical science, all such opinions as are fostered by ancient and popular belief, including those which assume Scriptural authority for their foundation. In their too hot zeal against dogmatical authority, we find some falling into the opposite rashness of lending their authority and favour to hasty and partial experimental deductions, or to doctrines still in their infancy, and checked or controverted by opposite opinions of equal weight. Let, then, the dangerous effects of gratifying too prevalent a taste for transcendental inquiries in science be duly marked and carefully avoided, regarding it as cause for gratitude and felicitation that they are corrected by the cotemporaneous activity of those philosophers who make experiment and strict deduction the sole measure and guides of their progress.

## ON SEEDS AND SAPLINGS OF FOREST TREES.

By Dr. J. D. HOOKER, F. R. S., etc.[*]

Forestry, a subject so utterly neglected in this country, that we are forced to send all candidates for forest appointments in India, to France or Germany for instruction both in theory and

[*] One of the Reports on the Paris Exhibition.

practice, holds on the continent an honourable, and even a distinguished place amongst the branches of a liberal education. In the estimation of an average Briton, forests are of infinitely less importance than the game they shelter, and it is not long since the wanton destruction of a fine young tree was considered a venial offence compared with the snaring of a pheasant or rabbit. Wherever the English rule extends, with the single exception of India, the same apathy, or at least inaction, prevails. In South Africa, according to the colonial botanist's reports, millions of acres have been made desert, and more are being made desert annually, through the destruction of the indigenous forests ; in Demarara the useful timber trees have all been removed from accessible regions, and no care or thought given to planting others ; from Trinidad we have the same story ; in New Zealand there is not a good Kandi Pine to be found near the coast, and I believe that the annals of almost every British colony would repeat the tale, of wilful, wanton waste and improvidence.

On the other hand, in France, Prussia, Switzerland, Austria, and Russia, the forests and waste lands are the subjects of devoted attention on the part of the Government, and colleges, provided with a complete staff of accomplished professors, train youths of good birth and education to the duties of state foresters. Nor, in the case of France, is this law confined to the mother country ; the Algerian forests are worked with scrupulous solicitude, and the collections of vegetable produce from the French colonies of New Caledonia, etc., contain specimens which, though not falling technically under Class 87, abound in evidence of their forest products being all diligently explored.

The collection exhibited by the Administration of Forests of France is by far the finest of its kind ever brought together; the enumeration of its contents alone fills an instructive pamphlet of 160 octavo pages, classified as follows, and which further contains a great deal of useful information on the geology of the forest regions, the growth, strength, and durability of timber, and many other matters concerning which no certain information is obtainable in this country. It consists of :—

1. Forest map of France, showing the relations between the distribution of the forests and the geology of the country.

2. A collection, in the shape of books, of the indigenous and naturalized woods. Each species is represented by several specimens, differing in their origin and qualities. The specimens, of which there are 1,300, are divided into two classes; namely, woods of ordinary leaf-bearing trees, and of conifers; these in each class are arranged alphabetically.

3. Collection of truncheons of the most important indigenous species; 223 specimens.

4. Experiments and observations on the density of woods, particularly with regard to age. Specimens exemplifying the opinions given.

5. Collection of seeds and fruit of indigenous and naturalized species.

6. Complete collection of corks of all ages and qualities, and of French production, furnished by the cork oak (*Quercus suber*) and the western oak (*Q. occidentalis*).

7. Barks and astringent substance suitable for tanning or dyeing.

8. Resins from the *Pinus maritima* and *P. Laricio*; methods of procuring them, and their various products.

9. Charcoals.

10. Different products resulting from the carbonization of wood.

11. Forest sawmills; three models.

12. Instruments for felling, prunning, etc., trees, and for collecting resin. A pusher for directing the fall of trees felled by uprooting. The 'Flamm' saw Rollers for the removal of logs from young plantations without injury to the latter.

13. Relievo of the valleys of Barr and Andlau (Lower Rhine), to show the arrangement of the forest roads established there. Sledge tracks with sledges, tramway with waggons, metalled roads.

14. Relievo of the perimeter of the plantations of Labouret, above Digne (Basses Alpes). Photographs of mountains to be laid down with grass or replanted.

15. Photographic forest herbarium, consisting of photographs of the branches with leaves, fruit, and flowers of the various forest trees, all of the natural size.

It only remains to add that the specimens are well selected and excellent, the method of ticketing leaves very little to be desired, and the arrangement is admirable.

With regard to the other collections, chiefly appertaining to Class 87, the reporter has little to say ; there was no English exhibitor, and up to the end of April, when the jurors were called together for the purpose of deciding upon the merits of the exhibitors, there were no collections of any importance ready for adjudication.

Further, various circumstances occurred that rendered it impossible to consider certain collections of plants, some of whose contents might be considered as referable to Class 87, from other cognate classes, and it hence became necessary to amalgamate the duties of Class 87 with those of other classes, including that class under which hardy conifers more naturally came, as objects of landscape gardening or ornamental planting, and not of forestry proper. Under this head comes the beautiful collection of hardy conifers of Messrs. Veitch & Sons, to which the first prize was awarded, with the full complement of marks ; and the same firm carried off the first prize for a collection of the rarest Coniferæ not yet in commerce.

The collection at Billancourt, which did not exist in April, was visited by Dr. Moore, F.L.S., associate juror, in August, and he found many very interesting plants suited for forest purposes amongst them, but they were not exhibited under Class 87, and I shall therefore allude to them here in reference to their being probably, at some future period, introduced into plantations in such considerable quantities as to be profitable as timber trees.

M. Accidin, nurseryman, Lisseux, was awarded the first prize for a collection of forest conifers, which consisted of the kinds usually selected for the same purpose in England, along with many rare species which are not yet sufficiently abundant for forest planting, though they may yet become suitable for that purpose when the prices at which they now sell are lowered at least ninety per cent. *Pinus grandis, P. nobilis, P. Nordmanniana, P. Benthamiana, P. Coulteri*, etc., all of which were in this collection, are not likely to be either moderate in price or plentiful for many years to come. There were equally rare Thujas and Cupressus in this collection, as well as other scarce coniferæ, which obviously cannot be considered under Class 87.

M. Accidin had also a large collection of trees generally used in forest planting, such as oaks, Juglans, willows, etc.

Among the oaks, *Quercus castunafolia*, *Q. ambigua*, *Q. aquatica* and *Q. haliphlæos* were fine foliaged kinds.

M. Rissot, Inspector of the Forests of the Bois de Boulogne, exhibited a good collection of conifers, more suitable in general for forest planting ; among which were some Mexican species of Pinus, which seemed hardy looking kinds. The same exhibitor had also a good general collection of forest trees.

A series of plants were also exhibited for the purpose of showing the effects of prunning by different methods, preparatory to planting in forests and in towns, as well as for ordinary ornamental purposes. This was not a successful exhibition, as many, in fact nearly all, the trees which had been brought for the purpose were dead, owing to their having been removed at a late period of the year.

---

# ON THE EXTRACTION OF COPPER FROM ITS ORES IN THE HUMID WAY.

## By THOMAS MACFARLANE.

In a former paper on this subject published some time ago in this Journal,\* I described a series of experiments, which had, for their object, the economical extraction of the copper contained in the poor pyritous ores of the Eastern Townships. The results of these experiments may be briefly stated here. It was shewn—1st, That it is impossible to remove from a very pyritous ore, by simple calcination with common salt, and lixiviation with water, more than a small proportion of its copper contents; 2nd, That by calcining such an ore with twice its weight of impure iron oxide, and the necessary quantity of common salt, it is possible to remove 95 per cent. of the copper; 3rd, That, if, in such an operation, a temperature much above redness be employed, copper is, to a considerable extent, volatilized; 4th, That in order to complete extraction it is necessary that the materials should remain undisturbed during calcination; 5th, That even with the use of a large quantity of iron oxide and salt, it is impossible to extract the whole of the

---

\* Vol. ii [2nd series], p. 219.

copper from ores containing purple copper or copper pyrites, without any admixture of iron pyrites. Although in some respects very successful, these experiments still left much to be wished for. Ores deficient in sulphur could not at all be efficiently treated. Even the pyritous ores required to be mixed with a large quantity of iron oxide in order to the complete removal of the copper. This, although favorable to the extraction, largely increased the bulk of material to be treated, and consequently the cost of calcining.

While visiting the Bruce and Wellington mines, on Lake Huron, last summer, I was forcibly reminded of the vital importance to them of an easy and economical process for extracting the copper of their ores, which consist, almost exclusively, of copper pyrites in a matrix of quartz. It may be safely assumed that one-fourth to one-third of the copper in these ores is lost in the present system of ore dressing. Of equal importance would such an economical humid process be to the Harvey Hill mines, in Megantic county, Quebec, where the ores are also too poor in sulphur to be advantageously treated by any known extraction process. It occurred to me that the difficulty, caused by the scarcity or absence of sulphur, might be overcome by furnishing the ore with sulphuric acid in the shape of calcined sulphate of iron, giving it at the same time the proper proportion of common salt, from the decomposition of which by the sulphate of iron chlorine might be developed for the formation of proto-chloride of copper. It next occurred to me that on precipitating the copper from the solution of the latter salt by metallic iron, a solution of proto-chloride of iron would result, which, on evaporation to dryness, would furnish an effective re-agent for treating fresh portions of ore. And, lastly, it appeared to me, that an easy method of procuring this proto-chloride of iron in the first instance would be to dissolve together equivalent quantities of green vitriol and common salt, crystallise out the sulphate of soda, and evaporate the mother liquor to dryness. The proto-chloride during evaporation might become partially oxidized, but this would not lessen its effectiveness in the proposed application.

At the first opportunity I proceeded to ascertain by experiment, in the laboratory, whether these ideas were capable of being applied successfully, and the following is an account of some of the trials made. Through the kindness of James Bennetts, Esq., Manager of the West Canada Company's works on Lake Huron, I had been furnished with various samples of ores from their mines. Slimes from the Wellington and Copper Bay mines were first operated on by calcining them with proto-chloride of iron in a muffle furnace at a dull red heat. Fumes of volatilized chlorides were abundantly developed, especially on stirring the mixture. The results obtained were very variable. With Wellington Mine slimes of 2.9 per cent., one experiment gave 0.5 per cent. copper soluble in water, 0.7 per cent. insoluble and 1.7 per cent. volatilized. In a second trial with the same slimes and a larger quantity of chloride, 1.5 per cent. were dissolved, 0.8 per cent. left insoluble, and 0.6 per cent. volatilized. In a third experiment with Copper Bay slimes of 2.1 per cent., the whole of the copper was rendered soluble. But such a result as the last mentioned was only attainable occasionally, and it became very evident that high temperature and unlimited access of air often combined to make the result unfavorable and at least uncertain. The temperature at which the sulphurets contained in the slimes oxidized, seemed to be so high as to cause a sublimation both of the chlorides of iron and copper. I therefore, in the subsequent experiments, calcined the one previous to treating it with chloride of iron.

The ore next operated on was an average sample of the crush-work at the Wellington Mine, as it comes from the crusher to the jiggers in the ore dressing works. On shaking it on a sieve having fifteen holes to the lineal inch, it was separated into a coarser and finer part, the former assaying 2.6 per cent. and the latter 5.2 per cent copper. On calcining and further pulverising the finer part, and sifting it on a finer sieve, it separated into one part, coarser in grain, and containing 4.41 per cent. and three parts finer containing 5.58 per cent. copper. The latter sort was heated over a spirit lamp, with one-fourth of its weight of proto-chloride of iron, in a retort through which a current of air had passage. In one experiment 3.9 per cent., and in another 4.3 per cent. of the copper contents were rendered soluble in water. In the first experiment water dissolved out proto-oxide of iron along with the copper, but in the second, which had been heated longer, all iron in the solution was present as peroxide.

Having observed in one of these experiments, that the air contained in the retort seemed sufficient for converting the proto-chloride of iron into perchloride and peroxide, ($6 \, Fe \, Cl + O_3 = Fe_2 \, O_3 + 2 \, Fe_2 \, Cl_3$), it occurred to me that the current of air passing through the retort might be dispensed with. Twelve grammes, calcined ore from the Wellington Mine, assaying 5.22 per cent., were intimately mixed with three grammes of the dry chloride, and heated over a spirit lamp in a common digesting flask for twenty or twenty-five minutes. These experiments resulted as follows :

|                              | I.   | II.  | III. |
|------------------------------|------|------|------|
| Dissolved by water per cent. | 4.76 | 4.76 | 4.96 |
| Remaining in residue    "    | .59  | .32  | .28  |

In II and III there were respectively extracted 91.18 and 95.02 per cent. of the copper contents. The residues contained respectively one-third and one-fourth of one per cent. copper. None of the solutions obtained in these experiments contained any protoxide of iron, but there was abundance of peroxide present. This proves that, although an excess of proto-chloride was used, all of it was decomposed as above explained. Little or none of the perchloride of iron was observed to sublime during the heating. It would therefore seem that, in these experiments, the protoxide of copper was converted into proto-chloride by simply exchanging its oxygen for the chlorine of the perchloride of iron ($3 \text{ Cu O} + \text{Fe}_2 \text{ Cl}_3 = 3 \text{ Cu Cl} + \text{Fe}_2 \text{ O}_3$).

Although the calcareous nature of the ores of Acton Mine gave little hope that experiments on them with this process would be successful, I nevertheless tried a few, but never obtained more than one per cent. of copper from an eight per cent. ore.

Ore of five per cent. from the Albert Mine, near Lennoxville, was next calcined and heated with one-fifth of its weight of chloride, as above described ; 90.2 per cent. of its copper was rendered soluble in water.

I next returned to experimenting with the slimes from Wellington Mine, which had been unsuccessfully treated by calcining them with the chloride in the muffle. They were first calcined, and then leached out with hot water, whereby some sulphate of copper formed in the calcination was removed. After drying they assayed 1.77 per cent. Ten grammes mixed with one gramme of the chloride and heated over the spirit lamp for fifteen minutes gave up 1.33 per cent. of its copper to water, while 0.44 per cent. remained in the residue. The same quantities heated for twenty minutes gave 1.55 per cent. soluble and 0.22 per cent. in the residue. Neither of the solutions contained protoxide of iron, and of peroxide, the solution from the first experiment gave more than that from the second.

The plan of using the chlorides of iron for the extraction of copper is not proposed here for the first time, but the manner of using it advantageously, as indicated by the above experiments, differs essentially from those heretofore proposed. The above experiments shew that direct calcination of a raw ore with the re-agent, under unlimited access of air, seldom leads to a successful or a reliable result. On the other hand, when the ore is previously calcined, the temperature kept low, and the current of air excluded, the application of the chloride becomes advantageous and practicable.

In the above trials, and others which have not been mentioned, the copper was sometimes determined volumetrically, and sometimes precipitated by iron and weighed. The residual solutions from the latter operation were evaporated to dryness, and the proto-chloride of iron recovered. The precipitated copper was easily compressed, in a diamond mortar, into little solid cakes readily fusible to buttons before the blow-pipe.

This process of extracting copper would seem to be capable of affording more reliable and more economical results than any hitherto proposed. Any ores, whether rich or deficient in sulphur, may be treated by it, except those containing carbonates of lime or magnesia. The exclusion of air, and the low temperature employed, render a decomposition or volatilization of the proto-chloride of copper, when once it is formed, impossible. There being no free acids in the solutions obtained, an equivalent quantity only of metallic iron is consumed. By evaporating the residual solutions, the re-agent is always recovered, and thus a further saving is effected. The amount of copper contained in the insoluble residues, is, in most cases, below, and never exceeds that of copper furnace slags, while the cost of the process will not exceed one-third of the expense of the ordinary method of producing copper from its ores by smelting.

With regard to applying it on the large scale, there would appear to be no grounds for anticipating any difficulty. The pulverisation of the ore would be most economically effected by wet stamps. If allowed to drain thoroughly, after being thrown out of the slime pits, it could then be completely dried and calcined, at the same time, in reverberatory or other furnaces. The roof of these furnaces might consist of cast iron plates which might form the hearths of chambers wherein the operation of heating the roasted ore with the chloride might be performed. The lixiviation is a matter of no difficulty, and with regard to precipitating the copper, it would be well to do this quickly, in vats heated by steam, in order to obtain a perfectly pure product. The evaporation of the waste solutions might be effected by waste heat from the calcining furnaces without any special expense for fuel. In short, there is nothing to prevent its economical application, and in all probability, an establishment for treating copper ores in this manner will shortly be established in connection with one of our Canadian Mines.

Actonvale, January 11th, 1869.

---

## ON THE ORGANISATION OF MOSSES.

By R. BRAITHWAITE, M.D., F.L.S. *

In former times many of the smaller cryptogamic plants were termed mosses, and although no order of plants is better defined or more readily recognized, the name is still vulgarly applied to lichens, as Iceland Moss, Cup Moss, and the shaggy forms growing on old trees; to algæ as Irish Moss; and even to some fungi. But the plants we have to consider are the mosses *par excellence*, Musci veri, or frondosi, as they have been termed, to distinguish them from the Musci hepatici, or liverworts.

By the ancients this group was but little regarded, for then plants were sought after on account of their real or supposed medicinal virtues; yet they had a Muscus cranii humani, or moss of a dead man's skull, which no doubt in the days of signature medicine was found of great service in head complaints. The first special work on the subject is the Historia Muscorum of Dillenius, published in 1741, remarkable for the excellence of its engravings, and containing also lichens and algæ.

Linnæus enumerates many mosses in his Species Plantarum, but he seems to have paid little attention to cryptogamic plants,

---

* Read before the Queckett Microscopical Club, June 28th, 1867, and cited from Science-Gossip.

and hence often confounded them.  His erroneous notion, that the capsule was an anther, and the spores pollen, led his followers astray, though we may chiefly attribute it to the want of sufficient optical assistance.

John Hedwig, however, now gave to the world those great works which have rendered his name immortal, and fully entitle him to rank as the founder of Bryology.  He was undoubtedly the first to discover the sexual organs in these plants, and his clear diagnosis of species is indicated by the great number which still bear the names he imposed.

These were followed by the valuable Bryologia Universa, and other works of the learned Bridel, whose critical eye greatly augmented the number of species; and in our day Wilson, and Mitten, and, lastly, Professor Schimper, have immensely extended our knowledge of them, the Bryologia Europæa, of the last named author, being the grandest contribution ever made to a single department of botanical study.

Bridel heads the first chapter of his Muscologia Recentiorum with the querry, " Quid sit muscus ?" (What may a moss be? , and this I hope you will be able to answer, after becoming acquainted with the details of their structure.

The mosses, to a cursory observer, may appear uninviting from their minuteness and apparent similarity, yet when we call the microscope to our aid, the exquisite beauty of their structure is at once apparent.  They are entirely cellular, and it is not surely a subject for admiration, that by mere diversity in form, arrangement, and construction of cells, we are able to characterize near 9,000 species in this one class of plants?

THE SEED OR SPORE—This is very minute, yet varying in diameter between $\frac{1}{4}$ and $\frac{1}{100}$ of a millimetre; in some minute mosses it is of large size, the capsule containing only ten or twenty spores; in others it is very minute and innumerable.  The spore is globose, of a yellow, rufous, or brown colour; its surface smooth or covered with rough points, and it consists of a mother cell, or primordial utricle, enveloped in an outer coat, or exospore, the contents being chlorophyl, starch, and oil globules, with mucus.

The first result of germination is the rupture of the outer coat, and protrusion of the primordial utricle or cell, which immediately commences division, the new cells repeating the process, until a dense felt of branched confervoid threads results, which we term the prothallium, and forming the green film we may often notice in spring coating damp walls and banks, and long mistaken for species of algæ (figs. 1, 2, 3).   From various cells of this, young

Fig. 1. Spore of
Funaria hygrometrica.

Fig. 2. Spore of Funaria hygrometrica
germinating.

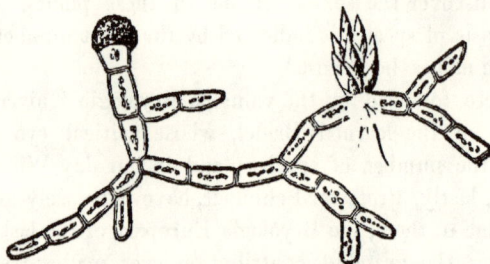

Fig. 3. Prothallium and young plant.

plants are developed, whose fine radicles penetrate the soil ; their leaves shoot up, and they become like the parent from which the spore emanated; and being now capable of maintaining an independent existence, the prothallium, no longer needed, dies away, except in a few minute annual mosses of delicate texture, where it is persistent during their whole life.   But some mosses rarely produce fruit ; yet it is necessary that their reproduction should be ensured, and we find prothallium also developed from tubercles on the roots, from gemmæ or buds occurring on the leaves, or even from the cell-tissue of leaves themselves; while in some mosses a portion of the leaves become altered into gemmæ, and clustered in a head on the top of a naked stalk called a pseudopodium, as in *Tetraphis pellucida* and in Aulacomnium (fig. 4).

Fig. 4. Pseudopodium of Aulacomnium androgynum,
with one of the gemmæ.

THE ROOTS.—These are slender fibrils, by which the plants are attached to their place of growth—the soil, crevices in the bark of trees, or rocks—and consist of a single series of cells, the septa between which are always oblique to the axis of the filament. Adventitious radicles or rhizinæ of a brown or purple colour also frequently occur on the stem, uniting the plants into a dense matted tuft, and like a sponge conveying water to every portion.

THE STEM.—Often simple, and sometimes so short as to appear wanting, it is in the terminal fruited mosses repeatedly forked, for on the cessation of each annual growth, a lateral bud is thrown off at the apex, producing an innovation or secondary stem; in the lateral fruited mosses, however, the stem is truly and repeatedly branched. It is of the same thickness throughout, for it grows only at the apex, or is acrogenic, and is composed of dense elongated cells, which thus render it firm and tough, those of the outer layer being often richly coloured.

THE LEAVES.—These are always sessile and simple, their form usually ovate or lanceolate, but varying in every degree between orbicular and awl-shaped. They are inserted spirally on the stem, though sometimes appearing to be distichous, or in two opposite rows; they may be erect, or spreading, or reflexed, or curled, and again they may be secund, or all turned to one side. The margin may be simple, or have a thickened border, entire or toothed, plane or wavy, involute or revolute.

The leaves may also be nerveless, but usually there is a central nerve, which may be short, or reach the apex, or be excurrent in a point, or long hair, and some mosses have two nerves. In the Polytricha, the nerve consists of a number of erect lamellæ, on its upper surface. The leaves consist of a single, sometimes of a double, or triple stratum of cells, the form and arrangement of which constitute the areolation, and afford characters of the greatest importance in the diagnosis of species, indeed used by some recent Bryologists, as Carl Müller and Hampe, for the chief divisions in classification.

In form, the cells are hexagonal, but varying to quadrate, rohmboidal, or linear, according to the density of their arrangement, and their surface may be smooth, or covered with minute papillæ. They contain granules of chlorophyl, which is often beautifully distinct, and the cause of the fine green colour, well seen in *Bryum capillare*, while in others it is expended on the growth of the cell, or the thickening of its walls, and thus in many mosses, while the cells in the upper part of the leaf retain their chlorophyl, those at the base are empty, hyaline, and elongated; in a few mosses the chlorophyl is wanting, and hence they have a white aspect, as in the family Leucobryaceæ.

Occasionally the basal wing of the leaf is occupied by cells, which differ from the rest, being enlarged or deeply coloured, and the presence or absence of these alar cells has been conveniently used by Prof. Schimper to divide the great genus Dicranum into two sections. When the cell-ends join by horizontal walls, they are termed Parenchymatous, and in one form of these, the cell walls are thickened, and the cell proper reduced to a mere point, producing the dotted areolations of Grimmiaceæ and others (figs. 5, 6). When the cell ends are pointed, we have rhombic areolæ,

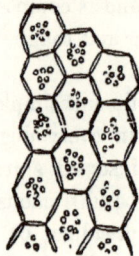

Fig. 5. Areolation of Pottia truncata.

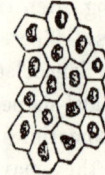

Fig. 6. Areolation of Grimmia apocarpa.

and these are termed Prosenchymatous, as in Bryum (figs. 7, 8). I must add that occasionally stipuliform organs occur intermixed with the stem leaves, as in *Hypnum molluscum;* these are named Paraphyllia.

An anomalous form of leaf occurs in the genus Fissidens, in which it appears to be vertical, and split into two laminæ for a part of its length. This split portion is, however, the true leaf, but the nerve and one wing have taken upon themselves extraordinary development, and there is also a lamina formed along the back of the nerve, these additional parts being named the apical and dorsal laminæ (fig. 9).

THE REPRODUCTIVE ORGANS.—It is now satisfactorily determined that these are of two kinds, male and female, and unless they occur near each other, the fruit is not produced; as an instance, I may refer to *Fissidens grandifrons,* of which male plants only have been found in Europe, female only in America, hence the fruit is unknown.

Hedwig was the first who pointed out the nature of these

minute organs, but his views were long opposed, for Roth and
Meese asserted that when sown, they produced young plants, and
hence were gemmæ or buds.

Fig. 7.   Areolation of
Bryum cæspiticium.

Fig. 8.   Areolation of
Hypnum rutabulum.

As in flowering plants, we find the sexual organs present three
modes of arrangement, and the species may be : —

Synoicous—when male and female organs are combined.

Monoicous—when they are separate, but on the same plant.

Dioicous—when separate, and on different plants.

The male or barren flowers are either terminal or lateral, and
consist of an involucre of minute leaves termed the perigonium ;

Fig. 9.   Leaf of Fissidens taxifolius.

these perigonial leaves vary in number, and in form and texture
differ considerably from those of the stem, becoming gradually
thinner and more delicate toward the centre.   Some mosses have
no perigone, but the male organs nestle in the axils of the stem
leaves ; in others the flower terminates the stem as a beautiful
disc or rosette, well seen in the coloured heads of Polytrichum ;
and again it may be gemmiform, or like a minute bud composed
of a few imbricated leaves, as in Hypnum.

Enclosed by the perigone are the antheridia, organs analogous
to the stamens of flowering plants; these vary in number, are

somewhat sausage-shaped, and usually intermixed with them are numerous jointed threads termed paraphyses, whose use no doubt, by the mucus they contain, is to keep moist and preserve the vitality of the antheridia, for in the open discoid flower they are most numerous, but in the closed gemmiform flower few or none (fig. 10). The antheridial sac contains the Spermatozoids, minute clavato-filiform bodies with two cilia, and coiled spirally, which on the rupture of the antheridium move about with great activity; they are most readily seen in the Polytricha (fig. 11).

Fig. 10. Two Antheridia and Paraphyses of Polytrichum.

The female or fertile flower, in a similar way, consists of leaves forming a perigynium, which enclose the archegonia, corresponding to the pistils of flowering plants; and so the oval base of an archegonium is named the germen, enclosing in its centre the germinal cell, and the tapering upper part the stylidium (fig. 12).

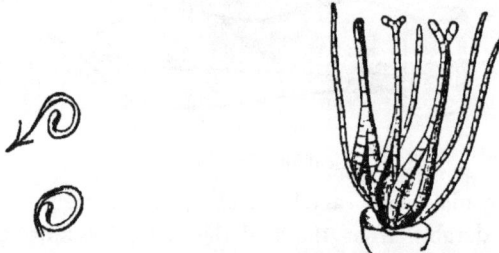

Fig. 11. Spermatozoids.

Fig. 12. Three Archegonia and Paraphyses of Bryum.

The inner leaves of the perigynium, as the fruit forms, become enlarged into a sheath round the base of the fruit stalk, forming what is called the perichætium, which is very distinct in Hypnaceæ.

Of the archegonia in each flower, seldom more than one is fertilized; sometimes, however, four or five may be, and we have as many fruits enclosed in one perichætium as in Mnium and *Dicranum majus.*

Having made you acquainted with the reproductive organs, we shall be prepared to follow out their functions. As stated, the antheridium at maturity bursts at the apex, and out pass the spermatozoids as a cloud of active particles; the archegonium equally prepares for their reception, the apex of the stylidium ruptures, the edges of the aperture roll back forming a trumpet-shaped orfice, from which we can trace a fine duct passing down to the germinal cell, and more evident now because it has acquired a reddish tinge. Both Hofmeister and Schimper have seen the spermatozoids within this canal.

The germinal cell, now fertilized, immediately commences its own proper development, first downward ; perforating the base of the archegonium, it fixes itself in the receptacle or apex of the stem, just as a stake is driven into the earth; then upward to form the seta or fruit stalk, and the contents of the archegonium being thus consumed, its delicate walls are ruptured, the lower part remaining attached to a process of the receptacle, as a little sheath—the vaginula (fig. 13) ; the upper carried aloft, becomes

Fig. 13. Young fruit of Orthotrichum crispum, showing Vaginula and hairy Calyptra.

Fig. 14. Mitriform calyptra of Encalypta.

the calyptra, or veil, and the seta, having attained its full length, begins to enlarge at the apex to form the capsule.

THE CALYPTRA OR VEIL envelops the young fruit, and is thin and membranous; it is sometimes torn irregularly, or it remains even at the base, when it is termed mitriform, or it is slit up on one side, when we call it cucullate or dimidiate; it is usually smooth, but sometimes densely hairy (figs 14, 15, 16).

THE THECA OR CAPSULE.—This presents an infinite variety of forms, but all of the greatest elegance; it may be globose, ovate, pear-shaped, or cylindric, straight or arched, erect or pendulous, smooth or furrowed. In some it is swollen all around at the base,

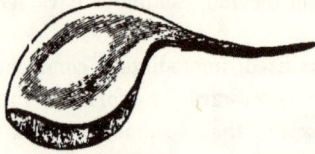

Fig. 15. Cucullate inflated Calyptra of Funaria.

Fig. 16. Cucullate conic Calyptra of Fissidens.

and this part is usually of a different colour, and is named the apophysis (fig. 17); in others it bulges out on one side of the base, and is then said to be strumose (fig. 18).

Fig. 17. Fruit of Splachnum ampullaceum with small conic lid, cylindric capsule, and obovate apophysis.

Fig. 18. Strumose capsule of Dicranum Starkii, with rostrate lid and annulus.

Closing the mouth of the capsule, we see a little cap—the operculum or lid, in shape flat, conical, or beaked; this, at maturity, is thrown off, either by the swelling of the contents or by the shrinking of a contractile ring of cells interposed between the lid and mouth of the capsule, which is named the annulus; well seen in the common Funaria. In the genus Andreæa there is no lid, and the capsule opens by splitting into four valves (fig. 19); and in another section there is also no lid, the capsule giving exit to the spores by breaking up from decay (fig. 20).  These

Fig 19. Schistocarpous fruit of
Andrœa.

Fig. 20. Cleistocarpous fruit of
Pleuridium subulatum.

characters enable us conveniently to arrange mosses in three
divisions :—

> Schistocarpi—the Split-fruited Mosses.
> Cleistocarpi—the Closed-fruited Mosses.
> Stegocarpi—the Lid-fruited Mosses.

The wall of the capsule consists of several layers of cells, the
outer of which becomes indurated at maturity, and often richly
coloured.

Enclosed with in the capsule is the Sporangium, or Spore-sac,
consisting of two strata of cells, the outer of which is contiguous
to the lining membrane of the capsule, or is suspended from it by
delicate threads; the inner is united to a pillar, occupying the

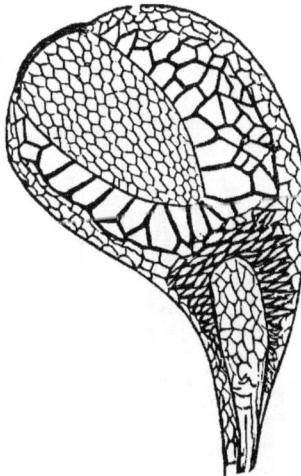

Fig. 20. Section of Fruit of Funaria, showing Sporangium suspended
by threads.

central axis of the capsule, and named the Columella, the apex of
which joins the lid, and sometimes falls away with it, though
occasionally we see the columella projecting from the mouth of
the capsule like a style (figs. 21, 22). The lid having fallen

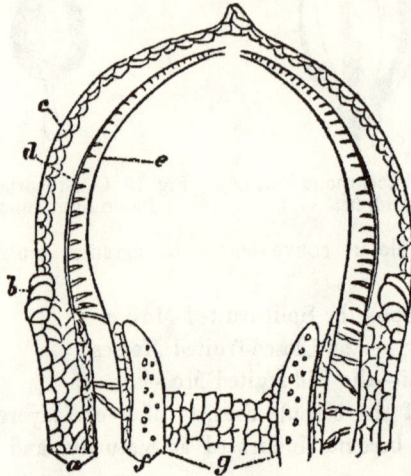

Fig. 21. Section of upper part of fruit of Mnium hornum, *a.* wall of
capsule, *b.* annulus, *c.* lid, *d.* tooth of outer peristome, *e.* tooth of
inner peristome, *f.* cavity of sporangium and spores, *g.* Columella.

away, the mouth of the capsule is seen, sometimes naked, when
it is termed gymnostomous, but usually adorned by the beautiful
appendage named the Peristome, consisting of curious hygroscopic
tooth-like processes in a single or double series.

The simple peristome, or the outer one when double, originates
from the lining membrane of the capsule; its teeth are always
constant in number, 4, 8, 16, 32, 64, and present an infinite variety

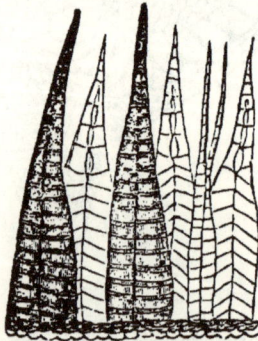

Fig. 22. Part of inner and outer peristomes of same.

of forms (figs. 24, 25, 26). They consist of two strata of cells, the outer in two rows, transversely jointed (trabeculate), richly coloured, and often separated for a part of their length, in the central or divisural line; the inner in one row, thin and hygroscopic, and projecting inward as transverse lamellæ (figs. 22, 23, 27). In the Polytrichaceæ, however, they are quite different,

Fig. 23. Transverse section of tooth of outer peristome.

and consist of a mass of agglutinated filaments, and Mr. Mitten uses this distinction to separate all mosses into two sections,

Fig. 24. Fruit of Tetraphis pellucida, peristome of four teeth.

Fig. 25. Splachnum sphæricum, with eight bigeminate teeth, and exserted columella.

Arthrodonti, those with jointed teeth, and Nematodonti, those with filamentous teeth. In the Polytricha, also, the top of the columella is dilated into a membrane, closing the mouth of the capsule, and joined to the points of the teeth; this expansion has been named the epiphragm or tympanum (fig. 27).

Fig. 26. Bifid tooth from peristome of Fissidens.

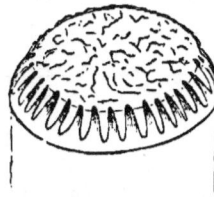

Fig. 27. Peristome and tympanum of Pogonatum aloïdes.

The inner peristome takes its origin from the outer wall of the spore sac, and is a thin plicate, or keeled membrane, divided into processes of cilia, which usually stand opposite the interspaces of the outer teeth, and occasionally one to three still finer ciliola, occur between the cilia (fig. 22).

The spores are formed from the cells, filling the spore sac, and are always free from the spiral threads found in the Hepaticæ.

In the above account I have not included the Sphagnina or Bog-mosses, as the views of recent writers tend to separate them as a distinct class, parallel with Mosses and Hepaticæ.

---

THE GENUS BOTRYCHIUM.—Dr. Milde has recently published an elaborate monograph of this genus, in which he recognizes the following species :—1. *B. Lunaria* Swartz. 2. *B. crassinervium* Ruprecht; a Siberian species. 3. *B. boreale* Milde; North Europe and said to be North American. 4. *B. matricariæfolium* A Braun. 5. *B. lanceolatum* Angstrom. 6. *B. simplex* Hitchcock. 7. *B. ternatum* (Thunberg). 8. *B. lanuginosum* Wallich. 9. *B. daucifolium* Wallich. 10. *B. Virginianum* Swartz. The first six species appear to be unduly numerous; Mr. Baker (very properly) condenses 2, 3, 4 and 5 into one, under the name *B. rutaceum* Swartz giving 5 the rank of a variety, but he recognizes 6 (which is hardly more than a variety of 1) to be a good species. The normal form of 7 is a plant of East Asia; the European *B. rutæfolium* A. Braun, and the American *B. lunarioides*, with its forms *obliquum* and *dissectum*, being reduced to varieties: the latter form is more of an accidental 'sport' than a botanical variety. Mr. Baker considers 8 to be a variety of 10; 8 and 9 are found only in East Asia. The normal form of 10, well known to Canadian botanists, is found throughout America from Canada to Brazil, and is widely dispersed in Europe and in Asia. D. A. W.

---

ERRATA.

On page 38, line 7, for 'ten miles daily,' read 'ten *inches* daily.'
On page 431, line 28, for '263,' read '268.'
On page 432, line 28, for '576,' read '575.'
On page 434, line 44, for '33,' read '53,' as the number of species in the Northern States referable to Caryophyllaceæ.

# INDEX.